Dirty, Sacred Rivers

# Dirty, Sacred Rivers

CONFRONTING SOUTH ASIA'S WATER CRISIS

CHERYL COLOPY

OXFORD
UNIVERSITY PRESS

# OXFORD
UNIVERSITY PRESS

Oxford University Press is a department of the University of Oxford.
It furthers the University's objective of excellence in research,
scholarship, and education by publishing worldwide.

Oxford   New York
Auckland   Cape Town   Dar es Salaam   Hong Kong   Karachi
Kuala Lumpur   Madrid   Melbourne   Mexico City   Nairobi
New Delhi   Shanghai   Taipei   Toronto

With offices in
Argentina   Austria   Brazil   Chile   Czech Republic   France   Greece
Guatemala   Hungary   Italy   Japan   Poland   Portugal   Singapore
South Korea   Switzerland   Thailand   Turkey   Ukraine   Vietnam

Published in the United States of America by Oxford University Press
198 Madison Avenue, New York, NY 10016

www.oup.com

Oxford is a registered trade mark of Oxford University Press in the UK and certain other countries.

Library of Congress Cataloging-in-Publication Data
Colopy, Cheryl Gene.
Dirty, sacred rivers : confronting South Asia's water crisis / Cheryl Colopy.
    p.   cm.
Includes bibliographical references and index.
ISBN 978-0-19-984501-9 (hardcover : alk. paper)
1. Ganges River (India and Bangladesh)—Environmental conditions.
2. Water—Pollution—Ganges River
(India and Bangladesh) 3. Water-supply—Ganges River
(India and Bangladesh) 4. Stream ecology—
Ganges River (India and Bangladesh) I. Title.
GB1340.G36C65   2012
333.91'620954—dc23      2012005796

9 8 7 6 5 4 3 2 1

Printed in the United States of America
on acid-free paper

Uma Devi Hamal places tika on the author during the festival of Dasain, 2007

# Contents

# List of Maps

# Acknowledgments

Like all writers, I have many people to thank. I doubt that I would have dared to undertake this book—let alone been able to finish it—without all sorts of help from dozens of people in several countries.

Ajaya Dixit and Dipak Gyawali of Nepal Water Conservation Foundation (NWCF) were my "water mentors" from early on. In 2005 Dipak encouraged me in what now seems an overly ambitious plan to try to tackle the water conundrums in the Ganges river basin. When I returned to Kathmandu in 2006 with a Fulbright grant based on the proposal Dipak had encouraged me to write, he and NWCF director Ajaya Dixit gave me advice, information, and support, and introduced me to an array of water experts throughout the region whose knowledge and contributions to this book have been invaluable.

For daily living during my years in Kathmandu, I relied on many friends. They helped me find necessities like apartments, printers, cell phones, yoga classes, computer experts, plane tickets, shoe repair, and a soft mattress.

I thank them for all that help but especially for their good cheer, good cooking, and special cups of tea—and for including me in their many festivities and holidays. Among these are Thakur Khanal, his family, and the staff at Royal Mountain Trekking; Arjun Adhikari and his family; Pushpa Hamal and his family; Shankar Tiwari and his family.

Special thanks to writer and friend Sushma Joshi for helpful tips and encouragement ever since we collaborated on workshops for young Nepali journalists in 2003; to Chrissie Gregory for Iyengar yoga and friendship; and to David and Haydi Sowerwine, who welcomed me into their home before I found an apartment in 2005 and introduced me to people in the expat community.

And a very special thanks to my favorite trekking buddy and guide Cholendra Karki. We trekked to places neither of us had ever been before; he made sure I got back.

In the years before I went to South Asia I worked with many people who inspired and encouraged me. Their support for my work doing enterprise environmental reporting in California and the northwest helped me develop skills that led to this book. The Society for Environmental Journalists has been a wonderful resource for me and other journalists over the years and provided a small fellowship that allowed me to do my first "regional" reporting, outside California.

Thanks to my former colleagues at KQED, the National Public Radio affiliate in San Francisco, who encouraged my work over the years. Thanks also to Frank Allen and the Institutes for Journalism and Natural Resources for their inspiring and illuminating institutes. During one of them, when I expressed some frustration about the strictures of six minutes of airtime, Frank said with a twinkle in his eye: "Well, you could write a book."

I am grateful to Rob Taylor and the International Center for Journalists in Washington DC, whose Ford Environmental Journalism Fellowship allowed me to go to South Asia for the first time as a reporter. And thanks to Peter Thomson of *The World*, formerly with *Living on Earth*, for advice on getting published and sensitive editing of radio features over the years.

The Fulbright Commission provided the grant that allowed me to begin this book project; its generous South Asia regional research fellowship made it possible for me to visit many of the places I explore in these chapters. Thanks to Mike Gill, former Fulbright director in Nepal, for suggesting I apply for that grant. And thanks to his successor, Peter Moran, now former Fulbright director in Nepal, for friendship, advice, and humor.

Many people in Nepal, India, and Bangladesh were valued informants for this book, and I thank them all for their patience and good humor. Some of the water experts and other informants cited here also read portions of the book, correcting inaccuracies and letting me know I was on the right track. Any and all errors that may remain in these pages are mine, not theirs.

I appreciate the professionalism and careful work of the editorial staff at Oxford University Press, including Jeremy Lewis, Hallie Stebbins, Marc Schneider, and Michelle Kelly.

Thanks to Kanchan Burathoki, who worked with diligence and good humor on the maps for this book; and to Ginna Allison and Eklavya Prasad for friendship and help with photos.

Finally, my undying gratitude to Bridget Connelly and Kitty Hughes, fellow writers and friends since graduate school, who have slogged through every chapter here more than once—from shaky first draft to final—encouraging me when I was uncertain and celebrating with me when things were going well. I can't imagine doing this without their support.

# Notes to the Reader

I use multiple names for the Ganges River in this book, according to the following rationale. When I talk about the Ganges watershed or Ganges river basin as a whole, I use Ganges, the name most commonly heard in the West for the great river. Ganga is the name for the river commonly used in South Asia, and I employ this name most often throughout the book, as it connotes the river revered by Hindus. When I am directly quoting a South Asian expert who has said Ganges instead of Ganga, I leave the quote as stated. In Bangladesh, the river becomes the Padma, and I try to reflect this. Experts in Bangladesh are more likely to say Ganges when referring to the entire length of the river, so I tend to follow their lead.

Sundarban is the term I've used to denote the mangrove forests on the Bangladesh coast. This is closer to the name in Bengali and Bangla, meaning beautiful forest. However, the name Sundarbans is the term used in most references and reports written in English. Both words denote the same region.

Measurements cited in the book are not consistent. Some are stated in metric form, many in American; and sometimes I used both if that seemed helpful. I have used many sources for the measurements cited in the book and have converted some to the American style if that seemed more helpful to the reader. On occasion it made more sense to keep the metric measurement instead of converting it. Sometimes I cite both. On quite a few occasions I am simply using my own observations and guesses about the size or depth or height of a place I describe.

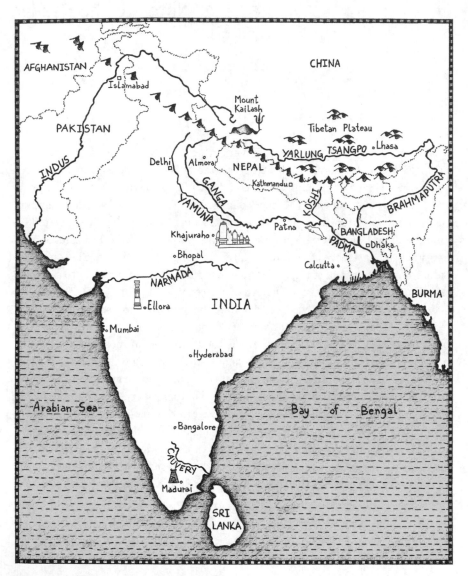

Map 1 South Asia—Major Rivers. Created by Kanchan Burathoki

# Dirty, Sacred Rivers

# Introduction

Red, hand-painted letters in Devanagari script inscribed on a yellow background tell visitors, in Hindi, "This water is as pure as the water from Ganga. Please keep it clean." The sign is painted on a wall near the entrance to a spring housed in a small temple in the Indian town of Almora. A shiva lingam painted red sits atop the tile roof that shelters a rectangular pool of clear water, embraced on three sides by white stucco walls below street level. Most of the people who come to this spring in the Himalayan foothills of eastern Uttarakhand obediently remove their shoes before descending the stairway to the stone pool.

The spring, called a *nola* in this part of India, is several hundred years old, locals say. Until fairly recently all the water used in this area came from hundreds of springs; some are small ponds like this one, others are spouts or *dhara* from which water flows. Now many of the springs are contaminated by trash and sewage. New construction destroyed some of them or blocked the sources that fed them.

The river that flows at the bottom of the valley below Almora does not have enough water both to support the region's agriculture and to supply household water for the city of more than forty thousand, where many people are now accustomed to water piped into their homes. Besides, it's expensive to pump water uphill into the town. Almora will soon have a full-blown water crisis. Already people go to the old springs that are still functioning. They need water because the supply in the city pipes sometimes dwindles; and many still prefer the taste and coldness of the spring water and believe it's good for their health.

The *nola* and *dhara* of Almora suggest some of the contradictions in South Asia's growing water crisis. Traditional systems have been neglected or abandoned, even abused, in favor of the promised convenience of modern ones. But those twentieth-century replacements have sometimes turned out to be unreliable and have left many people unserved. The population grows, construction continues, weather is uncertain, rivers and wells run dry, electricity for pumping is sporadic and expensive. And people have forgotten how to protect their water. The ponds that caught rain and replenished groundwater to feed the old springs were once carefully husbanded here in Uttarakhand. Now they are turned into playing fields or worse.

The admonition to treat the water in the *nola* as if it were the holy Ganga is sincere, but it's also ironic. The Ganges, called Ganga in India, is full of garbage, sewage, and industrial effluent these days. To respect the pretty little pool in Almora as if it were the contemporary Ganges could thus be to despoil it wantonly. The World Bank recently promised a billion dollars for cleaning up the Ganges. But that is not exactly what the admonition means; it means preserve the sanctity of the Ganges, honor the gift of water, and take care of it.

Dozens of thoughtful water experts I met in Nepal, India, and Bangladesh are trying to do just that; for a sense of the sacred is not sentimental or anachronistic. It embodies science, an understanding of how things work, and their limits. A huge price will be paid if that understanding continues unheeded.

The experts I've met and quote in these chapters think and do research and talk and write from many different angles, informing each other and trying to catch the ears of people in power in South Asia. They have a wealth of ideas that are too often heard only by fellow specialists. Yet their voices are part of worldwide conversation about resources and need to be heard.

Lalit Pande, my host when I visited Almora, asked me "what engages you here?" Why had I come all this way to wander around looking at rivers and old nolas?

I'm still not entirely sure why, but I know how it began. On a warm November afternoon, I sat on the worn seat of an old bus bumping its way down an unpaved rural road. We were descending rapidly on the way to Kathmandu. I leaned out the window to enjoy the last of the mountain air. I had already come down thousands of feet on foot.

A streak of blue-green water rippled down the middle of a chalky gray flood-plain to my left. The floodplain grew wider as we descended. It was no longer the mountain riverbank I had walked up two weeks earlier: steep, deep, and lush. This was dry, wide, and rocky, but the water flowed briskly.

Tired and a little feverish, I watched for hours as the river slowly widened and spread itself across a gravel floodplain; then as the riverbed widened, the blue-green river shrank inside it. People were in and around the river, mining gravel, watering animals, doing laundry. I was fascinated. As a reporter I had written about hydropower and riverbed gravel mining, about riverbeds dried up and fish populations devastated by dams in California and the Northwest, but I had never seen a river quite like this one: so wide, so much gravel, so many people. I had no idea what river this might be.

Some little voice, a small curiosity, asked: where are you going? The question was addressed to the river, but it could have applied to me as well.

This was the last day of my first trek in Nepal—something I had wanted to do for twenty years. I was a public radio reporter in San Francisco. I had not brought any recording equipment on this trip, just a camera. I wanted a complete vacation. The year was 2001.

I had waited in Los Angeles to connect with my Thai Airlines flight, uncharac-teristically delayed by an hour or so. The staff bustled around, serving even the economy passengers little sandwiches and fruit drinks to tide us over until our first meal on the long flight. Half an hour later I saw a big jetliner easing up to the boarding area. It was white, but the front was painted, like an amazing all-body tattoo. A winged dragon in golds and purples and reds was ready to take flight. The very airplane was magical, a mythical beast come to carry me away. My fatigue started to melt and excitement bubbled up in its place.

That first trek was not always fun. The altitude got to me; I was exhausted and lost my appetite until the Diamox kicked in. Then I came down with a flu, a sinus infection and fever. Yet I was in awe, finally being there in the high Himalaya; I didn't know where I was going and had to trust a stranger who spoke a different language to guide me. Nepal's mountains and rivers captivated me as they have climbers and naturalists for decades. The villages, the stupas and mani walls and yak-cow crossbreeds, the women in their long dresses—and of course the mountains themselves—caught me. I have gone back many times pursuing understanding of the watersheds, but mainly just to be there. Each time I ask myself what I am doing there, grubby and uncomfortable, too hot or too cold, slogging up impossible hills, no fresh vegetables for weeks sometimes. And each time I get back to urban comforts, I soon crave the mountains again.

The name of the river whose course we were following as the bus descended on the dusty road in November 2001 was the Indrawati. Wondering where it went started me on this journey, this attempt to understand the web of water connections in the Ganges basin. In time I would discover that the Indrawati, fed by the Melamchi River, joins with six others to become the Saptakoshi and flows into Bihar and then to the Ganga, like all Nepal's rivers.

I applied for a journalism fellowship and went back to Nepal two years after my first visit, this time with recording equipment, to teach workshops and gather material for stories. When I returned from Nepal, no longer getting support to do the kind of environment reporting I wanted to do in California, I resigned from the radio station.

Eventually, with help from a Fulbright research grant, I followed the rivers down to the villages that sip from them and to the megacities that suck them up and then ask them to carry away all imaginable refuse. In 2006 I set out to investigate the public policy debate, the natural forces, the human needs and the attempts to reengineer nature that are shaping conflicts over water in the far-flung parts of a hydrological system that sweeps from the Tibetan plateau to the Bay of Bengal.

This book chronicles my travels in Nepal, India, and Bangladesh, countries that are knit together by the Ganga and her various tributaries. I explain what I learned about glaciers melting in the mountains, sewage gluts and water shortages in the vast cities, and plans for engineering rivers that will have unknown consequences and perhaps limited benefits.

The book is a water policy travelogue through the greater Ganges basin; the title might be expanded to "Dirty, Sacred, Mismanaged and Dubiously Engineered

Rivers." Others have written travelogues, even environmentally focused ones, about the Ganges. They have chronicled trips down the mainstream of the great river from the Gangotri glacier to holy Varanasi and the Bay of Bengal. Instead, I zig and zag—down to Delhi, back up into the Nepal Himalaya, down into the Kathmandu Valley and Bihar and Bundelkhand—to show that all these are connected within the vast watershed the three countries share. As all roads led to Rome, all rivers in this region lead to Ganga. Though I am rarely on what would be called the mainstream of the Ganga, I am always traveling or trekking near some source, large or small, of the river.

My account is an homage to a mythically celebrated but very actual watershed. And it is also, perhaps, an elegy to these rivers, some already crippled, others in danger. People have asked me, when I tried to describe what this book is about, what my argument is. At first I was taken aback. I'm a journalist. I don't have an argument; that's not my job. But this is not entirely true. There is clearly a current in this book that is suspicious of modern engineering and celebrates the subcontinent's traditional water management.

There is no way I—a former medieval scholar turned environment reporter rather later in life—can claim to have answers to South Asia's water crisis, if there *are* right answers. So I give you many highly intelligent, trained, sane, and committed water experts from that region. These authorities more often argue for the lighter hand, the softer path: not no engineering at all, but less invasive engineering, and techniques that are localized, decentralized, and draw on traditional methods along with the almost-lost wisdom of local people.

I also present the arguments for more aggressive engineering, along with indicating the gray area, the uncertainty, the questions that can only be answered through trial and error. Which is perhaps why trying the soft path first, in earnest, makes sense. Lalit Pande told me, during my visit to Almora, that there was talk of bringing water from a glacier more than a hundred miles distant to slake the town's thirst since the local river was insufficient. Meanwhile some residents in Uttarakhand were trying to bring back the nearby sources of their water. For example, a retired army man created a pond on his land to store water for irrigating his vegetables. He can raise fish in the pond, and the water also percolates into groundwater. In another village, a pond that was turned into a playing field is being restored. Efforts like this are ongoing across the subcontinent as people begin to take charge of their own resources.

Ajaya Dixit and Dipak Gyawali, water experts in Kathmandu who are members of the extensive water intelligentsia in South Asia, advised me throughout this journey. They told me that my outsider's viewpoint could be useful even to readers in their part of the world. I can only hope they are right. I wrote these chapters chiefly for an American audience, hoping to spark interest in South Asia and show that problems of water management there hold a mirror up to the West and its choices. At times perhaps I have succeeded. At times I think I have inevitably done the opposite; the plight of Biharis perched on embankments during man-made floods or sleeping on sidewalks in Delhi in search of work because their fields are waterlogged may seem so remote to most Americans they won't identify with such struggles for survival.

Sometimes I have not gotten beyond my own cultural prejudices and expectations. I focus a lot on fouled waterways, and probably an implied judgment slips through, especially when I talk about the contrast between the idea of rivers as sacred and the fact of their filth. Except among American Indians, there is little sense of rivers as sacred in the United States. Laws and regulations have cleaned up American waterways and highways. From that evolved some sense of personal responsibility, or at least of habit and fear of fines. We also have a lot of land, and a lot of room for landfills: plenty of places to hide disposable diapers and endless food packaging. Out of sight, out of mind. Filth is just a word, an implied judgment on household garbage and sewage. We all have our messes to clean up.

A friend from Kathmandu who visited me in California some years ago kept exclaiming how clean everything was. He was the kind of guy who in Kathmandu kept the little street in front of his house and his neighbors' swept and washed. When I went to Kathmandu the first time I went around mentally agog all the time: cows wandering down busy streets, incredible temples, piles of garbage. I can't even remember now what struck me most because it's all become too familiar. And when I returned to the United States after my first fellowship, I did the same thing in reverse. California with its big clean, relatively quiet streets and huge supermarkets just looked weird: too much throwaway stuff on the planes, in hotels, in fast-food shops in the airport; toilets that flush automatically, according to some sort of clock.

Coming back to the United States I would also find unnatural disasters that made garbage and dirty rivers seem like a minor problem: a runaway gusher five thousand feet below the surface of the Gulf of Mexico, the first ocean I ever knew; an epic drought and fires in Texas; a cyclone in New England; fears that hydraulic fracturing or "fracking" will ruin groundwater.

I tended to compare tragedies, and to weigh their relative importance. A temporarily despoiled river in India seemed a very manageable problem compared to millions of gallons of loose oil. The oil spill is clearly more "important" than the sewage and the disregard that leaves the Yamuna and the Bagmati rivers such eyesores and insults to the nose. I saw how readily I snapped back to American-centrism, even after making good friends in South Asia, and looking at the world through their eyes, or trying to. Is the monumental BP oil disaster in the Gulf more important than what is happening to the South Asian rivers I have come to know in recent years? Is that the kind of problem that must be "solved" first? Or are they part of the same problem, on the growing list of nature's grievances, evidence of the great cost of continuing to ignore limits in South Asia and the United States alike?

Most discussions of resources get caught up in turf wars: nations and groups "own" the resources. But perhaps the rights we will have to learn to talk about are those of nature; and that language has not evolved enough yet to penetrate the halls of power. It's seen as fuzzy-headed, tree hugger talk. Yet maybe a river that can no longer flow freely has a grievance and rights of its own. Recognizing such rights will involve a far more nuanced understanding of cause and effect than is common on either side of the world.

As one of the experts I met, Rohan D'Souza, asks: is a river a resource or an endowment? If it's an endowment, it cannot be exploited beyond a certain point because it must be passed on intact to the future. Are there solutions for rivers in South Asia and the rest of the world that can bow to both sets of values, resource and endowment? Acknowledging the right of a river to flow will mean setting clear limits on how much reengineering of nature to tap resources is allowable before it's declared illegal. The voices in this book tell us eloquently why so much of the water management in South Asia has not worked. They also offer many ways to do it differently—to provide water for more than a billion people and still allow rivers to flow.

# 1

# Dirty, Sacred Rivers

The true Himalayas, miles above the sea, are still in the throes of
creation. The mountains are young, very young, adolescently
dinosaurian, terrible with ruthless youth.
—Han Suyin, *The Mountain Is Young*

I wanted to see the source of what we in the West call the Ganges. Here in South
Asia people call it Mother Ganga, Gangaji, the Great Ganga. At the edge of the icy
river that flows from the Gangotri glacier I scooped Gangajal—Ganges water—
into plastic soft drink bottles. I planned to take some of this water to friends in
Kathmandu, practicing Hindus for whom the drops of glacial melt would have
spiritual meaning. Along with its tremendous religious and ritual value, the water
of the Ganga has been shown to be both antimicrobial and richer in oxygen than
that of other rivers.

Revered beyond all others, this river is now abused in equal measure: harnessed
for hydropower near its holy mountain source, polluted with every imaginable waste
as it runs its course for more than 1,500 miles across the widest part of the Indian
subcontinent. One of the Ganga's main and equally sacred tributaries, the Yamuna,
flows through Delhi. Delhi, a city of more than fifteen million, owes its existence to
this river, which is now dead at its doorstep. Industrial effluents pour in upriver,
then Delhi adds its sewage.

During my first trip to Delhi in January 2007, I went down to the edge of the
Yamuna. I wanted to see just how bad the river's reputed pollution might be. First
I saw the barren ground along the riverside, strewn with rubble from the construction of a nearby bridge. There was little to tell me that this area was also the
site of regular religious practice where people come to do *puja*, take a little of the
water to splash on their heads, throw some flowers into the river.

Bunching up in the eddies under the bridge pylons were stray bits of colored
plastic and plastic shopping bags bloated with garbage, floating like sagging balloons half filled with air. They mingled with broken yellow marigolds scattered in
the water and bright red flowers set afloat in little cups by those who had come to
worship by the river.

*Map 2 Source of the Ganga. Created by Kanchan Burathoki*

A university student, a young woman named Parul whom I had hired to help me get around old Delhi that day, negotiated an appropriate fee with a man selling rowboat rides on the Yamuna. People who want to distribute the ashes of loved ones after a cremation often hire these boats. We climbed into his, its turquoise paint faded and peeling. The young man, dressed in worn jeans and a black jacket, steered the boat with long oars.

A light breeze of winter air brushed my cheeks. I was a little surprised, having been near the polluted river that flows through Kathmandu, that there was no overpowering odor that day. Nevertheless I came to have a literally bad taste in my mouth while we were on the water. As we pulled out from the shore I was afraid, though I'm a good swimmer. I peered over the side of the boat into the water. It was an eerie black, a dull mirror. At places there were bubbles coming up to the surface, making it look like a vast cauldron with the heat turned on simmer. At one point the boatman pulled up an oar; it was dripping with black muck, the sludge that covers the river bottom. The rowboat was in very little danger of cap-sizing in the still water; but the thought of touching, let alone of being submerged in the murk, froze me to my seat.

It was a quiet day; only one other boat was out on this stretch of water along with the blobs of garbage. Our boatman went downriver a ways, while a man poled himself upriver, riding upon what appeared to be an oversized grain bag that must have been stuffed with something lightweight and waterproof. As we neared the bridge, I saw a passing car send a bag of garbage over the railing into the water.

At the edge of an island in the middle of the river stood an abandoned structure; four square concrete pillars topped with rebar poked into the sky at the corners of a room-sized slab of cement. All four pillars of the unfinished creation were listing toward its center. Near one of the pillars a ragged monkey held court, glaring as we came closer. The boatman said the large rhesus had been here six or seven years and that people brought him food because he is an emanation of the monkey god Hanuman. He must have some magical powers, I thought, if he has been drinking from the Yamuna for that many years. To me he seemed like Cerberus, puffing his chest and baring his teeth, the river Yamuna a kind of Styx.

The boatman said he would drink from the river if he was really thirsty and did not have any other water available. But apparently he had not yet been driven to that. On the riverbank people continued their *puja*. Some just splashed their faces and heads with the water; others were immersing half of their bodies. A group of women was shielded by a U-shaped curtained shelter at the shore. One side was open to the river; they could remove some of their clothing in privacy for a ritual bath. The water was strangely clear when people cupped their hands and held it in their palms. A man doing his *puja* at the riverside told Parul it is not the water that is dirty, but our minds and hearts.

Though this stretch of the Yamuna offered some of the foulest water I have ever seen, the man's belief is well grounded in Hindu philosophy. Moving water is believed to be self cleansing. The Yamuna is moving water, even here, so it must be *choko* (pure), not *jutho* (polluted). Science confirms this belief: moving, oxygen-ated water can indeed eliminate impurities. Before these rivers were overwhelmed with sewage and effluent and modern refuse, they were in fact self-cleansing. For millennia these vast bodies could absorb and dilute all that a much smaller human population could put in them, including the bodies of animals and humans. Even though this is no longer true, the idea holds strong in many minds. The river is magic. It can take in everything, like a mother; hold it, kiss it, and make it better.

This may be one reason for the continued abuse of this and other rivers in South Asia. I spoke with political psychologist Ashis Nandy in Delhi: "People are used to presuming that rivers will maintain their old power," he says. "They have seen the majesty of the Ganges until thirty years ago. They are accustomed to the power of nature and that has not worn off for ordinary Indians as yet." He is optimistic that people will ultimately come to their senses and care for their rivers, but perhaps not soon enough to avoid irretrievable damage. "They have not seen the vulnerability of nature," he adds. "And that will take at least one or two generations. This crisis has come too early."

In Kathmandu one winter I was haunted by the face of an eight-year-old girl who drowned in the Bagmati River. Truly haunted, because when I tried to locate the photo of her I thought I had seen in a local newspaper, I could not. The silenced young face in my mind was streaked with the river's muck. The day after she died there was a photo on the front page of the local English-language paper of a Kathmandu policeman swimming out trying to find the girl. The photo showed him grimacing as he struggled to keep his head up while swimming through the murky charcoal gray water, garbage and debris bobbing on all sides of him. The little girl had already gone under. A bulldozer found her the next day. The policemen who tried to rescue the girl got skin and other infections after their Good Samaritan dips in the holy river.

A Nepali friend told me she had heard that the little girl fell into the Bagmati because it was so foul she didn't think she was on water. She thought she was stepping onto something solid.

The Bagmati is, for Hindus in Nepal, almost as holy as the distant Ganga, into which it eventually flows.

High in the mountains of the northern Indian state of Uttarakhand—where I traveled in September of 2007 and gathered Gangajal to take back to friends in Kathmandu—the Ganga flows free and clean. Its waters are also murky—kind of a dull khaki color—but this is due to the silt that the Gangotri glacier grinds from the earth as it moves down. The Himalaya, young compared to other big mountain ranges, still shed earth with abandon. This earth—transported by more than a dozen major and countless minor rivers—has made the Gangetic plain a fertile bread basket, its earth replenished yearly as torrential monsoon rains sweep away the top layers of the mountains and carry soil to the plains.

I had already climbed up to several sizable glaciers in Nepal, which can rightly also be considered sources of the Ganga, but I wanted to visit the vast glacier that carries its name: Gangotri, one of the largest glaciers in the Himalaya. It, like the others I've visited, is receding each year due to global warming. The Gangotri snout slides back about sixty feet each year. I can't call the visit a trek, because unlike my journeys to Nepal's glaciers I was only on foot for two days. And I can't call it a pilgrimage because I am not Hindu and did not walk up for devotional reasons. It was an excursion, a side trip on a quest to grasp what is happening to the waters of South Asia.

First I had to reach a town of twelve thousand called Uttarkashi, which I accomplished in a small car driven by a quiet young man who was a comfortingly safe driver. I had hired the car in eastern Uttarakhand, bordered by Tibet on the north and Nepal on the east. We traveled for two days, bumping along on roads that exist largely because of the Indo-Chinese border war of 1962. The roads the Indian army constructed led to development. Villages had become towns, towns had large hotels. Yet many roads remain one-lane and rutted, and in the morning they are crowded with water buffalos as villagers herd them to pasture.

Early on the second day of my ride to Uttarkashi, when the morning air was still cool, I glanced over my right shoulder through the driver's window as we wound down a hill. I glimpsed something bright and green, the pale green of algae or of spring. It was sparkling in the morning light, a translucent green haze. But something wasn't quite right. My brain tried to construct an explanation that would make sense: it's a mist, sunlight is reflecting on the intense green foliage and somehow turning the morning mist this pale, fuzzy green. But the next glance reoriented me. It was a lake, and a large one. The dull green mirror of glassy water was an engineered lake. The Tehri reservoir, which I had come to see on the way to Gangotri, had thus surprised me.

The 850-foot (260 meter) Tehri dam is a controversial structure planned in the days of Nehru, the beginning of the heyday of dam building in India. Construction began in 1978; Tehri started producing electricity in 2006. With a full reservoir behind Tehri dam, its generators would theoretically send more than two thousand megawatts of hydropower to the grid, along with 160 million gallons of water a day to Delhi, after costing well over a billion dollars.

Another surprise for me was that I did not immediately see the Tehri reservoir as anathema. It was just a flooded valley, rather beautiful in its eerie greenness, so different from the icy blue or slate gray of the Himalayan lakes I had seen. And in addition, I found myself thinking, it helps make electricity that will perhaps yield me a cold drink and warm shower to wash off the sweat from the hot car ride when I reach my destination in a few hours.

We wound down the road beside the brimming reservoir, where wooded hill-sides just slipped into the water with no division between the lake's edge and the wooded shore. An hour down the road I finally saw the raw wound of gouged-out, denuded hillsides. The sight made me think of the wounds on the stray dogs I have seen in South Asia: big, jagged, festering, and unlikely to heal.

Yet much of the disturbed earth I saw will "heal" in time. The reservoirs through-out the western United States are beloved lakes now, recreational destinations. If there is flora on their banks, it has adjusted to the yearly rise and fall of the water so there are no longer trees like those I could see at the edge of the Tehri reservoir, which looked surprised to find themselves where they weren't supposed to be. The disturbances of construction are long since gone, and in the less-populated North American continent, tens of thousands of people were not displaced to make way for the steep, sloping curves of the Hetch Hetchy or the Hoover.

Like those concrete American monuments to the will of politicians and the in-genuity of engineers, the Tehri dam has inundated a long valley. The waters of the Bhagirathi (the tributary that flows down from the Gangotri glacier and later takes the name Ganga) have backed up for twenty-five miles or so, drowning a couple of dozen villages and partially flooding many more. The old town of Tehri, which sat in the middle of the now flooded valley, has vanished. A hundred thousand acres of farmland were inundated, and a hundred thousand valley residents were displaced for the reservoir.

We continued to circle the reservoir, usually high on the ridges above it, winding around and up on a rutted dirt road for several more hours. The old road, which crossed the valley and would have taken us a fraction of the time, was under water. That's what my driver told me with his few words of English, pointing to a spit of land below us where a road ended abruptly at the water's edge.

I tried to take pictures of the dam as we circled up past it, but there was no place to stop. A sign said photography was forbidden in any case, this dam being a matter of national security for India. It seemed an unimpressive structure, with no particular grace or symmetry. I could not help but compare it to some of the rather handsome dams that dot the western United States, yielding irrigation water to make the desert green and electricity to make it glow. The Tehri Dam looks like a plug in the river; but it's tall, the largest earthen dam in Asia. It's also topped by a short service road covered in what appeared to be green gravel resembling Astroturf.

Why in this land of color and so much beauty did they not build a graceful dam here on a major tributary of the great mother Ganga? As I pondered this apparent incongruity, we went around a sharp bend in the road, and I saw a solitary rhesus

Flooded valley behind controversial Tehri Dam in Uttarakhand

monkey gazing almost pensively out over the reservoir from the shelter of a steel guardrail. The monkey was a common enough sight; but guardrails are few and far between in these mountains.

I found my hot shower and cold drink in Uttarkashi. The electric power did not come from the Tehri dam but from a smaller hydropower plant near the town itself. In the early evening I had a conversation with two people who were trying to restore the mountain ecology in this region.

Harshvanti Bisht teaches economics at the local college. In her spare time she is a mountaineer and has also established tree nurseries in an effort to reforest some of the high mountain areas that have been damaged by increased tourism. Harshvanti is a single career woman who had recently had an auto accident on a mountain road. An injury threatened the sight in one eye and her hopes of continued mountaineering. She came to meet me at my hotel after she had gone down to the river for *puja*, wearing a dark kurta. Her wavy hair was cut short, framing a long, oval face. She brought her friend Satish Dayal to our meeting because after nightfall she did not want to walk home alone, her night vision having been impaired by the injury.

Harshvanti told me about the array of hydropower plants to be built between Uttarkashi and the Gangotri glacier, my destination. Two "run of the river" plants—which do not store water behind high dams—were already under construction. The Indian government maintains that these run of the river projects do less damage than high dams, which are built to hoard the river's flow and release water as needed to generate electricity. Even though they don't submerge river valleys like the Tehri reservoir does, critics say the cumulative damage done in the course of building several small dams, along with roads to reach them in the unstable mountains, can still devastate the high mountain terrain. River advocates say the label "run of the river" is misleading, implying the river's flow continues unharmed. Instead they call these projects "diversion dams" because water is taken from the river and sent down a tunnel, thus drying up the stretch of the river between the diversion and the powerhouse far below.

"They are going to kill the Ganga," Harshvanti lamented, "the government is killing the Ganga." She was dismayed all the more by these efforts because she believes the plans have been made without regard to global warming. The mountain ecology is being disrupted for what may turn out to be no good reason. As the glaciers recede there will be less water for hydropower, according to some speculations, and some think the dams may be useless in a generation, either from lack of water or an accumulation of silt behind them.

Harshvanti has focused on reforesting the high mountain valleys, which have suffered from development, tourism, and an expanding population's cutting of trees for firewood. In the thin, dry air vegetation grows much more slowly than in the hot, wet oxygen-laden air of lower elevations. Harshvanti planted two tree nurseries. Since 1996 she says she has planted more than 12,000 trees and most of them are still alive. The government sued her for planting the tree seedlings on public land, protected by fences. She won the suit and had permission to continue for a few more months, after which she would have to get permission again. Though her mountaineering days may be over and her family wants her to move closer to the city, she says she loves her trees. When the foundation that was supporting her work ran out of money, she used her own salary to continue her efforts.

Her friend Satish had founded his own grassroots effort to discourage poor people from using the river as a toilet and garbage dump. Thin and wiry where Harshvanti is ample, he spoke quickly in a tumble of words that Harshvanti sometimes had to sort out for me in her more elegant English.

Satish wants to make the upper reaches of the Ganga watershed a world heritage site. "Ganga belongs to the whole world," he declares. "The water becomes rain and falls on everyone, everywhere." He says people are living closer to the river and they lack toilets as well. He acknowledges the river's power to clean itself but not when it is overloaded. India's increasingly consumerist society generates more waste, plastic and paper items having supplanted biodegradable predecessors like fast-food plates made of leaves and small clay cups from tea stands. Satish thinks the pursuit of money and consumer goods is making people oblivious to what they are doing. "They will have accounts in the bank but no clean water to drink and fresh air to breathe."

Satish, a technician at the polytechnic school in Uttarkashi, told me what had inspired him to begin his work. Once his mother had come to bathe on the bank of the Ganga, only to find that someone was using it as a toilet. He promised her he would make it cleaner by the time she visited the river again. He set about a public awareness campaign. In 1985 he started blowing a conch shell at five in the morning near the riverbank in Uttarkashi. With this horn, which is used in Hindu ritual, he wanted to waylay people in the neighborhood who might be thinking of using the river as a toilet that morning.

People call him Ganga man. All along the river between Uttarkashi and the glacier, Satish has tried to change the culture. He has worked to convince people in nearby villages they must not pollute their streams with feces or garbage because those waters in time reach the holy Ganga. He also pressured the government to build latrines. He says there are still not enough of them.

In time he became the father of four children and had less opportunity for predawn conch blowing, but he thinks perhaps his efforts have reached a quarter of the population.

"One hand throws a plastic bag in the river, the other hand reaches for water for *puja*," Harshvanti joined in as we talked about the river pollution. She thinks people know it's wrong to use the river as a dumping ground, but they do it anyway, lacking good alternatives like toilets or garbage cans.

Harshvanti declined my offer of tea. She said she had to go get ready for a trip to Delhi the next day. Satish and Harshvanti walked off into the evening, she leaning on him.

What is a river? It's a narrow body of water that is never the same—a process, a metaphor for change and impermanence. A river is also an object—defined by a line on a map—and it is an idea. For people in South Asia, rivers have personalities and take on divinity as gods and goddesses. But always a river is in motion. It may run swiftly or sluggishly, but if it is a river it must move.

The mountains that Himalayan rivers rush through are also in motion; they too are a process, though they seem the most stable of objects. For millions of years, continental drift pushed them up and created the precipitous channels the rivers follow as they plunge down, carrying eroded pieces of the very same steep mountains whose slope gives the rivers their force.

There are three layers of the Himalaya. The lowest hills have actually been formed by the silt that has been carried down from the highest ones. The high mountains rose from the thrust of the Indian subcontinental plate pushing against the Asian landmass. It first slammed into what is now Tibet and China about fifty million years ago.

Erosion from these youngest mountains in the world is massive. Erosion sounds like the sloughing off of a top layer, an epidermis; but no new skin is being created. In time the mountains will dissolve from the work of the rivers and rains and winds, the handmaidens that are delivering them back to the ocean from which they were once uplifted. The mountains and rivers are thus intertwined in the same process. The main difference is that the mountains are on a slow clock, the rivers on a fast one.

As I traveled across Uttarakhand on my way to Uttarkashi I saw how the rivers had sliced their way through the mountains and then merged, again and again. For eons the rivers of the Himalaya cut through the rocks, creating ever deeper canyons, following the gradient of the land and pulled by gravity. One day a pair of rivers might meet and capture one another, becoming one. These two merged rivers would travel on their conjoined way until they met some other pair of mingled rivers that had sliced through the mountains and merged. Then all four would become one.

For Hindus all these junctions or *prayags* are places of worship, auspicious sites. The power of water as an instrument of purification and a means to restore wholeness is even greater at these confluences. The Pindar and the Mandakini meet at Rudraprayag and become the Alaknanda. I spent the night at Rudraprayag on my way to Uttarkashi. My hotel was perched on a hill opposite the confluence. The rushing waters drowned out the sound of the noisy town nearby. Pilgrims descended the stairs from a small temple down to the water in the fresh, wet breeze. Butter lamps glowed in the near-dark; loud, recorded music played for the evening *puja* and reached me softened by the sound of rain and river.

The Alaknanda that I met in Rudraprayag captures the Bhagirathi at another special and well-visited prayag called Devprayag, the Bhagirathi having flowed down from the Gangotri glacier. But this is not a marriage of equals. One river is turbulent, brown, and frothy; the other sluggish and green. At first, standing barefoot on the river-washed steps that led right into the confluence, I did not understand why the rivers were so different.

A *baba* took my hand and led me down the stairs, put some purple flower petals in my cupped hand and then gradually poured water from a small brass pot on them, as he called out the names of gods and rivers in a loud voice. I only recognized a few of them, but he wanted me to repeat the names after him. When the brass pot was empty he scooped up more water from the river and continued the litany. I repeated the foreign sounds as best I could in a firm voice to match his near shout. Then I caught a mention of money between the sacred names. He wished long life for my deceased mother and father, then said something about four hundred rupees. Finally the ceremony ended and there was nothing but the bargaining left. I tossed the flowers into the swirling river and gave him a hundred rupees, a little more than two dollars.

After the ritual I went back up the stairs to a platform where a heifer was scratching herself against the corner of a concrete wall. I looked at the confluence again, and I realized that on my left was the free-flowing Alaknanda; on my right the sluggish Bhagirathi after it had been trapped behind the Tehri dam and slowly released. In front of me the two converged into the Ganga herself. The Alaknanda gave back to the shackled Bhagirathi some of her natural verve so that the Ganga flowing out of the hills to the plains had some energy.

All rivers that merge from here on down to Bangladesh will become part of Ganga as she takes the others into her arms. The Yamuna and Ganga descend from the same knot of glaciers in the Indian Himalaya and join each other far downstream on the hot Indian plain, the holy Yamuna taking her wealth of waste and symbolic freight into the great Ganga at Allahabad, an especially auspicious *prayag*, where millions of Hindu pilgrims come to bathe in the cold of winter every twelve years. Farther down, all the great rivers of Nepal also join: the Gandak and the Koshi merge with the Ganga in the state of Bihar, where the Buddha gained enlightenment.

In the second half of the twentieth century the Indian government began its own carving in these mountains, building roads and hydropower plants where rivers had been carving and merging for millennia. One of the many reasons people opposed building the Tehri dam was the danger of earthquakes, which are common in this tectonically active region—so active, in fact, that the Himalaya grow a little taller each year. The Bhagirathi was chosen as the site of multiple hydropower facilities, despite its holiness, because it has steep, narrow canyons and plenty of water—good for placing dams and good for getting a high velocity of water and thus more electricity.

The morning after meeting Harshvanti and Satish, I got an early ride in a "public taxi" up to the last town reachable by road, called Gangotri. The now diminished Gangotri glacier once came down as far as this town. Its Hindu temple remains the destination of many pilgrims who do not want to go farther on foot, chasing the receding glacier, whose terminus is now eleven miles higher up the valley.

The public taxi was a sturdy white Mahendra Max jeep, one of dozens that plied the mountain roads, carrying as many people as can be squeezed into them. On the way up to Gangotri I saw the hydropower plants Harshvanti had told me about: one complete, one under construction. I craned my neck for several hours trying to see out the window of the jeep, crowded with more than a dozen people. On the flat areas of muddy ground, fleets of earthmovers were parked. There were dozens of Komatsu and Volvo tractors, backhoes, and excavators. One piece of equipment in the well-funded construction effort was called a "Rocket Boomer," used for cutting through rock.

Higher up there was a flurry of activity at a large construction site. A new bridge over the narrow river gorge was half finished, its girders stretching out from both walls of the canyon, a wide gap still in the middle. Soon I saw the precisely contoured entrance to a tunnel, large enough to accommodate a lane of traffic. This was the lower part of the tunnel that would siphon water from upriver and carry it down to the new hydropower plant.

As we bumped along on the precarious one-lane road through the chaos and mud, I could see the cleanliness inside the tunnel. The road was open to the elements, to rains and landslides, while the carefully engineered tunnel was protected from them. Yet the difference struck me as in keeping with India's priorities now: throw up some sort of road so that sturdy vehicles (and hardy people) will have access, then get moving on engineering the hydropower that a power-hungry nascent superpower must have if it is to progress. Pavement, safety, and comfort can come later. That was surely the same sequence that other nations had followed as they industrialized.

It took close to an hour of winding up the mountain road to reach the other end of the fourteen-kilometer tunnel that was designed to tap some of the tremendous force of the upper Bhagirathi. One steel tube was already mounted in the top part of the tunnel; there was room inside the tunnel for three such penstocks to carry water down to the turbines.

Along the road were various small encampments of hastily constructed sheds made from corrugated metal and tarps with small cooking fires glowing outside. These were the homes of the itinerant workmen and their families. Officials lived in concrete houses on a hill above the worksite.

The trip up was uneventful and as fast as the rough roads would allow, which was sometimes faster than was wise. There were signs all along the road with clever warnings in English and Hindi to slow the traffic: *This is a highway not a runway. Road is hilly don't drive silly. No race or rally enjoy the beautiful valley.*

The return trip three days later showed me the precarious relationship between the man-made roads and the eroding mountains.

The South Asian monsoon is perhaps the single most important fact of life in this part of the world, and certainly the natural annual event upon which all life hinges. It typically extends from late June until September. In recent years it has sometimes had a halting start and extended into the autumn. Whether or not this is a long-term trend it's too soon to say. In September 2007 when I visited Uttarakhand, there were some heavy monsoon rains into the last days of the month.

On my way back down from a brief visit to the glacier, morning sprinkles turned into steady rain. After almost six hours of walking, the unprotected parts of my clothing and pack were soaked. I reached Gangotri again just as the light was fading, having walked in rain most of the day. Dry clothes awaited me in my duffel bag at an inexpensive lodge, along with some warm water and food and a long night's sleep.

It rained all night. The electricity was out. In the early morning, still in the rain, I waited inside another of the Mahindra Max public taxis. A fleet of them ferried people up and down the road between Uttarkashi and Gangotri. I waited for about an hour and a half—along with an Australian man named John who was also traveling alone, and an Indian husband and wife—for the vehicle to fill up.

Once underway we charged down the hill for about an hour. At 10:45 we passed the top end of the penstock tunnel I had seen two days before on the way up to Gangotri. Fifteen minutes later we saw a long line of vehicles pulled up. There was a landslide ahead. The Indian woman in the front seat advised against abandoning the vehicle and trying to walk to the other side of the obstruction. She said someone would come, maybe in an hour. I hoped her advice was based on experience and got back in the jeep to escape the light rain.

When the rain stopped I went down to see the landslide. Some of the people in other vehicles had left Gangotri an hour before we did, I learned, and had already been stopped for an hour and a half when we arrived. The landslide was an enormous pile of rocks filling the entire roadway. I could see sparks as some workmen drilled into boulders. Then there was a small dynamite explosion, apparently to break the boulders into smaller rocks. We all ran back up the road. No one was in charge of the crowd or the vehicles. Soon after the explosion, backhoes came and started chomping and biting and spitting out the rocks, tossing them over the edge of the road into the river below. In fifteen minutes they had cleared a lane.

At 12:10 our trip resumed. Probably the very lack of safety measures helped get us back on our way so quickly, only a little over an hour after we had arrived at the backup. And the nearby hydropower construction meant that the right equipment and manpower were on hand to clear the landslide. In Nepal a landslide might have stayed for weeks, with vehicles plying on either side of it and people clambering over the mudslide as I had done north of Kathmandu in the previous monsoon.

We whizzed down the mountain, but in fifteen minutes we encountered another landslide. I hopped out of the jeep and went down to take pictures, still in the same optimistic and amazed frame of mind as when we were sprung from the trap higher up the mountain.

But this second obstacle was different. It was not a pileup of medium-sized rocks but one huge boulder that sat right in the middle of the one-lane road, on a

Workmen try to remove landslide on road near Gangotri

sharp curve. It was the size of a hut. I couldn't imagine how it might be moved easily; a handful of eager young men were putting their full weight on it just then and it didn't budge. The really worrisome part, however, was the downhill side of the road. It was sliding away as I stood there, dark sodden earth slipping down into the river, slowly making the curve of the road narrower.

Anxious, I walked back and told John the Australian. He was the only other foreigner at this backup, because the few other tourists seemed to have gotten down the road just before this rock fell on it. I told him this one looked serious. He went to look and returned with a mirror image of my own fear on his face. We agreed that even if they cleared the road we were going to walk around the bend, not ride in a vehicle that could slip down the ravine in seconds.

We stood and waited, trying to imagine what our alternatives were at this point, far from our destination, far from just about anything that made any sense to us. Walk around the rock somehow? Call a helicopter from somewhere, perhaps? we joked.

Then a pair of bulldozers with backhoes appeared and started edging down the road toward the rock, approaching from both sides of the obstruction. On our side of the obstruction a round-faced man with a dark mustache, wearing a bright yellow helmet and rain slicker, came to a halt atop his backhoe. He could not get near the rock because several of the Mahindra drivers (not ours) blocked the way, getting into position for when the road was cleared, revving their engines. Fortunately the dozer driver on the other side of the rock was doing amazing things already, there being no cars between him and the obstruction. The driver aimed the bit of the bulldozer's shovel into the boulder, breaking it up and sending the pieces flying over the edge into the river. Even though it seemed a very good sign that the sliver of road held him, John and I still were not enthusiastic about getting back into our jeep to ride across.

The dozer driver on our side looked glum and seemed resigned to sitting this one out. But his face lit up in a big grin when I caught his eye in sympathy. He jerked his chin sideways in that peculiarly South Asian way, the quick tilt of the head a little like the shrug of a shoulder, saying "what can one do?"

The minute the last chunk of rock was thrown down the slope toward the river, the Mahindras revved and started moving forward, barely giving the downhill dozer, time to get out of the way. John and I stood watching as the eager drivers sped across the now open curve in the road, splashing through about a foot of muddy water. Not much fun to walk through, I was thinking, as I saw that our vehicle was on its way toward us. It paused to pick us up, and I got in very slowly, hoping thereby to create a small gap between our vehicle and the one ahead of us, just in case the road caved in at that point.

At 1:10, against all odds, we were on the road again, and half an hour later reached the bottom end of the penstock tunnel.

Safely back in Uttarkashi a few hours later, John found some welcome bottles of Indian beer in a government store. *Mountains are a pleasure if you drive with leisure. Life is a journey—complete it.* We laughed about the slogans we had seen on the way down and drank our way through four large bottles.

Compared to this dramatic ride from Gangotri back down to Uttarkashi, my walk up to the glacier that Hindus consider the source of the great Ganga had been less memorable. The glacier has long been the destination of *sadhus*—wandering holy men—and other Hindu pilgrims. But since the coming of the roads this destination is no longer a week's walk from the end of the road the British built, the last outpost of the former *raj* in the hills down below. It is accessible now to anyone willing to bump along for hours in an uncomfortable vehicle. And so twice a year (after the winter snow and before the monsoon, or just after the monsoon and before the snow) tourist-pilgrims flock here: to see and be seen, to pay homage, to please a spouse, to educate westernized teenagers in their faith, to take a vacation from the heat in the plains below, to refresh their minds and purify their souls.

I walked up to Gaumukh, the bottom of the Gangotri glacier, after the peak of the post-monsoon, autumn pilgrimage season; but still there were people on the trail. They ranged from middle-class Indians on horseback dressed in street clothes or *kurtas* to Europeans and Australians in trekking gear. Everyone coming down seemed more cheerful than the people going up—either because walking downhill is easier or the blessing gained from the pilgrimage reinvigorated them.

Along with the barefoot, solitary *sadhus* there was a colorful threesome of *babas* on horseback, all wearing lavender rain ponchos over their saffron robes, smiling and laughing with a gesture of blessing for the people on foot.

I walked up slowly until about four in the afternoon to an overnight stop, two hours below the glacier, called Bhojbasa. Groups of tourists and devotees were settling in for the night in the claustrophobic rooms of an ashram or in big, communal canvas tents. I chose the latter and positioned myself in a bed right by the tent flap where I could watch the moon and get fresh air.

It rained overnight. I started up toward the glacier early to finish the eleven-mile walk I had started the previous morning. The dry ground inclined gently, the trail winding around half-buried boulders. There was a small shrine just before the trail started up more steeply. A bell, some colorful flags and garlands, and several paintings of Hindu gods were arranged in a shelter of rocks. Soon the tall snout of ice, the tip end of the twenty-mile-long Gangotri glacier, came into view. This fifty-foot wall of ice is called Gaumukh or "cow's mouth" because some people think that's what it resembles. And the cow is auspicious, the emanation of Lakshmi, the Hindu goddess of wealth and light. I could not see the resemblance myself that day. I saw a jagged wall of pale blue ice, marbled and striped with gray glacial sediment, framed by snow-covered peaks aswirl with clouds.

This stunning sight was the source of the holy Ganga, the source—in rock and ice—of the goddess's gift to humankind. Various Hindu epics and scriptures tell the myth of the river's coming to the earth. Scholar Diana Eck has summarized the last part of the complicated tale of Ganga's birth.

> The Ganges is called the River of Heaven, flowing across the sky, white as the Milky Way. Long ago, she agreed to flow upon the earth as well. In her great mercy, she came to the aid of a king named Bhagiratha, who appealed to Lord Brahma to let the Ganges fall from heaven. Bhagiratha's ancestors, sixty thousand of them, had been burned to ash by the fierce glance of an angry ascetic, and only the funerary waters of the Ganges would raise them up again to dwell in peace in heaven. Having won from Brahma the boon of her descent, Bhagiratha persuaded Shiva to catch the Ganges in his hair as she fell, so that the earth would not be shattered by her torrential force. And so she plummeted down from heaven to the Himalayas, where she meandered in the tangled ascetic locks of Shiva before flowing out upon the plains of India.

Standing on a cliff above Gaumukh, at about thirteen thousand feet, I could see that two European tourists had approached from below and were admiring the ice face as they stood on the narrow verge of the river that poured from the bottom of the ice. An Indian man whistled at them trying to warn them away, because what they were doing was dangerous. A chunk of ice could fall on them or into the water and create a sudden wave. But they were too far away to hear him. They just waved like happy tourists, thinking he was saying hello. Several Indians were farther down the icy river where it was somewhat safer, doing *puja*, splashing themselves with the frigid water.

Above this point I could see only an intimidating pile of boulders and rain-threatening clouds, which convinced me to head down early, after scooping up some Gangajal. Both on the way up and on the cold wet return I posed the same question over lunch and tea as I chatted with some of the Indian tourist-pilgrims. How can the river considered most holy in all of India be so invaded and abused? Some simply registered the question with a somber look; others readily agreed it's an embarrassment, a tragedy.

One woman, keenly aware of the travesty of pollution, saw no conflict between the holy river's integrity and its potential hydropower. "We need the power for growth," she asserted firmly. Both the state government of Uttarakhand and the central government in Delhi are intent on harnessing the hydropower potential of the Indian Himalaya to develop the state's economy and send power to the national grid.

The hydropower project I saw under construction on my way up to Gangotri, however, was halted as it neared completion. The three dams that Harshvanti Bisht was most worried about the day I met her have been canceled. In 2010 G. D. Agrawal, a prominent engineer and environmentalist in his late seventies, began a fast unto death to stop them. His fast capped years of criticism from other river advocates. The short stretch of the upper Ganga between Uttarkashi and Gangotri is now protected; but twenty-six projects with diversion dams are going forward in this region, and hundreds more are likely under discussion.

In the town of Gangotri, on Harshvanti's recommendation, I sought out an aging yogi who was an environmentalist, mountaineer, and photographer. Originally from South India, Swami Sundaranand had been in the Himalaya since the 1950s, watching the changes. I posed my question about the abuse of sacred rivers to him, but he did not respond, at least not directly. Later he said: "This is a very bad time. Maybe in fifty years things will be better; people will listen and care for the resources here."

On the way down from Uttarkashi to Rishikesh, I rode in yet another Mahendra jeep. The driver was silent and unfriendly, and had perfected the unfortunate South Asian custom of honking at everything in and along the road. His honking was almost an involuntary tic. He honked at people, animals, and all other vehicles whether they were faster or slower, and would speed up just to honk at the rear of another vehicle. He didn't even spare groups of children on their way to school who were forced to share the road for lack of alternative. At one point he appeared to be honking at a pile of earth and rubble that had fallen into the roadway.

In the front seat a slender, young sari-clad woman sat crunched into the window, leaning out from time to time to throw up, either because of the jerky driving on the curving road, or because she was pregnant, or both.

Down, down the Bhagirathi channel we went. We retraced the route I had taken after we left the Tehri dam, but turned off onto another road before reaching it. Suddenly I noticed some trees near the river that looked as if they had waded in. The recent monsoon rains must have left the channel somewhat flooded, I thought. Then I caught myself more quickly this time. This was an arm of the reservoir, where the Bhagirathi stops being the free-flowing river that comes down from Gangotri and starts backing up. The plug of the dam was forcing water down a deep valley for twenty-five miles. The filling of the reservoir decimated the amount of water flowing in the Bhagirathi below Tehri dam.

As the Mahendra Max took us closer to Rishikesh, the road was straighter some of the time, the land less steep. I started to see enormous transmission towers dotting the landscape. They looked almost human, with legs sprouting from their steel waists, and short outstretched arms attached high on their long-legged bodies. The "arms" seemed to jut out from their heads, which were like a pair of rabbit ears with a wide mesh of steel brow in between. The great knobs of the transformers were clubby hands holding up the sagging high-tension wires. A single line-laden transformer sat in the middle of the brow like an eye. The towers marched out of the mountains and down to the plains like giant, alien soldiers. They reminded me of some early-twentieth-century French paintings that sought to capture the idea that humanity was becoming machinelike and machines were becoming human.

About half an hour from our destination, as we wound down a stretch of road in a deep forest, I saw a woman in a red sari, her arms bent, balancing a three-foot-high stack of fresh green fodder on her head. She was winding slowly, back straight, up the side of the road. The driver did not honk at her loudly, just a slight warning beep. The woman in her rich color and upright grace was an emblem of an older India: the enormous gray stick figures, the curiously human transmission towers with their stack of high-tension wires and transformers, were emblems of the new India.

Before I left Rishikesh I bought pale blue plastic bottles into which I planned to transfer the Gangajal once I got it safely back to Kathmandu. I wanted to present the water as a real gift to several people. The little blue plastic bottles, created for just such duty, had the Sanskrit character "Om" embossed on them. It's a graceful arch of a character, representing the single syllable that is understood to be the sound in which reverberate all the sounds and meanings in the universe. I wondered if I could tell the airline security people in Delhi it was Gangajal so they would let me take the bottles in my carry-on, where they would be less likely to leak. But I didn't want to risk having to pour the water out, unceremoniously, at the departure gate. So the three soft drink bottles of precious liquid went into my checked bag.

Back in Nepal, after decanting the water into the little blue bottles, I made a big point of telling my Hindu friends that this water was from high up in the mountains, near the glacier that was the source of the Ganga. The younger ones, with whom I had trekked in Nepal, seemed to appreciate this fact. Others seemed nonplussed.

Only later did I understand that it didn't matter. It was Gangajal. It would have meant just as much to them if I'd scooped it up far downstream where it would have already been subjected to the insults of human industry and urban life. It was the essence of purity to them no matter what humans might have put in it.

The Ganga is the sacred, silver thread that holds the Hindu world together. Far up in the landlocked reaches of Nepal—close enough as the crow flies but still a rugged or expensive distance to travel—and far down at the southern tip of India, thousands of miles from that silver thread, the river is magic in the minds of the people of South Asia. It's an emblem of life itself. If the Ganga dies—and there are plenty of dead and dying rivers in the world now—the river would surely live as a potent myth. But only that. The goddess who came down from heaven would have left the earth.

# 2

# The Real Poop

## How Rivers Become Sewers

> If I were making a country, I'd get the sewage pipes first, then the
> democracy, then I'd go about giving pamphlets and statues of
> Gandhi to other people, but what do I know? I'm just a murderer!
> —Aravind Adiga, *The White Tiger*

The train coasts into Agra, full of tourists coming to visit the newly scrubbed, sparkling white Taj Mahal. Through the window I see dozens of men squatting on the bare, flat ground between the railway tracks and a string of low buildings. Their backs are partially turned to the train tracks. Some squat alone; others hunch near each other in companionable duos. They gaze meditatively into the distance as they relieve themselves on the garbage-strewn ground.

I see this lineup for at least a mile as we glide toward the train station in the early morning. Some of the men carry little aluminum pots or plastic bottles as they approach the waste ground. The vessels sit on the ground next to men who are still engaged in their morning ritual. The water is for washing, for this is a land of washers, not wipers. I can't tell whether or not the train has surprised the men today by its punctuality. No one looks uncomfortable. No one is hurrying. Privacy seems irrelevant to them. And in any case the men's long white cotton tunics cover most of their bodies.

There are no women. Perhaps they came earlier, when it was darker, before there was a chance that the frequently delayed train would catch them.

I had not planned for a whole chapter in this book, let alone one right here at the beginning, to be about toilets and bodily functions. But there's really no avoiding it. Rivers pour clear and clean out of the mountains, then become sewers in the Gangetic plain. One of the biggest problems in South Asia is, simply, what to do with the shit of more than a billion people. Few of them have toilets as Westerners understand the word. Many have nothing at all—no outhouse, no pit latrine.

Globally, 1.2 billion people are still defecating in the open, approximately six hundred million of them in India. By 2010 more than half of India's billion people had access to a cell phone while only a third had some form of toilet to use.

From Delhi to Kathmandu to Dhaka, there is no way to talk about water in South Asia without talking about toilets, or the lack thereof, and about sewage treatment, or the lack thereof. That I hesitate, presuming the subject distasteful to readers, points to the problem it is for the whole of South Asia. It has been a largely hidden subject, a little like sex but with none of the fun. But the issue was a constant concern for me personally during my years in South Asia, and human waste remains such an enormous threat to health here as well as a major source of river pollution, I decided I had to address the subject right at the beginning.

The problem is staggering: The waste of a billion people in India, with and without toilets. Twenty-five million in Nepal. Upward of 150 million in Bangladesh. Half a billion people depend on the Ganga and contribute to its abuse in this most elemental way. Estimates of the amount of untreated raw sewage that enters the Ganga every day are hard to grasp: apparently something in the neighborhood of a billion liters. Much of it comes from homes that do have toilets, where relatively clean water is being flushed away and turned into sewage, which then turns rivers into sewers, a further loss of clean water.

A few heroes have tried to tackle this problem, but none with enough stature to kick off a sanitation initiative along the lines of what Nehru did to champion hydropower some decades ago. Most ambitious politicians in South Asia probably do not want their names sullied by association with such an undertaking, and it's certainly not an avenue for rising within government bureaucracies. Further, there's not much money to be made from the kind of toilets that are most needed, which are not pastel-hued porcelain ones with dual-speed flush. There being insufficient impetus to solve this omnipresent problem, it has simply persisted and gotten worse as the population grows.

One of the worst gifts the West gave South Asia may have been the water closet, the flush toilet. It will likely become a big liability in the United States and Europe as water supplies continue to be strained, but for sure in South Asia it's one reason rivers have become sewers.

How can this be? An instrument of sanitation the cause of fouling rivers? It's simply that adding water to shit makes sewage, and if there isn't an effective system to take all that sewage and render it relatively benign, as has developed in other parts of the world, it just sits there or goes to the rivers.

I'll venture that most people in the "first world" have not given this much thought. I had not. When I used the toilet in my own hundred-year-old bungalow in Oakland, California, where I lived for many years, I pressed the lever and went on with the day. When the opportunity arose I replaced an old toilet with one that used less water per flush, and like many ecology-minded northern Californians I did not flush for each pee but tried to conserve a little water, mindful of a drought in the 1970s when many of us began to become conscious that California's water had limits. But beyond those conservation measures, I gave no thought to where the "black water" went.

I learned at some point, because of leaks and work done in my basement, which pipe was the one for sewage from my single toilet: it was larger in diameter than the other pipes. I think I called a Roto-Rooter once. And once, while doing research for a radio feature on water quality, I even visited a local sewage treatment plant and didn't find it unpleasant. I felt complete confidence that this system worked, that all was well taken care of, flushed away into a sound system—one that would not leak barring a really catastrophic earthquake, and one that cleaned the water so it could be safely released into San Francisco Bay.

But the system I have known and used quite unconsciously ever since I was potty trained is nuts. In a world where the supply of clean drinking water is becoming an almost universal worry, why would anyone put clean water into a toilet? For South Asia, flush toilets and the treatment systems they require have become untenable long before they have become universal. For the Western world, the system as it is currently practiced will surely become untenable before too many more years have passed.

Raised in American cities, I have lived most of my life only a few footsteps away from abundant, safe water. The only times I was away from a tap with clean, clear water might have been trips to Mexico or if I ran out of water while I was hiking in the mountains. I have bathed and drunk and watered and wasted, coming up against the notion of scarcity only during occasional "droughts." And those years of minimal rain merely curbed certain behaviors, like car washing or lawn watering or compulsive toilet flushing.

During my years living in South Asia I became acutely conscious of how I was using water. Kathmandu was my home base for several years as I explored South Asia and its water issues. I had a lovely little apartment, in a fourplex built by a Nepali, chiefly to serve the Japanese diplomatic community. The flat lacked some of the comforts of the United States. I did not have an oven for cooking. And it was not safe for me to brush my teeth with the water in the bathroom or get water in my mouth in the shower. But by Kathmandu standards, I lived in luxury. Many people in the city did not have bathrooms of their own, or a reliable water supply even if they had a kitchen tap. I had an elegant bathroom with tile and marble surfaces, a tub and shower, both solar and on-demand gas options for hot water, and a clothes washing machine.

There was plenty of water at the tap in the bathroom and kitchen, for bathing and washing dishes that were then carefully dried. On occasion I almost used the water in the tap to rinse my mouth after brushing my teeth, or to rinse a carrot I was going to eat. Then I caught myself. In Kathmandu my drinking (and rinsing) water was delivered in large plastic jugs that had to be hoisted up and placed mouth side down in a water dispenser. They were too heavy for me. I timed my request for a new jug so that I would not actually run out of water, and I hoped two jugs would arrive so there would be a spare. And I had to be sure I asked someone in my landlord's employ who had the strength and coordination not to spill water all over the wall and kitchen to replace the jug for me. I read from time to time that even bottled water was not necessarily safe because tests often found some level of contamination.

The toilet in my apartment was far superior in make to other Western-style flush toilets I used in Kathmandu—let alone to the often perfectly acceptable squat toilets and outhouses that I frequently used in cities and rural areas of Nepal. Its seat did not slip off the rim, it had separate buttons for big flushes and little ones, and it always worked. The problem with this toilet was what happened after a successful flush. I was not sure, but I had my suspicions from the beginning. Every flush made me cringe a little with guilt. My landlord, who set up a sewage connection from the building, did not know where the dirty water went afterward. In time I came to understand that it did not go to a functioning sewage treatment facility because there really was none in Kathmandu.

About half a mile from my apartment building there was a small and smelly canal. Near it sat Saraswati Mandir, a temple honoring the goddess of knowledge. I crossed the canal and passed the temple as I walked to my yoga class twice a week. To avoid breathing the canal's stench, I took a breath before I reached the little bridge, then held it as I crossed, trying not to glance down to the litter-strewn bank where children often played and stray dogs lapped the foul water. In time I realized that this was surely the sewage drain for the entire quadrant of the city in which I lived. The canal originates in the hills to the north of Kathmandu. Actually, it is not a canal. Before it was so degraded, it was one of the small rivers that ancient inhabitants of Kathmandu revered, called the Tukucha Khola. More about it later.

I also spent time in Delhi, where municipal authorities are preoccupied with supplying water for more than fifteen million people. I frequently stayed in a small, pleasant guesthouse where the toilet in my private bathroom worked without fail, flushing a quantity of clear water, depending on which of the two buttons I pushed. The water in the porcelain bowl was not water I would drink, unlike the water in California toilets that I *could* drink if I had to. But the water in Delhi toilets would have been fine for drinking after a little boiling, and thus it was wasted. I wondered how much water might be going to criminal waste in the city's big, fancy hotels.

In Dhaka I stayed in a very nice hotel, much nicer than I really needed. It was a pleasant midrange hotel in Banani, one of the rich sections of the city. The Fulbright folks in the nearby American Center were being protective and recommended it. The staff was extremely helpful. The bed was big and soft. There was high-speed cable Internet and cable TV in the room, along with a little refrigerator, a plate of fresh fruit, and a glass of mango juice while I was registering.

And there was a fully functioning Western bathroom in pastel hues. I received what to me seemed an unconscionable amount of scalding hot water from a profligate showerhead, which poured on me with enormous pressure. Neither energy nor water-saving devices had yet been installed in this hotel, but the toilet seat was stable and the flush worked well every time. I reveled and cringed simultaneously as I enjoyed this bathroom. It was warm in Dhaka even in February or December, the months I visited, so I took a shower every day. I tried to cut them a little bit short and curtail toilet flushing when possible.

I also tried to convince the charming and well-spoken young man who cleaned my room not to change my sheets except maybe every few days. I had a king-sized bed to myself and barely disturbed his expertly folded corners when I slept. But he had other authorities to answer to: the hotel manager and his injunctions about how to cater to Westerners. I think he rarely did what I said on that score. He had a wife and two children and he had been away from them for several years, working in the Gulf where he could make more money than he could in Dhaka. What sort of "bathroom" did his family have, I wondered?

Bangladesh is virtually afloat for part of the year. It's the terminus of several great South Asian rivers that drain into the Bay of Bengal. The monsoon compounds their flow. Yet in the city of fourteen million that is the capital of this tiny burgeoning nation sometimes water shortages occur as the aquifers beneath the city are drained. How much water was I using every day in this Western-style hotel? The amount ten Bangladeshis might comfortably use for drinking, cooking, and washing? Maybe twenty?

Down the street from my hotel, perched on an open sidewalk, several families were encamped. They had built tent-like shelters against a fence on the sidewalk. Small children ran around, scantily clothed. Women in saris sat on the ground or squatted, knees elevated so that these slight women looked almost folded in half, scarves modestly pulled over their heads. How did they tend to their needs for water and hygiene? I wondered. There was no source of water nearby, and certainly no toilets. I walked by quickly, near the curb; I felt as if I were an intruder in their living space. They didn't ask for money. I would likely have given them some if they had. The men were perhaps out looking for work, or pedaling bicycle rickshaws if they were lucky.

In Nepal, where more than half the population may still lack functional latrines, some villagers look for a plastic bag before bedtime in case they need to defecate while it's still dark. They then carry the plastic bags to the river after daylight the next morning and dump the contents. They fear snakes as well as the darkness.

Bagging doggy poop in compostable plastic is becoming a common urban practice in the United States and Europe. Now something similar is being sanctioned as a way to deal with feces and urine during floods in North India when latrines may be inundated. They are called, appropriately enough, "peepoo bags." Developed in Sweden, they are being tested in India, Bangladesh, and Africa, in urban slums as well as rural areas. The bags are biodegradable, so they and their contents can be used as fertilizer. The makers say the bag works like a microsewage treatment system and can combat the spread of diarrhea infections. Until more permanent solutions are devised for South Asia, this seems like a good stopgap.

Ironically, the oldest toilet in the world may have been in the subcontinent. There is evidence in Mohenjodaro, in what is now Pakistan, of five-thousand-year-old sit-down toilets and bathrooms, along with a well-developed drainage system for wastewater. Perhaps the original idea for the water closet came from India and went West before returning and causing trouble. Other sites of the Indus Valley civilization show that drains from each house were covered by burnt clay bricks, creating enclosed sewer lines. These towns also used cesspools and soak pits to allow sewage to settle and wastewater to filter through the ground.

The most ancient Hindu texts, the Vedas, are almost as old as the Indus Valley civilization. Their hymns prescribe various modes of cleanliness. They forbid any pollution of water bodies, noting places where defecation and urination are permitted, and prescribe specific hygienic practices, including careful hand washing.

The history of toilets and sanitation throughout the world has been very up and down since Mohenjodaro. Well-off people usually managed something functional and aesthetic for their own needs, from Knossos to Rome. But poor people have generally had to make do with open fields or streets. In European cities the problem of disposal was once an enormous challenge, leading to filthy streets and outbreaks of cholera through the nineteenth century, killing tens of thousands in England alone. The Thames is said to have reeked even more than the Yamuna until sewers were built, dumping sewage downriver, away from the city.

In Delhi I visited the Sulabh Toilet Museum, established by a man who won the Stockholm Water Prize in 2009. The museum traces the history of the toilet from Mohenjodaro onward. It has some fascinating artifacts from the recent past: painted ceramic and porcelain toilets from the nineteenth century, elegant chamber pots, potties hidden inside wooden chests, one painted to look like a stack of books. But the toilet museum is just an amusing sideline for Dr. Bindeshwar Pathak, who—inspired by Gandhi—has spent much of his life trying to liberate India's hereditary sanitation workers.

The caste of sanitation workers or "scavengers," which is often called *bhangi*, was historically condemned to manually cleaning "dry" toilets, then carrying away the excreta to open fields outside of towns, often in buckets on their heads. The placing of such a load on the head was particularly horrifying to Gandhi and to Pathak as well, the head being the seat of the spirit and intellect. This practice persisted through the twentieth century and is still common in some parts of India. A similar sort of scavenging, without the caste implications and without buckets carried on the head, was practiced in the West into the nineteenth century, before water closets and sewers proliferated. Until then there was a market for "night soil," which was used by farmers.

Putting an end to this scavenging has been no simple matter, even though caste was officially abolished in India decades ago. Pathak had first to create and promulgate a cheap toilet alternative for millions of people in order to phase out the manual cleaning of toilets as a line of work. He wasn't trained as an engineer, but forty years ago he came up with the idea for what has come to be known as the two-pit pour-flush water-seal toilet.

The design is simpler than the name, which is nonetheless very descriptive of the system. The design involves two pits dug in the ground. Each has sturdy walls, which can be constructed of stone or brick or wood, depending on what materials are in abundance locally and how much money is available. The pits have earthen bottoms covered by sand and gravel. These filter the wastewater. When the pits are placed appropriately this usually protects groundwater from contamination while allowing moisture to drain away. A secure cover seals the top of the pit. Situated between the pits is a booth of brick or concrete or even bamboo; inside is a squat toilet set in a concrete platform. Sit-down toilets are not only uncommon in this part of the world, but are not even desirable to most people.

A drain attaches the bowl or "pan" of the toilet to one of the pits. A small amount of water is used to flush the waste: one or two liters at most, poured from a pitcher. Flush commodes use at least five times that much. The waste travels down into the pit through an angled drain; a "water trap" in the curved pipe below the bowl blocks odors. After some time—two years at a minimum, but in some cases far longer, depending on the size of the pit and the number of people using the toilet—the pit fills up and the drainage channel is connected to the empty pit. By the time the second pit is full, the waste in the first one has thoroughly dried and decomposed, rendering the pathogens harmless. The material is ready to be used as fertilizer. Switching back and forth between pits can continue indefinitely in an urban or rural setting. The whole construction can cost as little as fifteen dollars. The more expensive ones still cost under fifty dollars.

Similar toilets were already in use. Pathak's version perfected certain elements of those earlier designs to eliminate odors, leaks, and insects. And he stressed building two pits at the outset so that the toilet would always be functional and there would never arise a need to dump the pit's waste into a river or on the land. He did not patent his version because he was happy to see it copied anywhere in the world to promote elimination in a safe and relatively pleasant setting and provide a sanitary alternative for disposing of human excreta. It has been copied throughout South Asia and Africa.

Pathak's campaign began in 1970 in his native state of Bihar under the name "Sulabh," which means "easy to use" in Hindi. The Sulabh campaign is as much about changing people's behavior—both in terms of caste stigma and in terms of hygiene—as it is about building public toilets. Pathak persuaded the state government of Bihar to support his efforts. In the course of twenty years almost two hundred thousand two-pit toilets were built, freeing three thousand Bihari scavengers who were subsequently employed in menial but much cleaner work such as street sweeping.

As Pathak and his colleagues spread the Sulabh toilet—to villages, individual homes, and large urban communities in many parts of India—he could intervene in the lives of the scavengers. They were happy to be relieved of the shameful task of scraping latrines and carrying human shit. The work had condemned them to the very lowest rank of the "untouchables," now called Dalits or "scheduled castes." But it was the only way they could earn money because of the rigidity of the caste system as well as the stigma associated with their historical position in it. Over the years, Pathak and other activists working to abolish manual scavenging have almost succeeded. In 1970 there were still an estimated three million manual scavengers in India; now there are probably several hundred thousand still doing this work.

Pathak's own crusade began with a traumatic incident when he was six years old. He had been told he must not touch an "untouchable" person, but he was curious, and one day reached over to touch a woman from one of the so-called untouchable castes to see what would happen. His high-caste Hindu grandmother, horrified, forced him to swallow cow dung and urine as well as Gangajal—Ganges water—to undo the damage she felt was done to him by touching the low-caste woman. The holy cow's excretions and the holy Gangajal were meant to purify the boy.

Instead, the incident planted a seed in the small boy's consciousness. As a young man Pathak went to live, as Gandhi had, with a village of untouchables. There he saw a young woman being forced into scavenging after marriage. He tried to intervene on her behalf, but he had no alternative to offer her at that time. Another day he saw a young man, badly injured, left lying in the road because no one would take him to the hospital. No one wanted to touch him because he was an untouchable. Pathak took him to the hospital but it was too late; the boy died. Pathak vowed to spend his life trying to eradicate the stigma that led to these tragedies.

In addition to the two-pit toilets, Pathak, who earned a PhD in sociology in the course of his work, created centers where the bhangis could learn new skills for earning their livings. During my visit to the Sulabh campus in Delhi, I met a group of women who had been scavengers. They were all from the same town in Rajasthan a few hours west of Delhi. I talked with them through a Sulabh staff member who spoke English.

Some said they used to feel too sick to eat after doing their day's work. I felt sick after merely looking at pictures of women doing this work, scooping shit with a couple of pieces of flat metal into a basket or bucket. People had defecated in public latrines onto bare stone, sometimes inside rudimentary "booths." As I sat with these remarkably strong women, all dressed in blue saris, I could feel myself recoiling at the thought of what they had had to do. I even imagined a vague odor of feces. One older woman was quiet and looked sad, as if still weighed down by more than thirty years of cleaning toilets and carrying shit. But the other women were lively; even those who were shy were eager to talk about their lives. They smiled and told me about their children.

The group had elected Usha Chaumar as their leader. She said the other women probably chose her because she is confident and makes the rest of them laugh. She was eager to speak some English with me. Usha was thirty-three when I met her. She had started scavenging when she was seven, going out to help her mother, whose family had been doing this work for generations. She never went to school. She says her family had to live outside of town, which they could enter only very early in the morning to do their work. They had to wear bells around their necks or hit the ground with a stick to announce their presence. They were not even allowed to spit on the road; instead they wore little pots hanging from their necks so that they could spit after breathing the foul air of the latrines. They could not touch utensils, or enter houses, and the coins they used to pay for food had to be put in a bucket of water so that a servant could clean them before the shopkeeper touched them.

Usha told me these details without self-pity. After she had been doing this demeaning work for twenty years, Dr. Pathak came to her neighborhood outside the town of Alwar, Rajasthan. He asked her and other women whether they would quit scavenging if he gave them other work. Usha, with support from her husband, leapt at the chance. Pathak set up a center to train the women, where they learned to make snack foods to sell. They learned to sew, or work as beauticians. Fortunately, Usha's children had never had to clean toilets. They were too young when she was still scavenging.

Another of the former scavengers, a pretty woman in her early thirties named Lalita Nanda, said she used to feel like the dirt she carried on her head. Lalita was not born into the scavenging caste but married into it as a teenager. She cried and begged not to do the work, but finally gave in to pressure from her mother-in-law and did the work for eleven years until she heard about Dr. Pathak's training center. Now she works in a beauty parlor. I saw a shy, even perhaps a wry smile as she told me that customers from upper castes came to the beauty parlor. She said she no longer felt discrimination. She and the other women were being invited to weddings and into people's homes. At least in the community where Lalita lived, the stigma of scavenging seems to have been superficial, one that could be banished with a good bath, the donning of a bright clean sari, and discovering a talent for hairstyling.

After other women told me similar stories, the group sang a melancholy song about their lives and the betrayal they had felt during the years of doing demeaning work and being shunned for it. Then we all went to the dining hall, where we ate a hearty lunch at an enormous oval table. The ladies told me about flying to New York to visit the UN. They were representing the Sulabh campaign and helping to promote the UN millennium goals of safe sanitation for all.

Thinking about it later, I am struck by how much the shunning of these women shows a keen, age-old sense of germs and disease translated into the rigidity of the caste structure, creating a human tragedy. A distorted understanding of infectious disease helped maintain a system that provided some people with other people to clean their toilets. Ideas about contamination were transferred from the contaminant to a human being; and within the rigid caste system, contamination was deemed permanent instead of transitory.

As I toured the Sulabh campus in west Delhi I saw some of the other achievements of Pathak's campaign. On the road just outside there is a Sulabh community facility with many toilets and showers and a place to do laundry. The women's facility—and I assume the men's as well—was bright and spotless and there was no odor. I would happily use these toilets. Instead of being attached to underground soak pits, the toilets drain into an airtight tank—a biogas plant where an anaerobic process digests the waste and catches the methane gas produced. The gas is used for cooking stoves at the Sulabh campus.

The people who use such community facilities, which the Sulabh foundation has built throughout India, pay a modest fee. The money covers upkeep and the salaries of the Sulabh workers who keep the facility clean. Often these workers are former scavengers. Pathak's campaign has been criticized because so many of the liberated scavengers continue to do this sort of work, disdained even by Dalits of a slightly higher status. I heard from people who had used such facilities in other parts of India that they were not always maintained to the high standards of those situated next door to the Sulabh campus in Delhi.

At the campus I also toured the school Pathak founded to give the children of scavengers and other low-income families a chance at better jobs. The girls at the school were eager to tell me about what they were learning: computers and dress design as well as beauty salon skills. Perhaps this next generation will have some choice in the work they perform, none of it having to do with toilets.

Gandhi was so horrified by the work the *bhangis* did that he decided he and the followers at his ashram had to take care of their own excreta. Sheel Raj Shetty thinks Gandhi had the right idea: make everyone responsible for his or her own shit.

Sheel, a young architect, works for a company in Bangalore that promotes ecological architecture and rainwater harvesting. I met him when he was on assignment in North India after a flood in an area that sorely needed toilets—the kind that could both survive a flood and not contaminate floodwaters.

"Everybody needs to shit," he began when I asked him to tell me about eco-san (ecological and sanitary) toilets. "First of all, the concept of flushing your fecal matter is a Western concept. Who says we need to flush with water? We put it together and flush it together. Why?

"If you excrete in a field, the sunlight takes care of it, dries it up. Bacteria in the soil works on it, and it decomposes back to soil." But this natural method doesn't work so well anymore in the waste ground near railroad tracks, or in the fields around villages. There are far too many people and way too much shit. "Nature can't take care of so much shit in one place," Sheel said. "You have to spread it out."

Sheel noted that the average person excretes 150 to 250 grams a day, about half a pound. Eighty to ninety percent of this is water. If you remove the water, "what's left could be contained in a small matchbox." Feces smell more when water is added. Starve the bacteria by not using water and there is no stink. "What we say is: 'Don't flush your shit. Don't put water in.'"

Urine is also mostly water, so mixing it with feces and water as is done in toilets and sewer systems creates a soup for pathogens to feed on. In the absence of good sewage treatment, it all just goes into the rivers and other water bodies. All the pathogens are in the feces. Urine is sterile, with very little chance of carrying disease. What it does have is nutrients that are being sent to the wrong place; they create algae blooms in water bodies when they could be valuable for agriculture. While those nutrients are being thrown away, Indian farmers buy expensive chemical fertilizers, which pollute rivers and stress soil.

"So we want to sanitize it at the source, *and* make everyone responsible for his own shit," says Sheel. The solution may seem radical right now, but it's simple and it's likely the future. The eco-san toilet gives the user separate receptacles for different excretions. The solid waste goes into a simple chamber, often just a tin or plastic bucket with a removable lid. After the waste is sprinkled with some ash, there is no odor. Adding some dried leaves of the neem tree further absorbs moisture and kills germs. A chamber about two feet by two feet can last a family of five for a year. The dust-like residue is not dangerous, and it has some minimal value as fertilizer. "With eco-san everybody gets a toilet and," Sheel concludes, "gets to shit in dignity."

What to do with the urine? Ideally, collect it and sell it to farmers, unless one has a garden that can use a few liters a day. Human pee, according to some studies, makes gardens bloom better than cow urine or chemical fertilizer. How to transport urine to the fields? Urine-diverting toilets would have to be linked up to some sort of transportation system. That logistical problem combined with the challenge of changing people's habits are the chief obstacles to urine-diverting eco san toilets.

First, people would have to learn that pee and shit do not belong together, and that pee is golden in more ways than one. Human pee captured through urine-diverting toilets may turn out to be a good way to recover phosphate, a vital nutrient for agricultural production and a nonrenewable resource that is now mined from rock. Supplies of rock phosphate are running out, but urine contains plenty of as yet unmined phosphate.

Recently a pilot project in Nepal worked on transforming human urine into a dry, easily transported form. Like most development projects, it has its own acronym: STUN, or "struvite recovery from urine in Nepal." Struvite is similar in composition to kidney stones. Its crystals contain three nutrients: magnesium, phosphorus, and nitrogen.

I learned of this project from Marijn Zandee, a Dutch researcher collaborating with UN Habitat in Nepal. In his office he handed me a little packet of struvite pellets, each about the size of a mustard seed. Struvite powder had been made into pellets for ease of use; it tends to blow away in its powdered form. The pellets had no odor.

The STUN project started in a village called Siddhipur just south of Kathmandu where some farmers were already using urine to fertilize their vegetable crops. They diluted the urine from the eco-san toilets in their homes with plenty of water; urine typically contains too much salt to be used straight.

The Siddhipur farmers claimed that combined with compost, urine produced better crops. They also said they used fewer pesticides than with inorganic fertilizers.

Eco-san toilet in Bihar, inside and outside (photos courtesy of Eklavya Prasad)

Marijn says there's no scientific evidence for such pest resistance yet. He speculates that the plants perhaps get all the nutrients they need and are healthier, so they can better resist pests.

The Siddhipur experiment aimed to make a local business of struvite, attempting to create a model that could be copied in similar settings. Jerry cans filled with urine from villagers' toilets were carried by bicycle to a central location where the urine was mixed with magnesium and the liquid drained away, leaving struvite powder. The idea was for the sale of struvite to pay for the collection and processing.

"But," says Marijn, "we were too optimistic about collection efficiency and volume." The collectors were to be paid by volume, but there was not enough urine available for collection. And it was hard to find collectors for the waste: the experiment ran into status and caste issues in these Hindu villages.

But people's aversion to participating in urine collection is being overcome in some agricultural communities in Africa where farmers can't afford fertilizer. In India such a practice might be propagated. The ancient yogic and ayurvedic practice of auto-urine therapy, or *shivambu*, which involves consuming one's own, sterile urine as medicine would seem to embody an acceptance of this excretion. Marijn says that based on his experience in Nepal, "it is much easier to convince people to use their own urine than other people's. But I think some farmers, who really see the benefits, will in time have no qualms about using other people's urine."

The Siddhipur experiment in commercial use of urine taught the researchers a lot even though it did not succeed in the way they had hoped. The trick, says Marijn, would be to scale up. Collecting urine from individual households was too time-consuming for the amount it yielded. Five hundred liters of urine would yield only a kilogram of struvite. But such a project might work if there were a tank for several households and the urine were collected once a week.

"The nutrient value of just plain urine is low because it's so diluted. So a centralized collection system starts to look necessary." And the remaining liquid is still high in nitrogen, says Marijn. So the next step is to find a way to recover it. "The problem is how to reuse the effluent or recover more nitrogen. If we could collect from all of Kathmandu, the scale would probably be enough to produce fertilizer in a cost-effective way." Marijn says urine from about twenty thousand households would be sufficient to start a business, provided transportation costs were minimal.

Research is ongoing to find a way to reclaim all the nutrients from urine, but the technology doesn't exist yet. "Phosphate is the most acute problem because it's finite. We have to find a way to recover it. In Europe they are starting to recover it from mixed wastewater through a process developed in Canada," Marijn says, "and in Japan they recover it from swine waste."

"Kitchen gardens are an ideal use for urine. Nearby fields are also fine. But much beyond that it doesn't work until it's really large scale because of the relatively low nutrient concentration of urine. For now, the value of urine is the extra food you can grow. Empower people in remote communities to use urine," Marijn says—areas where there is no chemical fertilizer even available.

The farmers of Siddhipur are still using urine to fertilize their fields, and research on recovering all of urine's nutrients continues. Marijn's organization Eawag is starting a project in Durban, South Africa, where ninety thousand urine-diversion toilets offer an opportunity to experiment with scaling up the process.

Sheel Raj Shetty says eco-san toilets could be an answer to several of India's key problems: the shortage of toilets, the shortage of water, the dangers to health from water pollution, and the high cost of sewers. The possible source of fertilizer to promote food security is an added boon.

This seems a fine solution for rural India, where a farm or kitchen garden could use both dry fertilizer and urine. The eco-san toilets being built in flood-prone areas of rural India resemble Sulabh structures: a simple booth of brick or bamboo holds the squat toilet whose bowl can be cement or plastic. The difference is that there are no pits. Under the elevated platform are small receptacles for the feces; one is a spare for when the first fills up. After six months the feces—which have been sprinkled with ash—are dry manure. Urine goes into a separate receptacle and can be collected, diluted, and used as fertilizer.

Having realized that water pollution and water-borne illness will be a drag on India's continuing economic vitality, the government has relaunched a campaign to end "open defecation." Sulabh toilets are being built for many of those six hundred million people who still have no choice but the open fields or land near the railroad tracks. Excellent as these toilets may be, using a toilet of any kind is a lifestyle change for people who have defecated in the open all their lives. Some rural women who have gone to the open fields since childhood maintain they are constipated for days after they start going into those little rooms. They feel claustrophobic in the small enclosures.

A young man who was spearheading the sanitation campaign in one Indian state told me that many people he had worked with were reluctant to switch to toilets. Women in particular missed the social opportunity the trip to the fields as a group had always afforded them. This reminded me of a story I had heard in another part of India, in Uttarakhand, where women were not always happy to have piped water in their homes because they missed the walk to the spring to fetch water. The walk allowed them to chat and catch up on family affairs. The walk to the fields or the spring was also a break from household chores, a time of conviviality. Promoting enclosed toilets often requires a well-organized campaign with sensitive "motivators" from the community in order to alter ingrained behavior. Otherwise the toilets built during these well-meaning campaigns fall into disuse, the money is wasted, and the opportunity for sanitation perhaps lost for another generation.

According to one water specialist I spoke with, the lack of toilets and the consequent pollution of rivers and groundwater was a bigger issue in India's rural areas—where the bulk of the population still lives—than in urban ones. Even though the problem of the polluted Yamuna looked overwhelming, the avenues of pollution—usually storm drains—were visible and could easily be identified. All the technology for dealing with that kind of sewage is available if governments take the initiative. Rural pollution is far more diffuse and thus harder to control.

A good place to start an experiment in urine-diverting eco-san toilets might well be India's train system. Not only is the land near train tracks a frequent site for informal public toilets, as I saw outside of Agra, but the tracks themselves are the recipients of excreta. The latrines in Indian trains open directly to the tracks. Passengers are requested not to use the toilet when the train is in the station, but not all of them can comply and some simply forget, which means train stations, even in Delhi, can be highly odiferous places.

All of us coasting by the lineup of bottoms visible from the window as the train pulled into Agra that morning were thus potentially contributing our share to the pollution problem, whether we were in first-, second-, or third-class cars. The trains of India are rolling pee and shit machines, you might say. At any given moment there are millions of people on the move in India's trains. Most Indians use the trains, so this would also be a way to get the public comfortable with the idea of separate receptacles and waterless toilets that are odorless and pleasant to use, as well as with the idea of excreta being a resource. For rural people this may be natural. For urban ones it's a "lifestyle change."

After my first visit to the banks of the Yamuna, I spoke with water and sanitation specialist Catherine Revels about the sad state of sanitation in Delhi. Revels then worked at the Delhi office of the World Bank as a regional team leader for the water and sanitation program. I asked her how long she thought the city could function without some kind of radical improvement.

"Sadly, I think it could go on for a long time," she replied. She said people were still unaware of the risks that could be caused by one drop of water coming out of a system operating so badly. She told me a story about a colleague who was sick for three weeks with typhoid. The colleague's doctor guessed that a drop of water left in a restaurant glass infected her—though there are many possible ways of contracting typhoid. "So she can do everything right at home and still there's a health risk." And the poor people are suffering most, Revels added, since they are the chief victims of the forty-five million cases of diarrhea each year in India. According to UNICEF, a thousand young children in India were dying each day from untreated diarrhea.

"In the US, I remember in the 70s we had a few problems with polluted rivers," Revels continued. "In my own city I knew someone who had fallen into the river when he was a young boy and it had eaten up his sinus passages. This is the Saint John's in Jacksonville, Florida. That was because sewage was going straight into the river and industrial pollutants too. When they decided to clean this up it was in response to Environmental Protection Agency pronouncements that it had to be done."

Revels mentioned some other egregious examples of river pollution in the United States. These rivers were cleaned up when the national government took charge:

On the one hand getting serious about regulations, on the other providing financing for cities to straighten out the problem. A lot of money
was spent and those rivers were cleaned up and we don't have that problem anymore. And it was very fast, it happened probably within ten
years. There's an awful lot of money floating around in India being spent
on public works. With the amount of money that's being spent, if there
were also serious monitoring and control—actually having service outcomes and environmental outcomes—you could do it very quickly here.
There are sewage treatment plants here—there are plants designed to
clean this up—so it just takes them being serious about it.

Courts have ordered Delhi to clean up the Yamuna, which means the city must
make its sewage system work. Revels said she believed the Indian government
was seriously trying to tackle the problem, but "once you order something to be
done it takes a little bit of time to *get* it done. I think you'll see a difference here,"
she said. That was in 2007. Four years later things had not yet improved.

One problem Revels noted was that money was often lavished on sewage treatment plants that didn't work.

They're either underutilized or unutilized completely, or they don't get
the flows they're supposed to have. So along with the money you spend
on the plants you have to spend commensurate time and effort on the
soft side of this, which is policy—policies for getting people connected to
the system. Sometimes you have to subsidize the connections and you
have to make people abandon systems that aren't working. That requires
some changes to plumbing in housing and in people's yards. And it comes
down to political will. It requires a lot of communication, consultation,
and awareness raising.

Governments shy away from controversy, and trying to solve infrastructure problems in a city as big, diverse, and vocal as Delhi will lead to controversy.

There's a lot of controversy over sewage collection, about doing something in the slums. There's controversy any time you dig up streets, and
controversy over who is going to pay for it. People have self-provided—
septic systems, various things. Do they want to pay again for a connection? We faced this in the US. People with septic systems didn't want to
connect to sewage. They would have to pay again, so they have to be convinced it's a good idea. But in India, if they don't connect, there won't be
the necessary flows into the treatment plant and the plant won't operate
effectively. That's happening with almost every treatment plant that's
built in this region.

"It's a waste of public money if you look at it that way. The money needs to be spent, but you need to do the controversial things. So it's not that easy, none of it," she concluded.

At the time I spoke with Revels I had not come to question the idea that sewage treatment in Delhi should mirror the method used in the United States. I could not imagine any other way of doing things. Now this dilemma strikes me as an excellent opportunity to approach the problem from a completely different angle, to treat the shit closer to the source. After reading more and talking to people like Sheel Raj Shetty, I could imagine an entirely different scenario for Delhi.

Indian officials, at least as quoted by the Indian press, are fond of saying that they are tackling particular problems "on a war footing." What would cleaning up the Yamuna "on a war footing" look like? San Francisco recently implemented a law requiring all residents and businesses to separate food waste for composting so it would no longer have to go to a landfill. The program seems to be working well. What if Delhi's mayor were to say to Delhi residents: you have one year to start dealing with your family's sewage at the source. Sewage service in Delhi, such as it is, will end in a year.

This would apply, of course, only to people who have flush toilets and a supply of piped water. As an incentive to get the system set up faster, households and high rises would receive a rebate in their water bill if they were ready to suspend sewage service sooner. Delhi would have to make eco-san toilets available for a reasonable price, and would have technicians on hand to assist in the setup. Many of the nonprofit organizations that are currently working on these issues would be enlisted to help in the effort.

Delhiites should have alternatives. They could go the septic tank route; some people in Delhi have them already, so they could continue using them provided they were being maintained and emptied appropriately. That would still qualify as taking care of sewage at the source; but residents should be encouraged to switch to biogas plants to make use of the resource.

In the meantime, the city government would have to supply good communal toilets in the slums and poorer sections of the city so that they too could comply with the injunction to treat their own waste—something along the lines of the Sulabh facility I saw at Bindeshwar Pathak's campus in West Delhi with its biogas digester. Everyone would need to have access to a toilet whose waste did not go to the Yamuna.

This could certainly be called tackling sewage in Delhi "on a war footing." Could it work?

I talked to Sheel Raj Shetty again to see what he thought of this scenario. "Someday soon there won't be much choice but to implement something along these lines," he said. Theoretically it's possible, but problems lie in how to create alternative home solutions that would work for such a huge and diverse city, then how to tackle the solutions technically and get people to implement them.

South Asia is diverse even in sanitation terms, said Sheel. "Squatters and sitters, wipers and washers, able and disabled, old and young, aware and unaware, sensitive and insensitive, haves and have-nots, high-rise inhabitants and individual housing. So do we have a solution for everybody's need? Maybe we do. As always technology is simple. Getting people to use it is the challenge."

I wondered if there would be demonstrations and riots in Delhi if this sort of solution were forced on its busy inhabitants. On the other hand, I have also wondered why people are *not* marching in the streets over the state of the holy rivers that flow through Delhi and Kathmandu.

Sheel continued: "So what happens to the human waste after it's separated at the source? OK, let's start calling it a resource; it's not waste anymore. Every person excretes five hundred liters of urine and fifty kilograms of fecal matter every year. Delhi's population is more than fifteen million. Imagine the amount of resource we have! Now where do we store it or use it? Do we work on a plumbing system to ship all this urine to the farms around Delhi? Remember, urine has all the fertilizer and feces has all the pathogens."

Some waterless urinals were built in Delhi as part of a pre–Commonwealth Games beautification project. If these were to proliferate in the city, the next step would be to hook them up to a storage and transport system and process the liquid into fertilizer.

"Do we start thinking in terms of urban agriculture?" Sheel continued. "Let's assume we achieve that also: we have roof gardens, public park or community vegetable gardens. Now we are talking about food security. We've already saved a lot of water if each person saves 40 liters of water per day by not flushing." Water supply and sanitation are so intertwined, serious innovation in sanitation would save water and alter the ongoing debate about how much water a city like Delhi really needs. And it might reduce the large amounts of electricity needed for sewage treatment.

But then Sheel sounded discouraged. This would entail an enormous lifestyle change. People—middle-class ones at any rate—would have to get past a mental block, fearing that the lack of abundant water to flush waste away would mean smells or accidents, an encounter with unpleasantness. "Try talking to one of the higher castes, say a Brahmin man, about being responsible for his own shit. Over his dead body! Caste, religion, beliefs, gender, status. Imagine what we are up against!"

It *is* daunting, any way you look at it. Turn the whole thing upside down and do something radical like this, or get a functioning "Western" sewage system for Delhi. Both seem overwhelmingly difficult to achieve. Maybe that's why the Yamuna is still a sewer.

But, as Sheel said, someday soon something like this will have to happen in Delhi and Dhaka, Calcutta and Kathmandu. There's simply not enough water in South Asia to go the Western route: a porcelain bowl full of drinking water, flushable after every act of excretion. The very lack of toilets and functioning sewage treatment in India and other parts of South Asia is an opportunity to create a sustainable system. And the solutions can so easily be an alternative to the fresh-water hogs Western toilets are. The entire West and the middle class in Asia are now going in a direction that is not sustainable, however desirable it may seem compared to walking into a field or using a river or a smelly latrine.

Dr. Roshan Raj Shrestha is a pioneer, showing the way to an eco-san future and making it seem normal. He practices precisely what he preaches in his role as chief technical advisor to UN Habitat's Water for Asian Cities program, for which he travels throughout the subcontinent and other parts of Asia to promulgate ways to deal with water shortages and sewage gluts.

Since 2002 he's had a home system that does many of the things that may be common in the subcontinental middle-class home someday. It's a multifaceted approach to water conservation, widespread use of which would offer an economical alternative to the increasing water shortages predicted for cities like Kathmandu and Delhi.

Roshan's home is on one of Kathmandu's many hills, just down the slope from the ancient stupa of Soyambhunath, believed to be the oldest holy site in the Kathmandu Valley. The view is lovely, but being up high means the area often receives little water from the strained city supply; the water that does come is intermittent and requires someone to be home to turn on a pump and suck it up. I know other people who live in the same part of the city where the Shresthas live; already some of them simply never shower at home, instead having to go to a club or to a friend's house because they don't have enough water. Either that or they pay for tankers to deliver water.

Roshan's home, dubbed the "eco-home," is not even connected to Kathmandu's water or sewage lines, such as they are. Instead his family relies on water from the sky; two roof terraces are fitted with rain gutters and drains that catch the yearly monsoon, which is then stored in an underground tank. The water the family stores lasts for eight months of the year. The rest of the year they use their well, which they also recharge with rainwater each year.

The Shrestha family shepherds its water carefully, using half as much as other families might because they don't flush everything away. The Shresthas recycle gray water from showers and sinks to use in the garden, or in their one flush toilet, or for washing the car. Roshan and his wife use an eco-san toilet near their bedroom. There is one regular flush commode downstairs, which guests and the children use. He says he installed that one at the outset, since he had not yet tested the eco-san toilet and did not know what a success it would be.

The solid and liquid wastes from the eco-san toilet help fertilize the small gardens on the home's ground level and terraces. The dried, decontaminated feces are added to the household compost pile.

Roshan says the water harvesting and storing systems amount to approximately 1 percent of total costs when builders install them during construction. Retrofitting can cost more, chiefly to build the underground tank. But he says it's still cheaper over time than buying water from private tankers, the only option for most people whose supply from the city is meager. He said his family uses six thousand liters a month and needs about seventy thousand for a year. Since he collects two hundred thousand liters, he's recharging more than a hundred thousand back into groundwater through the dug well. Families who recycle their gray water can get by with smaller, cheaper tanks.

Roshan acknowledges that a number of aspects of this system offend Hindu sensibilities about water that is *jutho*—polluted in a ritual and religious sense. Hindus believe they cannot, for example, water flowers they will offer to the gods with water that has been used already. He says this is not an issue of sanitation since some people still bathe in the horribly polluted Ganga or Bagmati. But still water that has been used, say in the kitchen—even if it is then processed through his reed bed filtering system and stored in one of the household tanks to use for watering the garden or washing the car—would be deemed ritually impure.

He says his mother wouldn't use the recycled water for a year after they moved into the new house. Then she came to appreciate the system and the recycled water, acknowledging that this was a way of honoring the value of water. Roshan thinks the water recycling he is practicing will require even more time and education to be accepted in South Asia than in the West because of ritual prohibitions. Westerners just want to know that water is safely sanitized.

Pioneers can lead the way, but governments have to follow to institute new systems. "I started this practice in 2002. It's already 2009. I demonstrated that it can be done, but still very few people are doing it. The ideas are not incorporated in any sort of planning. If Kathmandu did this we could save more than 50 percent of our water. We're always behind big projects instead of the small ones. The two should go together."

I asked Roshan if my fantasy of switching Delhi to eco-san toilets could ever work. "Probably not," he said. But cities like Kathmandu and Delhi will have to set up onsite sanitation, a decentralized sewage system. That way people could keep their water toilets but would greatly minimize the amount of water used. The toilet would use recycled gray water that, after it was flushed and became black water, would go to a septic tank or be used in a biogas system.

This will still require an investment from each household, along with breaking old habits. How to motivate the people to accept the changes? "Water scarcity will demand change," Roshan said. People here will realize they can save money, save time, and even avoid the stress created by Kathmandu's chronic water shortages. "And then other values have to be brought out," he said. "People have to be proud they aren't polluting their environment, polluting their rivers." But all this has to be supported through regulations, policies, and incentives. "Maybe the first generation will need rewards and punishments and education. The next generation will do it automatically." Roshan said he does what he does because he's a scientist; but for others there has to be an organized effort.

Changing attitudes about ritually polluted water will be challenging enough in this part of the world. But everywhere in the world our feelings about feces may have to change. "The way we treat our shit—that it is useless, and should not be touched—we need to switch that mentality," says Roshan. "Shit and urine are useful, like people used to think at one time." Maybe it would be possible to motivate people in water-stressed cities to deal with their own sewage by assuring them of an adequate water supply. Make people happy with a regular, predictable supply and then maybe they will be more cooperative about sewage.

When I first started asking questions about water supply and sewers in South Asia a few years ago, I did so on the assumption that aiming for what is frequently called a "24/7 system" made sense. It had never occurred to me that there was any alternative. A "24/7" system simply means there is water in the tap all the time, under pressure, as American city dwellers expect. But only in rare instances is water in South Asia available in the lines all the time.

Throughout South Asia, water may arrive perhaps once a day or once every few days. A resident must be at home to open the tap and fill up his or her storage tank when the water is sent down the pipes to a particular neighborhood. Tanks are typically round, made of heavy reinforced plastic, and hold about a hundred gallons. Such tanks are generally on the roof of the house. The homeowner uses an electric pump to get the water up to the roof, then gravity takes over and supplies water to the bathrooms and kitchen until the tank is depleted. Thus the house itself has a 24/7 supply, but only as long as there is water in the tank.

The reason I had water in my tap in Kathmandu was that my landlord made sure his underground and roof tanks were filled regularly, through a combination of city supply and water tankers. The water the tankers delivered may have come from springs in the nearby hills or it might have been pilfered from the city system or through an illegal groundwater pump. But my landlord paid someone who had fetched it from someplace to bring it to the building. I was puzzled when I first lived in Kathmandu and saw my landlord advertising twenty-four-hour water, hot and cold. In such an attractive house wouldn't twenty-four-hour water be a given? Eventually I realized why he made it a selling point. And then I came to understand why South Asia should not even aim for a 24/7 water supply: there simply isn't enough water for a Western 24/7 system, with flowing water in city pipes at all times. And as leakage is anywhere from 30 to 70 percent, according to some guesses, you wouldn't want to be sending water down pipes all the time in any case.

Another Western assumption that is finally being questioned, even in the West, is that water in the pipes must be drinking quality. When such water will be flushed down toilets to pollute a river, the concept starts to look absurd. So what is the best solution for this part of the world? Is it different from that in the United States and Western Europe? Don't bring the water up to such a high quality? Perfect sewage treatment, then improve water quality? But if water is short already, it makes no sense to aim for a system that will send any of it down a toilet.

Sustainability-minded progressives in California and other places are treating the clean water in their taps carefully and recycling some of it. Their low-flush toilets are still using drinking water, but water from kitchen sinks and washing machines is sometimes diverted to the garden. This water could be used in toilets, as in Roshan Raj Shrestha's eco-home. Gray water systems in homes are often not legal yet in the United States because regulators are a little behind on this agenda. But that will change in time. New buildings are putting in double sets of pipes so that black water from toilets is not mixed into a big soup that all has to go to sewage treatment. Instead, gray water can be used again with a little filtering. And the black water can, after treatment, perhaps water golf courses.

As water shortages in the United States increase, systems like this will also. But Americans are never going to want to give up the water in the tap that is safe for drinking, for washing vegetables they can eat raw, or for showers where they don't have to fear a little water slipping into their mouths. This American certainly does not. It's what I looked forward to on my summer visits back to California.

Tap water that some of my friends don't consider quite up to their standards without a device to filter out tiny traces of heavy metals and chlorine is to me ambrosia. I head for the tap soon after my arrival and drink deeply. I make salad after salad in minutes, without having to soak lettuce in iodine for half an hour then wash it again with bottled or boiled water. I won't go on about the glories of this system, which Americans have, for generations, considered a God-given right. Wonderful as it is, it will have to change. The fact that our water is already drinking quality and sewage systems work—so that for the most part water bodies are not polluted— does not mean it's acceptable to continue to send Sierra snowpack into toilets.

Throughout South Asia, governments are still aiming to emulate the centralized solutions for water supply, sewage, and electricity that have so far worked in the industrialized world. But there is a buzz in South Asia about other ways of doing things. I find a lively debate on the subject when I peer into my e-mail every day. Decentralized sewage treatment, or as Sheel Raj Shetty says, finding ways to make "everyone responsible for his own shit," is the model there. Horrifying, you may say. Okay for Indians who have no toilets to begin with, you may say, but not for me. I want to flush it all away and send it to the experts, the modern bhangis who get well compensated for taking care of it.

But it need not be unpleasant. Roshan Shrestha and his family aren't suffering in their eco-home in Kathmandu. Perhaps homes in all settings may someday be much more self-sufficient than they are now, more like rural homes once were with their own wells and outhouses with soak pits. I was surprised at the number of organizations in the world working to spread some form of eco-san technology worldwide. There's an upscale version in South Africa, for example, that looks just like a sit-down flush toilet in a modern bathroom. Nothing about it would lead the "Western" user to feel ill at ease, and the householder doesn't have to interact with the mechanized process until months later when there is a bag of dry, odorless compost outside, at the end of the system. I was even more surprised to discover that in my own former hometown, Oakland, California, a few residents are experimenting with urine-diverting toilets.

Many forward-thinking environmentalists in India are proposing and experimenting with a decentralized model for both water supply and sewage treatment. These do not follow the American and European models and they avoid the expensive infrastructure and maintenance that the types of projects supported by the World Bank require. With the right kind of push from governments, new systems could be perfected and instituted in Asia long before they are in the West. Even though the thrust in Delhi is still how to make the Western water and sewage model work, in other parts of India there is more innovation.

Many urban people in South Asia have become accustomed to using Western toilets, and that may be pleasant for the people inside the clean bathroom, but not when that same person crosses the Yamuna River to the other side of Delhi. How can it make sense to build another dam in the Himalaya to supply water for toilets in Delhi that will then pollute the Yamuna and the Ganga? Clearly, imitating the West and using clean water in toilets is a terrible mistake in South Asia, even if they were all low-flow toilets—which at the moment they are not. Eco-san may be a stretch now, but when water becomes too expensive to waste in toilets, maybe it will become one solution.

People managing their own waste at the household or neighborhood level starts to look like the only sensible solution if sewer systems are not working in this part of the world, and more water is being dammed and diverted and shunted from far away to cities like Delhi where it basically contributes to the problem. Better still, governments have to decide that sewage is a resource and get busy finding ways to use it and save their rivers. The Yamuna and the Ganga will never be clean until this problem is solved. And if it is solved, that would likely mean there is the will to tackle industrial effluent and garbage too, which contribute greatly to river pollution.

Why pump raw sewage for miles to a sewage treatment plant, especially if the plant isn't working, the lines are leaking, and it will require tearing up a whole city's infrastructure? Is that the best way to do things in the subcontinent just because it has so far worked to do that in Tokyo or London or San Francisco? It may continue to be a workable system in some parts of the world. Right now in the United States we have both the technology and the money for sewage management, but there are a mere three hundred million souls pooping every day.

The water and sewage infrastructure in the United States is aging; sewage overflows have happened after extreme storm events in the United States. The public money that was available to build sewers and sewage treatment plants may not be available to replace them. And not all parts of the United States and Europe have even attained the kind of tertiary treatment that is considered a safe standard for releasing water back into rivers. Climate change may soon mean even less water is available in an already overextended system.

Perhaps the West and Asia will meet, striving toward the same middle ground where eco-san toilets, biogas units, and phosphate recovery from urine have become the norm for everyone.

# 3

# Delhi's Yamuna

"There is no water in our flat this morning, Mrs Puri."
"No, Mr William, and I am telling you why."
"Why, Mrs Puri?"
"You are having guests, Mr William. And always they are going
to the lavatory."
"But why should that affect the water supply?"
"Last night I counted seven flushes," said Mrs Puri, rapping her
stick on the floor. "So I have cut off the water as protest."
She paused to let the enormity of our crime sink in.
"Is there any wonder that there is water shortage in our India
when you people are making seven flushes in one night?"
—William Dalrymple, *City of Djinns*

Hell, no. I can't stand politicians. A politician is someone who
promises you a bridge, even when there's no river.
—Gregory David Roberts, *Shantaram*

As challenging as I often found living in Kathmandu, my home base as I explored South Asia, I felt as if I had left a safe retreat and entered a maelstrom when I flew down to Delhi. By January of 2010 I had visited several times and had learned to negotiate the immense city to some extent. Still, the morning after I arrived, just the process of obtaining a SIM card for my phone sapped my energy. The tall man at the tiny general store near my guesthouse was quite pleasant. He apologized several times for all the forms I had to fill out to purchase a prepaid SIM card: the government was requiring even more documentation since the 2008 attacks in Mumbai. I had to go back to my room for a photo, then find a shop to make photocopies of my passport and visa. Then the store owner had to call the guesthouse owner, whom he knew, to vouch for me. I made a small mistake on one of the forms and had to start over.

Yet this was only part of what taxed my equanimity. As I stood on the sidewalk at the shop counter negotiating the details, the store owner dexterously handled a couple of dozen other customers who crowded up on both sides and behind me to demand soap, toothpaste, recharge cards for their cell phones, potato chips, or little plastic pouches of milk. The man yelled for one of the boys who worked for him to bring a desired item from a shelf in back and bag it; he fanned through little stacks of recharge cards to find the right one for a customer; he took money and made change and still kept the process of my SIM card moving along, however slowly. Most of the other customers were yelling too, some of them quite close to my ear. I don't understand much Hindi, but these were not angry exchanges. It was just business, just the pace of Delhi. I felt as if I were in a casino, everyone clamoring to get in the game, get the dealer's attention. "Chill," I mumbled to myself. "It's just milk. It will still be here a couple of minutes from now." But this is Delhi.

Many Indian leaders want this to be a world-class city. "World-class" probably needs definition, but let's assume we all know what the term means. One thing it means for Delhi is a metro system, which was in the planning for several decades before construction began in the 1990s. As I ride around the city in three-wheeler auto-rickshaws, forever confused about where I am even though I've ridden on the same streets many times now, I see the construction of the new metro. The signs at road diversions read Larsen and Toubro, a company that was founded by Danes but is an Indian company from Mumbai. The first stage of the metro came in on budget and before schedule, an impressive feat.

Early in 2010 the city's usual frenzy was heightened by the intensity of preparations for the Commonwealth Games, which Delhi would host in October. India snagged the games, a kind of Olympics that Great Britain and all its former colonies hold every four years, by offering $100,000 to participating countries, along with airfare and lodging for their athletes. The Delhi Development Authority built a "village" of high-rises to house those athletes on the east bank of the Yamuna River, near the Nizamuddin Bridge.

Several years before the games, I first traveled over this bridge to East Delhi to meet Manoj Mishra, director of Yamuna Jiye Abhiyan, the Yamuna Forever Campaign. Manoj retired in 2001 from the Indian Forest Service and helped form a small environmental group. In 2006 he and his colleagues saw that Delhi's lifeline, the Yamuna, urgently needed attention. Its sad state was not improving, even though it had been the focus of several rescue efforts. Manoj and his group wanted to make some visible progress in restoring the river by the time of the 2010 Commonwealth Games. For them, a world-class city should boast a clean and protected river. "I did not know at that time," he says, "that these very games would become our enemy number one. Not our enemy, but the enemy of the river."

When he decided to take up the challenge of the Yamuna, Manoj began to educate himself about rivers and floodplains. He soon came to believe that focusing on the water that flows in Indian rivers was, ironically, at the heart of the country's failure to protect them. "This is the biggest tragedy of rivers all over the world. We are looking at our rivers only as channels of flowing water. Which they are, but the rivers are much more than that. And the single-minded obsession of everybody—including the state, the courts, activists, everybody—has been the flowing water in the river."

Manoj has a chiseled, intelligent face with high cheekbones and an aquiline nose. A trim man, who looks as if he has not spent all his time behind a desk, he has an air of gentle, if sometimes weary, authority. He has piercing deep-set eyes behind square glasses. His well-trimmed beard and mustache are almost white, while his wiry hair, a thick shock of which sits above his wide brow, is still salt and pepper.

He told me what he had learned in his trips up the Yamuna. "The river is a very dynamic system. It's not a single entity from where it starts to where it meets the sea. The river in the hills, the river in Delhi, in the plains—it's very different. And until we can appreciate this difference, that this is the river in summer, and in winter, and during floods . . ." As he lists these, he moves his hands from one foot to two and then to five feet apart. "Which river are we talking about?"

Manoj doubts that efforts to revive the Yamuna in Delhi will get anywhere without an emphasis on the river as an ecosystem. And that means, among other efforts, keeping cities and industry from encroaching on the floodplain. While almost everyone else was talking about the sewage in the Yamuna, Mishra's campaign was focused on the alarming amount of construction going on in her floodplain. "You see there is a technology to clean the water, but there is no technology if the river as an ecosystem is gone."

Manoj obtained documents under India's recently enacted right to information laws and came to believe that the construction of the village for the Commonwealth Games was illegal. His group started writing letters to the government, pleading for the site to be changed. "Please try to understand," they exhorted officials, "do not kill this river in the name of a ten-day event." The village of high-rises built for Commonwealth Games athletes was not the only intruder on the flood-plain. Delhi's new metro situated a station and maintenance yard on low-lying land near the river. Yamuna Jiye Abhiyan protested this construction too. The group made a presentation to the government of Delhi—a zone like the District of Columbia, called the National Capital Territory. "They have said we are right but no one was willing to do anything," said Manoj.

Thus for several years the shrinking Yamuna floodplain became the center of a political battle over how to develop India's capital, a city of close to sixteen million that has been growing by half a million a year. The advocates of development seem to have won the battle, but Manoj Mishra and other opponents spent three years in court trying to halt the construction of the athletes' village. The case was moving through the courts when I first visited him. Yamuna Jiye Abhiyan later won the case in the Delhi High Court, but India's Supreme Court overruled the judgment on technical grounds. By then, construction of the high-rises had already begun.

Manoj said the lieutenant governor of Delhi's National Capital Territory later promised him there would be no further construction in the floodplain. By then the metro and games facilities were almost complete. High-ranking government officials wanted the project to proceed, not just because of the prestige of hosting the Commonwealth Games and presenting Delhi as a world-class city, but because developing the Yamuna riverside is lucrative and takes advantage of prime land near the center of Delhi.

Delhi, like that subtropical American city, Los Angeles, has a lot of cars, asphalt, and palm trees. It's a burgeoning city in a semi-arid terrain, sprawling into far-flung suburbs as people seek space to live and work. The Yamuna-side development is one of the city's few opportunities for infill, and for what would be, were the Yamuna not a cesspool, a pleasant riverside location. The only other opportunity for development in the heart of the sprawling metropolis is to evict people from slums, which has happened at the Commonwealth Games' site and in other parts of Delhi.

In response to outcries over development of the Yamuna's floodplain, Delhi's chief minister, Sheila Dikshit, countered that all major cities have used their riversides: "Show me another city in the world which has not developed its riverbanks. Development has to take place." Critics point out that the Yamuna is not like the Seine or the Thames. It's a South Asian monsoon river, prone to spectacular and unpredictable floods.

Manoj Mishra warned that developers were inviting disaster, noting a flood in 1995 that inundated the very spot where the village was under construction. Sheila Dikshit said that the Yamuna does not flood anymore. But she added that in any case, "we have a system. The moment the river rises above a danger point, we evacuate people immediately."

Evacuation wasn't much of an issue in 1995 and earlier. This was all agricultural land. South of the site where the athletes' lodging was under construction, and just below the Nizamuddin Bridge, a small group of farmers was encamped when I first visited Manoj in early 2008. They were six months into what would become a two-year protest, registering their opposition to the construction of the athletes' village through an around-the-clock sit in.

We parked on the side of the road a few yards from vehicles speeding toward the bridge to go into central Delhi, then walked down the road embankment onto the floodplain. Six men and one woman, most with graying hair, were seated on rugs laid on the ground under posters with grainy black-and-white photos of Gandhi, the original nonviolent protestor. These middle-aged activists grew up in villages scattered around what has become urban East Delhi. All their lives they had farmed land in the floodplain—land that at the time of independence India's new government leased to their families. The lease agreement stipulated that Delhi could reclaim the land for public use; otherwise it would remain farmland.

The farmers told me, through Manoj, that the nearby Commonwealth Games construction had caught them by surprise. They believed the government would never allow this kind of construction in the floodplain. Baljit Singh, one of the farmers, said: "We are all villagers. We are the population of Old Delhi." Singh is one of the few who are still farming here, on the other side of a nearby embankment. "My land is there. We grow vegetables, wheat, fodder. We sell it in the market near here." But other farmers were evicted to make way for the sprawling games village.

Singh said the farmers in the Yamuna floodplain had agreed to cede the land at any time for public use. When the Nizamuddin Bridge was built, the farmers readily surrendered the land. Some years later, when a large Hindu temple called Akshardham was to be built, the government did not talk to the farmers first. It simply removed them. Singh and the others believe the temple was not a public use of the land; nor is the Commonwealth Games housing, which was slated to become expensive condos for urban dwellers after the games.

When the farmers started their protest, construction of the Commonwealth Games village had not begun. By the time they quit their protest two years later, a cluster of modern ten-story buildings rose from the floodplain, towering above the remaining fields of vegetables and flowers. The 1,160 units in these high-rises built for the athletes went on the market more than a year before the games were to begin. The web promo, with animated cartoons of people strolling on the grounds and photos of elegant interiors, promised that residents would be "insulated from the noise and frantic pace" of Delhi. It's a "landmark address in the heart of Delhi." The "wonderful riverfront" is another selling point. Whether this last bit of marketing was hopeful or ironic or blatant misrepresentation, I can't say. Prices for the condos, which have from two to five bedrooms, started at three hundred thousand dollars.

Manoj said the temple called Akshardham was no less an enemy of the Yamuna than the secular invaders. Both the temple and the Commonwealth Games village are private enterprises. But they can be billed as quasi-public uses, Manoj said, which makes opposing them much more difficult than it would be to oppose a simple, for-profit development project. Akshardham opened the door. After that it was impossible to stop the Commonwealth Games village because it could no longer be maintained that the land was meant only for farming.

Crossing the Nizamuddin Bridge, I had seen the Akshardham Temple rising from the floodplain. An enormous structure, bigger than a coliseum and unlike the elegant Hindu temples I had seen in Nepal and other parts of India, it had become a tourist site. I decided to see this controversial intruder for myself—even though my preference was for ancient structures that would tell me about the past. Akshardham seemed like one that might tell me something useful about the present, about New Delhi.

Bal Kishan, the driver I had come to rely on when I visited Delhi, drove me to Akshardham late one Sunday afternoon. The father of two arrived at my guesthouse a few minutes early, as he always did. A slight man in his thirties, his hair is cut close to his head and he has a thin mustache. His large, round eyes showed worry as they always did, but he smiled quickly in greeting before he got down to business threading his way through Delhi's congested streets.

It was a busy day at the temple—a work holiday for people in Delhi, who tend to work half days on Saturdays but have Sunday off. We went east over the Nizamuddin

Protestors camp out to oppose construction in floodplain of Yamuna River

Bridge, past a new metro station. The temple's parking lot was larger than baseball stadium lots I had seen at home. At four in the afternoon people were still pouring in on foot and cars were lined up at the gate. Bal Kishan patiently eased the little silver car toward the lot. Nearby on the other side of a tall fence, work was underway, perhaps for another parking lot. Signs announced: *Parsvath Metro Mall. Opening Shortly.* A yellow crane bent and grabbed a mouthful of Delhi's red earth, then disgorged it onto a dump truck. Red dust rose like smoke. I thought I smelled clay.

Entrance to the temple was free. I joined the hundreds waiting in line to become part of the thousands milling around the acres of grounds. We were pressed between steel guardrails, women on the left, men on the right. We shuffled up toward the head of this first line for about half an hour.

Women and men waited separately, like at Asian airports, for a security check. There were six lines of each gender. Everyone was polite at first, then as we got closer to the head of the line the crowding started. Ten women squeezed past me. At the top of the line, two men held ropes in front and behind a cluster of people, pinning in the next group that would be let loose to hurry to the security building and line up again. Finally we moved through the building where the security staff checked bags.

The main temple was closed that day because some sort of work was going on inside. But there was plenty to see outside. An elaborate frieze wrapped all around the base of the temple. Hundreds of elephants about the size of bulls and other animals closer to their normal life size enlivened the bas-relief. Fables and parables and words of wisdom were inscribed on signs beneath the tableaux. I tried to remember some of them when I came back to the car, because we weren't even allowed to write inside the temple, let alone take photos or carry a water bottle. No phones, no cameras, not much of anything was allowed inside except for wallets. I had a little notebook in my passport purse and the lady guard made me promise I would not write in it if she let me keep it. A writer without my notebook: I would not have gone in if I had had to surrender it. I promised not to write anything.

One of the stories depicted told of a hare inviting an elephant to its house, but the little creature worried the elephant was too big for his dwelling. The moral: a heartfelt emotion reaches God. Another story about a goat or sheep that gets a ride on an elephant demonstrated how we can all join in greatness if we're sincere.

After I circumambulated the temple, I wandered through the grounds. Series of ornate arches encircled fountains and courtyards. At the back of the compound was a food court that was itself as large as a good-sized temple. It too had carvings and decorations, though not as elaborate as the temple's.

The temple is made of sandstone indigenous to Delhi, like the materials used in the city's Mughal landmarks, mosques, and forts. I found myself wondering if somehow this modern-day temple was meant to rival, even supplant, the Taj Mahal down the road in Agra, built by a Mughal emperor, a Muslim. Indians celebrate the Taj Mahal as a national treasure, its most renowned monument.

Akshardham was built when India's highly nationalistic and Hindu-fundamentalist-leaning party, the BJP, held power. Akshardham's size, shape, and grandeur, along with the expanse of green, watered lawns surrounding the buildings, assert power, wealth, and solidity. But Akshardham lacks the grace and harmony of the famous Muslim king's shrine for his wife. I thought of the far distant Ellora caves in Maharashtra, also grand and Hindu. They have elegance and proportion and they required astounding feats of engineering and sculpting. But they are not so accessible to most Indian Hindus. This Hindu theme park is.

Manoj Mishra is a Hindu, and he believes his work on behalf of the river is not his alone. "I have this firm conviction that somebody is using me as a medium, and if I think that I am doing something I am mistaken. It is some power somewhere who has chosen." But he thoroughly disdains Akshardham for invading the floodplain and taking the farmers' land in defiance of traditions established at independence. Manoj said he had never been inside the temple. He felt its construction was calculated to open the door to real development, to the high-rise housing where Delhiites with money will live when the games are finished.

"The land was given by the state. It was a private trust strongly supported by the then head of the state, the BJP government. So the BJP government started the process of colonization of the riverbed through creating this so-called temple. And the current government is taking that as an excuse for creating this residential colony. Akshardham. This so-called temple has become the excuse for everything that has happened. I will not go."

Manoj also warns it's at risk. "This area where the so-called temple stands was under deep water in the 1995 floods. These constructions have not yet seen a de-cadal flood, that's why they think they can get away with it. Ultimately it is the river that will decide."

The day after I arrived in Delhi in January 2010 and managed to get my SIM card happened to be a holiday, Republic Day.

The center of the city near India Gate was virtually shut down, cordoned off to all but VIP vehicles. This holiday was different from India's Independence Day, which celebrates India's freedom from British dominance. Republic Day memori-alizes the establishment of the state and its constitution.

I thought of going to see the parade. Sandeep, the owner of the guesthouse where I stayed, told me it would be almost impossible to see anything; better to watch it on TV, he said, since I didn't have a ticket and the roads would be mobbed anyway. If it were a bright sunny morning I might have risked it, if only to see the crowds themselves. But it was foggy and cold, so I stayed in my room with WiFi and cable TV and hot tea.

The dignitaries on the TV screen were wrapped in coats, scarves, and hats as they watched the precision marchers in their red turbans, green coats, and white trousers, swinging their left arms hard and fast, a full 180 degrees. India's military might was on display. Shiny green tanks rolled down the well-paved street. March-ing soldiers followed, then members of the navy, a band in red coats. India Gate loomed above the marchers in the fog.

After the military display came the floats. The children in the crowd bright-ened, and the dignitaries looked more interested. Floats showed daily life: eating, cooking, indigenous dancers from all the states. One float featured bamboo pipes being used for irrigation: sustainable water use, right up my alley.

"Muscle flexing," observed one Indian commentator during the display of mil-itary might. I thought of American Independence Day—the nation's birthday, which just happens to be my own birthday. The USA has plenty of other holidays for its muscle flexing, so July 4 is more for county fairs and fireworks, a kind of patriotism, an assertion of softer American traditions. It's a good time for a back-yard party with salads and grilled chicken.

For many Americans it's a day for patriotism; but for many others, it's a day off, as it seemed to be for the young people I later saw strolling around Delhi nibbling snacks. For others, in both countries, it's just another workday. I don't know if Bal Kishan was paid for the morning of work he missed. By two that afternoon the roads were open again and Bal Kishan arrived to take me back over the Nizamuddin Bridge for my second visit to Manoj Mishra at his office in the colony known as Mayur Vihar, a name that means the dwelling of the peacock.

We drove by a new metro station on our way to the condo that Yamuna Jiye Abhiyan uses for its offices. Since my previous visit the metro had extended a new line, which passed just behind the group's building. Outside on the landing I caught the smell of sewage from a drainage channel that also passed near the cluster of apartment buildings. Manoj said this channel was an offshoot of the thirty-foot-wide Shahdara drain, one of the main culprits in the pollution of the Yamuna.

Indian engineers were planning to intercept the sewage that passed into the river from this drain and another large one on the other side of the river. Some maintained that capturing and treating these two sources of sewage would take care of half the pollution that now reaches the Yamuna. The drainage canals are storm drains, not intended for sewage at all. Until last year the one by the building did not have sewage in it, Manoj said. He thought the channel was blocked now because of the construction for parking lots at the new metro stop.

Yamuna Jiye Abhiyan's address was East Delhi 91, 178F Pocket 4, Mayur Vihar Phase I. The address says something about Delhi's vastness, and even with all those details Bal Kishan had to stop and ask around to find it. As we drove into the block of buildings, I lost count of how many virtually identical concrete block, four-story buildings were in the enclave. Each block had four numbers, each numbered building had eight units.

I didn't see any of the area's surviving peacocks or peahens, but pigeons, the emblems of big cities, took brief flight and scrabbled around for perches on the buildings. Laundry was draped outside some of the windows. Small cars and motorbikes clustered on the pavements; there were air conditioners in some of the windows. To some people it might look like a big-city tenement, but it's not. It's a middle-income colony. Each apartment has a different style of window, door, and gate, indicating different owners.

In the two years since I had met him, Manoj had shifted the emphasis of his campaign to focus on the fresh water taken out of the Yamuna before it reaches Delhi. "What we have finally come to is unless this river has its own water flowing around the year there is no revival for it. They may do all the sewage treatment that they want, but they cannot revive the river until and unless there is a flow."

Far too much water is taken out upstream from Delhi, Manoj says, mostly for agriculture in the states of Uttar Pradesh and Haryana, which surround the city-state of Delhi. About 150 miles upstream there is still plenty of water in the river.

The depleted river is then so overloaded with the city's waste it becomes a sewer. "Sewage treatment is part of the solution," he continued, "but we have not even halfway accomplished that. If sewage were the problem the river could take care of itself with full flows. When you rob the river of all its water why are you calling it a river at all? It's not a river, that's the tragedy. The only secret to river survival is fresh flow."

Gaining approval for, funding, and building interceptors to capture sewage instead of letting it flow into storm drains is no small task in Delhi's water and

sanitation bureaucracy. But restoring the Yamuna's flow is an even thornier political issue because it involves the two powerful Indian states that encircle the National Capital Territory.

Water from the Himalaya passes through Haryana before it reaches Delhi. The
canal that siphons water to send to Uttar Pradesh's farms is also upstream, in
Haryana. Different parties are often in power in these adjacent states of Uttar
Pradesh and Haryana, complicating agreements over water. But Haryana also
shares the Yamuna after Delhi's pollution has overwhelmed it, because the state
surrounds Delhi on three sides, including downstream.

"If Haryana's government sees any additional water as just drinking water for
greedy Delhi, it will always be reluctant to give it. But if it's to revive a sacred
river, which its people share and revere, they are bound to help," Manoj believes.
"But there needs to be political will, and it has to come from the top." Sonia Gandhi—the chair of the Congress Party and mother to India's "crowned prince,"
Rahul Gandhi—should be at the table. And with her the chief ministers of Delhi,
and of Haryana, along with India's minister of water resources and its environment minister, said Manoj. If such powers were assembled to thrash out the issue
of the Yamuna, then something might happen—if they could find a way to do it
without upsetting any one of the parties involved in the water dispute too much.

"Slowly, slowly," said Manoj, "we are getting the people in power to understand
this issue."

When I visited the graying protestors on the floodplain in 2008, the only woman
in the group said: "Yamuna is like a mother to me. I eat what she gives." As Bal
Kishan drove us back across the bridge two years later I thought of her as I saw
bright, fresh vegetables—red and white radish, green cabbages—for sale near the
bridge. Perhaps they were produced by farmers still growing crops on the east side
of the river.

Crossing the floodplain—if I did not turn to my right to see the Commonwealth
Games' "work in progress" signs—I would not know I was in Delhi. Green fields
scattered with trees spread out before me, the same landscape I might see from a
train traveling across the Gangetic plain anywhere between here and Bangladesh.

Turning back, I see little clusters of corrugated metal lean-tos—hastily assembled dwellings for workers building the new New Delhi. A half dozen concrete
mixers sit idle on the dusty red earth of an empty lot nearby, their rough white
bellies still instead of revolving.

A new water treatment plant was built for the high-rises in which athletes were
to live during the games and Delhiites afterward. The water for the 1,160 condos
will come from deep under the Yamuna riverbed. But the water under the high-rises
might not last many more years. This very encroachment into the floodplain shrinks
the recharge zone for the city's groundwater. Then the water table drops, requiring
deeper and more expensive drilling of wells and pumping of water. Ultimately, the

more the groundwater is sucked up and recharge zones are paved over, the more Delhi will depend on water shunted down the Ganga Canal from the Himalaya. Right now it's about 50-50, half from the Himalaya, half from the Yamuna and groundwater.

Baljit Singh and the other farmers I met near the beginning of their two-year protest said that even though they were losing their right to farm on the land, they wanted to protect the river. So much construction would compromise the already shrinking groundwater supply for the city. "It's more important to protect the river, and that's why we are staying here. We have been protecting this land. We have been keeping it green. This is not land, this is water body," said Singh. "We are here to safeguard the riverbed."

Singh and the other protestors who lived with the Yamuna all their lives have a sense of responsibility toward the river that has become rare on the subcontinent, perhaps especially in a city like Delhi. The battered Yamuna floodplain is the most damning evidence of this. And scattered throughout Delhi are the remnants of a water system whose functional and aesthetic potential have been at best ignored, at worst demolished.

Nitya Jacob—a journalist who writes about water and oversees an online forum for progressive ideas about water management—showed me some of them. One day we took a walk around Hauz Khas, a body of water that covers more than a square kilometer of busy south Delhi. The lake was in walking distance of the guesthouse where I stayed during many of my visits to the city. I had seen signs for Hauz Khas, but not knowing the name means "royal pond" in Urdu, I didn't visit it until after I had read Nitya's book, *Jalyatra*. His book is a "water pilgrimage," chronicling Nitya's visits throughout India to find what remains of traditional ways of harvesting the monsoon and distributing it through the year.

A reservoir was first constructed at the site in 1295. It was a source of water for irrigation, a source of fish and fowl, and a place for the nobility who built the two-story open pavilion at one end of the lake to escape summer heat. On a summer day the lake can still cool the hot breezes that blow in Delhi. Muslim rulers created many such large artificial ponds or tanks called *hauz* in the region.

In the past Hauz Khas would have been fed by small seasonal streams pouring into it during the monsoon; now it's fed by a sewage treatment plant several miles away. The sewage is close to raw when it leaves the plant; the journey via an open channel aerates and cleans the water to some extent, but, as Nitya observes, the water we see as we stroll beside the lake is a "bilious green." Still, it's a reuse of water, and provides a lake for an urban park.

We looped around the lake, following its meandering curves. Surprisingly few people were enjoying the peace of the tree-filled park the day we visited, even though it is such a welcome refuge from the hubbub of Delhi. A few young couples escaped the prying eyes of parents; a large flock of geese honked noisily.

Nitya says the bottom used to be paved with stones that would keep water in the lake yet allow it to percolate through the ground, cleaning the water and feeding the groundwater table. But now the bottom has been cemented so the water remains stagnant, draining into the Yamuna during monsoon overflows. If the wastewater were treated better, it could be allowed to recharge groundwater even now, and be usable again.

Archaeological evidence shows there was a city near the Yamuna as early as 300 BCE. Evidence of water harvesting and management reaches back at least to 700 CE. Many of Delhi's ponds and reservoirs were natural depressions that filled up in the monsoon, Nitya tells me. Small dams were all that was needed to manage the water and make it available year-round from ponds and wells. Until Delhi's rapid expansion in recent years, this region was full of ponds that stored excess monsoon water and whose seepage kept the groundwater level high so that shallow wells were enough to supply residents with drinking water.

Delhi had seemed brutally dry to me except when rain was falling. But Nitya says the area is not really dry. "There used to be forests in abundance. It's become dry because of the amount of concrete in the city now. Delhi isn't short of water. It's more a question of distribution—and waste," he adds.

Serious rainwater harvesting from the roofs of all the buildings that now carpet Delhi's landscape could bring the water table up to boost the city's supply. But so far there's little talk and less action in this regard. Nitya notes that in the ample monsoon of 2010 the groundwater table came up thirty feet. With the right kind of conservation efforts, Delhi might depend on its groundwater instead of needing more water from the Himalaya.

What would push the city toward action? I ask. Nitya chuckles. "An upper-middle-class crisis, where they don't get water. Or if they have to pay ten times as much as they do now."

Another day I took the new metro to the outskirts of Delhi to meet Nitya in Mehrauli. Once a village, Mehrauli has now been swallowed up by Delhi's sprawl, though it doesn't yet enjoy any city services. There's a scattering of trees here, but I can't visualize this unplanned urban mess as the swath of mango orchards it once was.

The graceful Mughal-era pleasure palace Jahaz Mahal remains; it was once surrounded by a cooling moat. Nearby is a water tank called Hauz-i-Shamsi, fenced in now to prevent its use as a toilet. The rectangular tank stays full of rainwater year-round, Nitya tells me, though it's not as large as it once was. Some of the land it covered has been claimed for buildings, the same fate that has overtaken hundreds of Delhi's old water tanks. The fence doesn't keep garbage out, it seems. One end of the tank is a pea-green soup of algae and refuse. We see more garbage, along with a dead dog that has been dumped here, as we walk around the tank outside the fence.

People in urbanized villages like Mehrauli don't seem to want to pay for garbage collection. Nitya tells me that developers may even encourage locals to use these ponds as dumping grounds. After a few years of that abuse, it's easy to dismiss water bodies as garbage dumps and proceed to improve them by building on them. In the past, social codes kept these areas clean for the sake of all who used the water source. Now wastewater from the nearby apartment building drains into Hauz-i-Shamsi.

I ask Nitya why people abuse water bodies like this so badly. He pauses. It's a question he has wrestled with; one theory, he says, is that after government appropriated the delivery of water—which began with the British and continued after independence—the people who had carefully tended to their own water sources for generations lost interest. And they seem to have lost the ability to connect the dots—that keeping ponds like this clean would improve their chances of a clean and regular water supply.

Whatever caused the sense of sacredness or of public responsibility to deteriorate, Nitya blames India's increasing water crisis on the loss of five thousand years of traditional water management. Nitya says he felt humbled on his *jalyatra* by the knowledge ancient Indians had about managing water in this relatively arid but paradoxically water- and monsoon-rich land.

Traditional water management in India was a sophisticated, decentralized system that began to erode under British rule and eventually collapsed all around the country as alien methods were imported. There was such a wealth of methods of water management in India—adapted to specific locales—that centralized government could not easily take over and manage them efficiently. Now pumps and pipes, legal and illegal, drain India's once abundant underground stores of water and threaten the nation's continued economic progress.

Before I left Delhi I wanted to see one of its few intact stepwells. Stepwells, called *baoli* in this part of India, are just what they sound like: wells you can walk into. Different styles of stepwells can be seen all over the subcontinent.

Ugrasen ki Baoli was right across the street from the Fulbright office, but I never knew it was there until Nitya told me. Even then I had to hunt around for the right path that would lead me to where it was hidden behind the well-guarded gates of office buildings and homes on Hailey Road. When I climbed a short flight of stairs from the path to the top of the stepwell, I saw a stunning view: a subterranean construction of stone descending five stories lay at my feet, about sixty feet wide and two hundred feet long.

Full of monsoon rain, it would look like a long pond with a row of archways stretching along both sides. Only as the water level began to drop after the rains ended would the layers of lower archways gradually become visible. The stone stairway would take visitors down to the water as the level dropped.

A fringe of trees at ground level surrounded the stepwell; behind the open arches at the back of the structure I could see high-rises in the distance. I descended the wide staircase all the way to the bottom, since there was no water in evidence. Half a dozen young Indians were visiting the site too. A good sign, I thought. They are appreciating some of their heritage, though I guessed the magnificent fourteenth-century

Ugrasen ki Baoli, stepwell in Delhi

stepwell might seem an oddity to them. The walls on both sides at each level had open arcades, the archways opening into small rooms. These were places where one could bathe using a bucket of water from the central pond. Private bathrooms are the norm in Delhi now, not such a communal style of using water. But in the Indian country-side, partially clothed men and women still bathe at public water sources.

I peered down into the round well at the back of the edifice, meant to hold clean drinking water. The day I visited, in winter, I could see only a little filthy water at the bottom of the well.

Back up at the top, a young man who was serving as caretaker showed me a photo in his cell phone taken some years ago; it captured the stepwell filled almost to the top after the monsoon. But water doesn't stay for long anymore. Underground water competes with the new metro, so its builders have pumped out the cool subterranean store of water that once meant water security for Delhi.

Madhu Bhaduri is, like Manoj Mishra, a retired civil servant turned activist. Her graying hair adds a touch of elegance to her round face and gravitas to her youthful energy. In 2004 she had recently retired from the Indian Foreign Service, having served as ambassador to Portugal. She was doing volunteer work for a donor-funded organization called Parivartan, whose name means "metamorphosis." The group was chiefly concerned with exposing government corruption and helping Delhi's citizens combat it.

Madhu read a newspaper article about an effort the World Bank had launched in partnership with the Delhi Jal Board—the city's water authority—to reform part of the water supply system. According to the article, the World Bank had assessed the assets of the Delhi Jal Board—its pipe system, water treatment plants, and other facilities—at a figure that she called "shockingly low." She was alarmed because she thought this assessment might be the first step toward selling and privatizing the utility.

Seeking information from the Delhi Jal Board, she was at first refused on the grounds that the official files were not for public scrutiny. She took the issue to court. After five months she got a "mountain of papers." Another Parivartan member, a retired income tax commissioner, spent two months scrutinizing them. He found items that both confused and alarmed the group's members.

Madhu was dismayed to see no mention in the documents of sanitation. "So it wasn't going to deal with wastewater," she exclaimed. "The real financial burden of water is not in supplying the water but in dealing with the wastewater." Nor did she see any mention of the great inequities in Delhi's water supply, in which pipes in some areas of the city run dry while others have plenty. The neighborhoods where diplomats and government bureaucrats live, along with those that house the Indian Army, never have water shortages.

The agreement between the World Bank and the Delhi Jal Board stipulated that only multinational water companies, not Indian ones, might be selected to provide services. Consultants were to be paid $24,000 a month each—a figure that is not atypical for foreign consultants and engineers, especially in the start-up stages of large projects. In addition, the work proposed did not address the complete lack of water pipes in parts of the city where very poor people lived.

Parivartan also found out that various large, corporate, and government users owed large sums to the Delhi Jal Board—including the police, building contractors, and even companies that bottled the water for sale. "I've been trying to figure out about these water tankers," said Madhu. "Actually this is Jal Board water being sold by private companies. It's been treated; otherwise they could not sell it."

No wonder the public utility was floundering with such hefty uncollected bills, Madhu concluded. She discovered one customer had an outstanding bill for thirty million rupees, well over half a million dollars. Many owed more than a million rupees, more than twenty thousand dollars. "It's not the poor in Delhi who are cheating the taxpayers, but the rich!" she exclaimed.

Parivartan came to doubt that just hiring a foreigner—whose goal at least in part would be to make money—was going to correct the problems in Delhi's water sector. Then there was the problem of leaks. The DJB itself had estimated more than 40 percent of the water was lost this way.

I joined Suchi Pande one day outside the office of the Central Information Commission. A researcher and Parivartan volunteer, Suchi had gathered information about the various multinational corporations with which the World Bank was doing business.

Tall cotton trees studded with their strikingly red blooms mesmerized me as I waited for Suchi on a warm March day in Delhi. She and other volunteers had set up a table outside the Commission to help citizens who had come seeking information from public agencies under the Right to Information Act. They wanted to know why they hadn't been promoted, or what had happened to their pensions. The volunteers were here to make sure the petitioners understood their rights and to help them appeal if the Commission had not secured the information they sought.

Suchi finished talking to a petitioner, then started telling me why Parivartan had been concerned about the World Bank's methods. She said the bank seemed to have made it impossible for all but their chosen company to compete. Beyond that, the group's members were dubious about what any chosen multinational company would actually do.

"We asked them: 'OK, if this is a management contract what are the responsibilities of the contractors?' We found out it was not their responsibility to arrange for water to meet the 24/7 promise. They were not making any investments, they were not going to add to any infrastructure. Providing water was the Delhi government's responsibility," said Suchi. "The only thing they had to ensure was 24/7 availability at the district metering area. Whether water reached each individual household was neither their responsibility nor Delhi government's. That didn't seem like an improvement to us. It just seemed like they were telling them 'do what you want—you're the experts.'"

The Parivartan members were also bothered by the inequity of trying to provide 24/7 water for some neighborhoods when there were "areas in Delhi that don't even get two hours. Have you tried first seeing if people want to go from two to six?" asked Suchi.

Suchi said that Parivartan's objections to the project were not simply a blanket rejection of any species of "privatization" but were based on the inadequacies of the plan under consideration. "It was more on technical grounds—that this project was so flawed. And the fact that it hoped to address the water needs of the citizens of Delhi but they had been left out of the consultation process. We kept asking the government for a consultation, for time to talk about all this, and we couldn't get it."

In November 2005, about a year after Madhu Bhaduri began her campaign, she read another newspaper article. The government of India had withdrawn its loan application to the World Bank, possibly at the suggestion of the bank. Parivartan members were relieved and happy, and proud that Indian democracy had allowed them to prevail where in other countries water privatization efforts had succeeded. "I think what has been revealed is that in a democratic country it cannot be done," said Madhu Bhaduri. "It's one thing to do in a country without democracy, another thing in a country like India."

I spoke with Catherine Revels, the World Bank's regional team leader for the water and sanitation program, a couple of years after the Government of India's change of mind. Many critics still referred to the effort as an attempt to privatize Delhi's water system.

"Privatization means selling off the assets," Revels began. "It would mean the private sector owns the water assets." That was the prospect that alarmed Madhu Bhaduri when she read that the assets of the Delhi Jal Board had been appraised. "That rarely happens in the world," Revels continued. "The bank was working with the Delhi government on something called 'private sector participation.' So the government would continue to own the water assets and be responsible for water delivery. It would contract the private sector in to help.

"The term 'privatization' is misused intentionally by a lot of people who are anti–private sector," said Revels, "because it carries the connotation that the government is giving over responsibility for water to someone who is only focused on profit, and that would never be the case."

She added that the effort in 2004 was intended as a pilot program, focusing on only two of Delhi's twenty-four zones. The zones were in South Delhi, generally considered one of the more upscale areas of the city, but with both rich and poor neighborhoods, according to Revels. South Delhi, where I had been staying when I visited, frequently had water shortages. But there was never a shortage at my guesthouse because the owner of the converted family home had storage tanks and pumps so that he could store enough for his guests.

"They were going to try to improve service delivery by reducing leaks and instituting metering to get some accountability and equity about how much water was being used or wasted—and, frankly, stolen," Revels said. "People tap into the main transmission lines and do illegal connections. I have some interesting pictures that show illegal connections. They look like spaghetti coming off the main line." She said such illegal tapping not only gives some people free water, it promotes leakage.

Revels agreed with Madhu Bhaduri about illegal vendors siphoning water out of the DJB system and then reselling it. "What's happened is the service has not been provided through the formal provider, so you've got a lot of people who have stepped into the gap and they're making a lot of money off this. People are paying a middleman for something they would otherwise be paying into the kitty that could help provide a better level of service for everybody. A homeowner pays the city anyway. Then pays again."

Thus, between the industries that have grown up to fill the gaps in public water service and the omnipresent plastic bottles of so-called mineral water for sale everywhere, water seems to have been partially privatized here with or without a World Bank loan and a foreign management company.

Water equity in Delhi has been complicated by its rapid, unplanned growth. Slums and illegal settlements created "fundamental issues around water service and land tenure," said Revels. "Some of these things go way beyond what you think of as the water sector. How on earth do you get a handle on that?"

Under the project proposed for the World Bank loan, the city was going to try to begin to get a handle on some of it. It would "start to legalize what has been handled illegally. There would be a formal level of service for a fair price to people who sign up for the service." People who were making money illegally from the current system would not want to see that happen, said Revels. That combined with a widespread mistrust of government and misrepresenting the plan as "privatization" undermined the effort, according to Revels. "So it wasn't hard for certain groups to raise a lot of doubts about what was being proposed. You can make a nice negative campaign of something that is quite reasonable when you look at it on paper."

Explaining the requirement that the company chosen for the management contract be international, Revels said at that time no Indian company had ever provided "24/7 water"—continuous water supply service with meters and accountability. But by now Indian companies could probably qualify to handle the job because it would not require experience serving fifteen million people. "Delhi doesn't have 100,000 problems; it has a hundred problems a thousand times over," she said. "And that was the purpose of this pilot: to break it down into nice chunks and then roll it out, learning the lessons from the pilot so you could roll it out into the whole city.

"I think if the public sector can deliver, excellent," said Revels. "And if there are ways to improve public sector delivery, we work on that. But if the public sector isn't delivering and there seem to be too many obstacles to correcting its inadequacies—as there seemed in Delhi—people who are working in the public sector can be managed to deliver a better level of service. It's all about incentives. It's always the same people. You're not bringing in a different group of people to deliver the water service. Just giving the people already on the ground the right incentives."

Revels said the astronomical cost Parivartan estimated was erroneous. "They made a really funny calculation. If a management contract cost X amount per month for two zones in Delhi, then by the time you roll it out to 24 zones there would be millions of dollars going to the private contractor. But we know that's not how it works. There is a cost for bringing in experts. But it goes down over time as people get trained. As you bring on more business you don't bring on more experts from outside. In this way they were able to spread misinformation and scare everybody.

"What you see is that middle- and upper-income households who typically influence decision makers here are getting an OK level of service. They are paying more for it than they should, but they're self-providing; they're making sure they have underground storage tanks, overground storage tanks, pumps, filters, everything they need to get water. In my house, I take a shower every morning; it's not a problem. And middle- and upper-income families across India are doing the same thing. So they don't need to see a change."

One of the water experts I spoke with believed the World Bank and the Delhi Jal Board had not managed the public relations campaign for this project very well. They had been too secretive about what they were trying to do. For a project in another Indian city, the World Bank made hiring a public relations firm a first step.

During a later visit I stayed with friends in East Delhi, a complete change of scene from my South Delhi guesthouse. My friends lived in a typical East Delhi colony where residences consisted largely of separate, gated co-operative housing enclaves. Each co-op typically had several ten-story buildings each, with parking on the ground level and a small expanse of lawn. These co-ops, like the new high-rises built for the Commonwealth Games, were sucking up water from the deep aquifers near the Yamuna floodplain.

The ride back to central Delhi crossed the Shahdara storm drain, one of the odiferous channels carrying sewage to the Yamuna. The road passed through an expanse of barren ground. On a flat strip of land no more than twenty-five feet wide and about half a mile long were congregated an array of animals: dogs, cows, water buffalo. The enormous buffalo were lying on the ground surrounded by garbage. I could see no water for them to drink, and though there was a lot of garbage it would not have been enough to sustain them. These huge bovines must have been refugees from the remnants of rural life that still survived on the lowlands near the Yamuna. I had even seen dung cakes drying on a ledge, the blue roof of a new metro station gleaming nearby. The dung cakes would fuel a cooking fire. But how did the water buffalo reach this barren strip of land, cut off from the remaining fields by a canal full of sewage and streets with speeding vehicles? Were they waiting to be taken to a slaughterhouse?

I passed down the same road one night, thinking the animals might be gone. But they were still there, dark humps in the dim light. This was some shadow land, it seemed, not part of the world-class city some people wanted Delhi to be, but no longer part of the orderly agrarian life promised at independence for its outskirts.

Emerging from a daytime transit through this same area, Bal Kishan steered the car back onto a main road into Delhi. Roadside vendors sold flowers and motorbike helmets at the intersection. After a half hour in frenetic traffic Bal Kishan deposited me in South Delhi in a posh area called Greater Kailash.

Here I was to meet Deya Roy, a young Delhi woman who was working on a PhD in science policy at Jawaharlal Nehru University. She had recently completed a master's project on Delhi's water system, a project she chose because she wanted to understand why there was so much inequity in Delhi's distribution of water, and whether there might be a more rational way of distributing costs throughout the system to make it work better.

Deya steered me toward a restaurant where she ordered delicious kebabs and curry. A pretty young woman with a heart-shaped face and smooth shiny hair, she wore a stylish kurta. One might not expect an elegant young woman to be so passionately interested in a subject like the Delhi Jal Board. She was also passionate about dancing. Living in Delhi, which she loved, not only gave her the opportunity to study some thorny problems in public policy but to pursue dance in a multitude of forms, from classical Indian to Latin.

Deya grew up in an upscale area of South Delhi, but for the past twenty-five years her family's house had received water for only half an hour each day. Then the new Sonia Vihar water plant went into operation. It treats water sent down to Delhi from the reservoir behind the Tehri dam that I saw on my way to Gangotri. Now water is available to the Roys an hour each morning and another in the evening.

"Here in Delhi everyone thinks the rich get water and the poor don't," Deya said. "But in North Delhi, many areas that are actually very poor are getting water twenty-four hours a day because they are near the main line. While here in South Delhi, farther away from the main line, people get less." Illegal connections, like those the World Bank's Catherine Revels told me about—along with low pressure in the pipes—exacerbated the shortage in South Delhi. "People think the rich can fend for themselves so it doesn't matter—they'll buy water from tankers." Deya said her family had installed pumps and tanks and an Aquaguard—a high-tech water-purification device.

In the course of the research for her master's thesis, Deya visited the Delhi Jal Board's water treatment plants and found that they successfully treated the water to drinking standards. Customers near the main line in North Delhi could safely drink water from the tap because the pressure in the pipes kept it clean. The problem, she said, came in the old supply lines, which frequently leaked. Some of the water system in the city dated from the British Raj and would be very difficult to replace. Pipes were so deep it would be an immense challenge to reach them, and in any case the layers of the new New Delhi—Internet lines and auto flyovers—would make the task even harder.

Replacing the old treatment plants or the aging water lines would not only be enormously expensive but would invite even worse water shortages for a while, and thus would be difficult to manage politically. Nonetheless the city was attempting to replace some of the old pipes as it prepared for the Commonwealth Games.

"So is some form of 'privatization' the answer?" I asked Deya. "Outsourcing the management of Delhi's system to a company that specializes in urban water management?" "I don't know," she said. "If you take better care of the present system it can work quite fine. And I don't know that we require a 24-hour water supply. India is a water-intensive society—especially north India. People would wash their cars if they got extra water and probably give their balconies a proper wash every day. Twenty-four hours would not be a good idea unless the water is priced high."

Deya thinks the Delhi Jal Board charges too little, less than it actually costs to produce the water. Politicians seem to be afraid of making people angry if they raise prices, she said. She and other researchers have found that even the poor are willing to pay for water. In fact, she said, they are paying already for small plastic packets of drinking water because they know that the water from pipes in their homes or standpipes in common areas isn't clean enough. "If you give them a clean water supply for a few hours a day they really would *like* to pay. And we would like to pay," she adds. "The cumulative cost to my family for an aquaguard and other pumping and storing adds up."

Finally, the Delhi Jal Board needs reform. Its system for promotions is not based on professionalism, says Deya. She had seen that engineers and scientists were not receiving promotions while office staff got regular, automatic ones. So the incentives to do the job well were not rational.

When I was interviewing water experts in Delhi in 2007, I tried for several months to gain an interview with the Delhi Jal Board's CEO, Arun Mathur. I got many polite put-offs from his assistant and no interview. I later learned he had left the job under a cloud. The Delhi High Court ordered a two-week jail term for Mathur in November 2008 along with two of its head engineers. They were convicted for failing to prevent sewage from flowing into the Yamuna two years after assuring the court they would stem the flow. A law dating from 1974 prohibits untreated effluent in the river, a law so flouted it may as well not exist.

The angry justice criticized the DJB for its endemic corruption. The *Times of India* described the step as "extraordinary" and applauded it. The order for Mathur was stayed the following month. Later he became the director of the Finance Department for Uttar Pradesh, one of the states bordering the National Capital Territory.

In my quest to understand more about Delhi's water system—especially why so much sewage was still going into the Yamuna—I went to meet with an executive engineer in the West Delhi division of the Delhi Jal Board. Deya Roy had told me that he was part of a new generation of DJB staff, ethical and hardworking, which gave her hope the system could be sorted out and eventually provide Delhi with a safe and equitable water supply.

Dalbir Singh was delayed for our appointment by a meeting in another part of Delhi. I sat in his office, where an aide brought me tea and biscuits as the winter day faded into dusk and the noisy rush-hour traffic outside competed with music from a nearby temple. The office was in an old, one-story building, its floors bare and walls in need of paint. A thick slab of glass covered the big desk, which faced a few short rows of folding chairs, as if lined up for morning briefings. The insignia of the Delhi Jal Board was etched in the glass and a small Indian flag stood on it. A map of the West Delhi sector covered part of the wall behind the desk, along with a file cabinet piled with folders and binders. Singh hurried in and told his aide to bring me more tea. He apologized for being late; I reassured him I was not in a hurry.

Dalbir Singh is a handsome forty-year-old man, broad-shouldered, with a dark beard and mustache. He wore a turban, as most Sikhs do; his was a deep red that matched a dark red tie, both of which contrasted strikingly with his black pin-striped suit and white shirt. He is married to a teacher and they have one child.

Singh was professional and unpretentious and seemed comfortable with his authority. He was gracious and friendly to me, but had to decline my request to record him, otherwise he would have had to get permission through the public affairs office to continue the interview. But he answered all my questions without hesitation.

Singh had been working with DJB for ten years. He had trained as a civil engineer, then pursued an MBA in finance. He helped me understand how the provision of public services had gotten out of control in this vast city, where about a third of its fifteen million citizens lived in thousands of "unauthorized colonies."

Colonies in Delhi are roughly equivalent to what in the United States might be called neighborhoods or subdivisions. As Delhi's population mushroomed in recent decades, the government could not keep up with the growth. People were pouring in, looking for work. What Singh referred to as the "land mafia" saw an opportunity to make money by setting up housing on private agricultural land or government land. Though these neighborhoods may not have been legal according to the procedures of the Delhi government, they were filled with people whose votes mattered to politicians.

Politicians helped the citizens in unauthorized colonies get water, at first from water tankers. Then the local politician might approach the Delhi Jal Board asking it to connect the colony to its water pipes. Eventually the politician might ask the government of Delhi to "regularize" the neighborhood, to make it legal and eligible for all city services. Some of these areas might be slums, but more likely they are not. Many are filled with Delhi's middle class. And many have in fact been regularized, though thousands more remain illegal. Recently hundreds were regularized, which simultaneously gave the government the power to exclude and even tear down any further illegal construction in those neighborhoods.

This is just one aspect of the challenge the Delhi Jal Board has to meet. The challenges of DJB are the very same challenges facing Delhi: there is no way to control the population, there is no functioning master plan for the city, and there is increasing dependence on neighboring states for water. "Despite so many difficulties we are maintaining the system," Singh said with some pride.

And echoing what Deya Roy told me about the lack of incentives for workers, Singh indicated that the level of effort in a civil service organization like the DJB is inconsistent. "In government it is my choice whether to work. People don't have incentives—no rewards, no punishment." Twenty-two thousand DJB employees served Delhi's 16.5 million citizens, but not all of them did a good job. He said he had eight hundred people working just for him. Some of them came to work, some of them did not; some of them worked when they showed up, some of them did not. It could take years to fire an employee, because labor unions and civil service rules would protect a laggard or a good employee equally.

Meanwhile the good workers simply pick up the work others should be doing. "But our generation is trying. I feel personally that Indians have the capacity to do what's needed." That applied to the issue of bringing in foreign or any other external manager to correct DJB's inadequacies. Instead of hiring someone again and again to come fix things, Singh thinks DJB personnel should learn to do it themselves.

A customer once gave him flowers to thank him for his help, Singh told me. But he said he had to be honest with customers and tell them what he could and could not—would and would not—do for them. A religious man, Singh said he saw God as responsible for his having a good job. "Getting a paycheck for giving people water makes me happy."

I asked Singh about Delhi's sewage dilemma. The unauthorized colonies created problems here as well. Tankers might come and deliver water to them, but no tankers came and took their sewage away. Even after the DJB put water pipes in these areas, there were no pipes to take the sewage away. Providing water in the unauthorized colonies created sewage that had no place to go. So it went into storm drains and on into the Yamuna.

Singh said the unauthorized colonies accounted for about a quarter of Delhi's sewage. I heard from other experts that about half of Delhi's sewage—that is, the sewage from the areas with sewage lines—did not go to sewage treatment plants but into storm water drains and thence to the Yamuna. Then there was the 50 percent of the people in Delhi who lived in unsewered slums and neighborhoods. That waste, on the ground, was likely contaminating groundwater, Delhi's shrinking aquifer.

Delhi has seventeen sewage treatment plants, some built with aid from Japan, which has been very involved in Delhi's sewage for some years now—without much progress, according to some critics. If the needed connections were made, those plants could deal with about three quarters of Delhi's sewage. But currently there is unused capacity in them, because the sewage is going to the storm drains and the river, not into the treatment plants.

Singh said there had been a plan at DJB ever since he had arrived ten years earlier to build an interceptor that would capture the untreated waste going into the Najafgarh drain. Along with the Shahdara in East Delhi, the Najafgarh on the other side of the Yamuna is one of the largest sources of sewage. The waste would then be transported to sewage treatment plants. The plan was finally to be implemented in 2010 and would take two years to complete. In a gently ironic tone, Singh said that it could take a long time for the government to make a decision.

Critics of the plan say that even successfully intercepting the sewage from the large drains won't bring the Yamuna up to "bathing quality," as ordered by India's Supreme Court. Manoj Mishra thinks plans for sewage treatment as currently conceived probably won't succeed because the DJB is starting from a false premise, underestimating the size of the job it has to do. "They are saying that the total sewage the city is producing is 2,000 million liters per day; whereas the pollution control board says 3,800—almost twice as much. The DJB is making the calculation based on the water it supplies. But there is a huge amount of water consumed in the city from groundwater which is not supplied by DJB, and it too becomes sewage," says Manoj. "There is no regulation. They have no clue how many bore wells are in the city. If you are designing something based on a gross underestimate, then you are bound to fail."

A recent government pronouncement called for no sewage in the Ganga by 2020. That would have to mean no sewage in the Yamuna as well. "Can you wait ten years?" I asked Manoj. "The river cannot," he replied.

Work was underway twenty-four hours a day during the year before the Commonwealth Games—technicians and laborers were working on the metro, working on the electric supply lines. Power went off periodically.

One morning during my visit I saw an article in the morning paper about a homeless woman sleeping on a sidewalk with her toddler and two-month-old baby. They had been hurt by an out-of-control driver during the night. He was arrested for possible drunk driving. He had been coming home from a call center with a friend after finishing work. The homeless mother had come to Delhi to get work. The woman seems to have simply collapsed on the sidewalk in her exhaustion.

In many parts of Delhi in 2010—especially the central ones that tourists and visitors were likely to frequent during the Games—there were countless sidewalk homes. Working families had constructed little tents by throwing tarps over a set of U-shaped, bamboo supports. Other dwellings were even more humble, just tarps hung from the tops of an iron fence. In the daytime the "roofs" hung loose against the wall or fence; at night their owners pulled them out taut and sheltered beneath them next to the wall.

Nearby were signs: "Work in Progress, CWG—Commonwealth Games." Other signs were painted with big letters: PWD, for the Public Works Department. One sidewalk home had used a PWD sign for a wall. Some of these workers were digging up sidewalk pavement and spreading new concrete slabs during the day, then living on the sidewalks at night.

These workers were needed if the breakneck pace to get ready for the games was to keep up. What would happen to them when they were no longer needed? Would they be thrown off the very sidewalks they were beautifying so that visitors could walk down them and admire the gracious and green parts of Delhi that radiate outward from Connaught Circle?

I told Manoj Mishra and others I spoke with in Delhi how impressed I was by the political will being summoned for the Commonwealth Games effort—both the metro and the other work going on around the city. In response I would see an ironic smile, a knowing look on faces. "Urgency brings opaqueness," Manoj cautioned about the speed of the Games preparations. Regulations could be set aside, controls relaxed, and quality could suffer. "When regulation and accountability are set aside, corruption can walk in."

Still, if something as complex as the new metro, which involved retrofitting much of Delhi—tearing up streets, redoing the electrical grid—was coming along nicely, just how was it that sewage could be such an issue and the Yamuna still a cesspool?

Senior water specialist Ramaswamy Iyer smiled wryly. "It's just not a priority," he said. Delhi's Chief Minister Sheila Dikshit oversaw the building of the metro system and the switch from petrol to compressed natural gas for public vehicles to clean up Delhi's air pollution. Delhi's citizens would have to tell her they want real action on the Yamuna.

In 2010 the World Bank and the Government of India announced a new restoration plan for the Ganga, which the Yamuna joins at the sacred *prayag* Allahabad about 350 miles downstream from Delhi. The plan included a billion-dollar loan from the bank.

Two years before, India had declared the Ganga its "national river." Environmental researcher Gopal Krishna saw this as an alarming development. "I'm scared because when India declares something its national whatever, it will kill it!" This is a familiar phenomenon to Americans. The bald eagle, the symbol of the United States, was only recently removed from the endangered list. The grizzly bear was declared California's state animal only after all the bears in the state had been shot. Landmarks throughout California took Indian—Native American—names after most of the tribes had been massacred.

Gopal pointed out the contradictions in the World Bank's involvement in India's holiest river. Upstream it was funding dams, downstream it was funding cleanup. Even more ominously, he said it was too late to test the water in the Ganga, using stations all along the river as suggested in a recent proposal. For people who live on the banks of the river and have been exposed to years of industrial effluent and other pollution, "it's time to test their blood."

"I think rivers are showing a mirror to the country," says Gopal. "The claim to being religious and spiritual is fake, bogus. If the people were really spiritual they would not let the rivers get polluted.

"These are flowing temples, flowing rivers. There can be no better symbol of divinity than flowing water. Yet the government in power, irrespective of its political color, lets diversions of rivers happen with no protest from religious institutions.

"A sacrilege near a temple causes outrage," says Gopal. "Religious leaders stand exposed because they have not taken a stand with regard to safeguarding rivers and they have the power to influence public opinion and see that steps are taken to address pollution."

Another water specialist, Himanshu Thakkar, thinks all the attempts to clean up the Yamuna River have failed because there's been no attempt to ensure that the real stakeholders have a role. There should be a government committee for every ten-kilometer stretch of the river, he said, with half the members from the local community, half from government. I thought about the dedicated, graying protestors who could no longer farm near the river but still tried to speak out.

"Until this is fixed I don't see how there can be a clean river. Delhi's population has no connection with the Yamuna today." When the river was alive there was a fishery, farmers like those I met at the Yamuna Jiye Abhiyan protest, average citizens who drank from the Yamuna, bathed in her clean water. "Today people in Delhi have lost this relationship," says Himanshu. "They have no role in governance, so they are insensitive to the state of the river."

Even all that hydropower development I saw up in Uttarakhand on my way to Gangotri is connected to how the Delhi Jal Board functions, Himanshu pointed out. "If the public utility is functioning properly, I don't need a pump. I will be using less power, and the whole system will require less power. Right now Delhi is using the maximum amount of power. Imagine if everyone's pump were removed! There would be so much less electricity consumption in Delhi, maybe we would be able to forgo a couple of dams in the north."

During my years researching and writing this book I took long summer breaks back in California, visiting friends and house-sitting. During the summer of 2010 I periodically checked Indian news sources to see how preparations for the games were coming along. By late August there was trouble. The venues were not ready. The Yamuna was rising, getting close to the athletes' village high-rises as Manoj Mishra had warned. There were twice as many cases of dengue fever as usual.

Dengue isn't like some strains of malaria, which can lead to recurrences. But it's not something you want to be exposed to if you can avoid it. Apparently the increase was not just because of flooding along the Yamuna but because the construction sites had created a lot of little breeding grounds of standing water for the mosquitoes that carry the virus.

A month later, just before the first athletes would arrive, stories penetrated Western media. Advance teams for the athletes reported that the well-appointed apartments in the village were dirty. Toilets were filthy. There were muddy paw prints on bedspreads. I could just imagine it. Workmen had been using the toilets, most likely. Then stray dogs tracked in monsoon mud when doors were left open. All 1,160 apartments could not be in this condition. Still, I cringed at the bad press. The incident provoked a largely irrelevant debate about "different standards of hygiene" between East and West.

I found myself hoping this would blow over quickly. Knowing Western tendencies to stereotype and dismiss and be smug, I started to feel protective. I did not want to see India embarrassed by this venture, however ill advised it might be.

A footbridge collapsed near one of the game venues and workmen were injured. Some tourists were shot in old Delhi. The stories marched through my regular news sources. The BBC was first with stories of the trouble, then *The New York Times* had articles on its inside pages. Finally there it was on the front page; a color photo showed a flooded stadium, scaffolding askew, and a lone Indian wading in knee-deep water trying to gather up floating construction materials. India was getting plenty of attention, but not at all the kind the world-class label seekers wanted.

NPR aired a light story about the Indian police recruiting langur monkeys to scare off the hundreds of rhesus who might harass visitors, but Indian reports weren't so cheerful. They told of government-dispatched bulldozers arriving at a slum school, giving the students and teachers three hours to vacate before they would demolish the school, located five hundred meters from the games village, as a security risk.

It started to look as if athletes might not come to Delhi, as a few had already announced, citing dangers to their health: one from New Zealand, then one from Africa. "Toughen up, guys," I muttered after reading the article. "Someone will clean the toilet before you get there. Don't worry. It won't bite you." Then I thought of the dengue.

After one of my visits, I left Delhi early in October. A few days after my return to Kathmandu I attended a rare chamber music concert. Four young Dutch musicians had come to Kathmandu. I thought I might go to all of their concerts, just to drink in the sound of classical European string music. Halfway through I started to feel cold, even though in early October the evenings in Kathmundu are still summery. I pulled my cotton scarf tightly around my neck and shoulders and sank back into the music. By the end of the concert I felt I was coming down with some thing. It was well past dinnertime, but I wasn't hungry. I was beginning to feel achy and feverish. I gratefully accepted a ride home from an acquaintance at the concert, though normally I would have walked.

I collapsed in my apartment with a splitting headache after taking a painkiller. Several days into my collapse—having twice been taken to a clinic by friends—and after one misdiagnosis of typhoid, a doctor announced I had dengue fever. The day-biting mosquitoes that carry the dengue virus are common in Delhi that time of year. They had not yet reached as far north and as high as Kathmandu—though some public health experts thought that would come in time with global warming—so it was most likely I was bitten before I left Delhi. The incubation period was right, the symptoms were right.

For two weeks I moved from bed to couch and back to bed again, and only left my apartment to go to the clinic. The only good thing was the loss of a few pounds. A month later I still got tired after normal exertion.

October is the worst time for dengue. "Why on earth did they schedule the games for now anyway?" I exclaimed in exasperation. The weather is fine in Delhi until December. An athlete could be flattened for weeks with a case of dengue.

Then, the day before the games start, my heart swells with relief for the whole country. Things look like they're back on track. A tough Brit says everything in Delhi is fine for him. Some athletes may be pulling out—but because of injuries, not fears of dengue, terrorism, or toilets not up to Western standards.

They're going to pull it off, like people had been saying, like a Bengali wedding or a Punjabi wedding. It looks like a mess, then somehow it all works out at the last minute. Many would continue, as they should, to decry the skewed priorities that led to inequities, blunders, wastes, and outright corruption—to billions of dollars spent that might have bought Delhi a healthy river.

But for these ten days, I thought, let them pull it off somehow so the rest of the world can applaud this rich and fascinating *masala* of a nation for what it has done in six decades of nationhood. Maybe soon, that young democracy, that amalgam of ancient cultures, will martial the same energy that went into the Commonwealth Games to make its sacred rivers a priority.

# 4

# Melting Ice Rivers

It is a wild and a little known valley and tales of the legendary
yeti, or abominable snowman, have poured from the mouths of
the handful of Sherpas who live there.

In Tibetan, rolwa means "furrow" and ling means "place"—and
local people believe that the straight narrow valley of Rolwaling
was plowed out more than 2,000 years ago by Padmasambhava, or
Guru Rinpoche, the Indian yogi who introduced Buddhism to
Tibet. After creating the valley, Padmasambhava meditated here
for three months.
—From websites advertising treks to Nepal's Rolwaling Valley

From a remote outpost of global warming, a summons crackles over a two-way
radio several times a week:

*Kathmandu, Tsho Rolpa! Babar Mahal, Tsho Rolpa! Kathmandu, Tsho Rolpa!
Babar Mahal, Tsho Rolpa!*

In a little brick building on the lip of a frigid gray lake fifteen thousand feet above
sea level, Ram Bahadur Khadka tries to rouse someone at Nepal's Department of
Hydrology and Meteorology in the Babar Mahal district of Kathmandu far below.
When he finally succeeds and a voice crackles back to him, he reads off a series of
measurements: lake levels, amounts of precipitation.

A father and a farmer, Ram Bahadur is up here at this frigid outpost because the
world is getting warmer. He and two colleagues rotate duty; usually two of them live
here at any given time, in unkempt bachelor quarters near the roof of the world.
Mount Everest is three valleys to the east, only about twenty miles as the crow flies.
The Tibetan plateau is just over the mountains to the north. The men stay for four
months at a stretch before walking down several days to reach a road and board a
bus to go home and visit their families. For the past six years each has received five
thousand rupees per month from the government—about $70—for his labors.

The cold, murky lake some fifty yards away from the post used to be solid ice.
Called Tsho Rolpa, it's at the bottom of the Trakarding Glacier on the border

Map 3 Nepal Himalaya—Rolwaling and Khumbu Valleys. Created by Kanchan Burathoki

between Tibet and Nepal. The Trakarding has been receding since at least 1960, leaving the lake at its foot. It's retreating about 200 feet each year. Tsho Rolpa was once just a pond atop the glacier. Now it's half a kilometer wide and three and a half kilometers long; upward of a hundred million cubic meters of icy water are trapped behind a heap of rock the glacier deposited as it flowed down and then retreated.

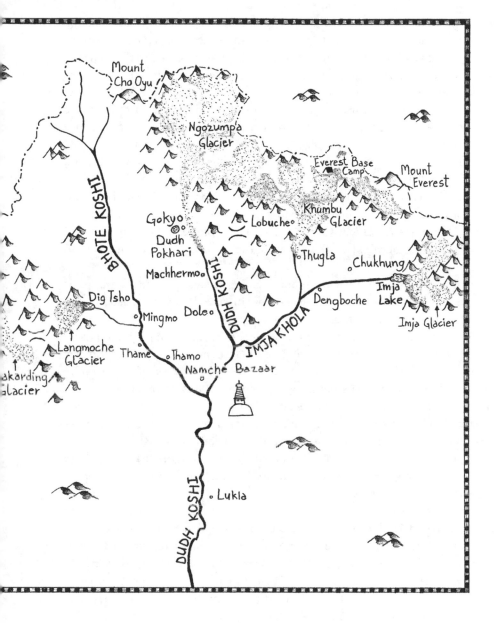

The Netherlands helped Nepal carve out a trench through that heap of rock to allow some of the lake's water to drain into the Rolwaling River. Construction teams finished the work in June 2000. This effort aimed to relieve pressure on the glacial moraine where Ram Bahadur was living when I joined him for three days in the summer of 2006.

It takes a long time to rouse someone in the slow-paced offices of the Department of Hydrology and Meteorology, down in the monsoon warmth of the Kathmandu

Valley. By the time he establishes contact, Ram Bahadur is a little frustrated. The man at the other end can't seem to hear the numbers. Ram yells them repeatedly in both English and Nepali. The numbers are perhaps abstractions to the man who takes them down, and to a great extent to Ram Bahadur himself, whom I accompanied when he went out to check the water level on a stationary pole near the shore of the lake. The measurements have meaning for the hydrologists, but there will have to be a lot more such data before they can interpret them and venture a guess about how fast this glacier might melt and Tsho Rolpa expand.

Glaciers themselves might be abstractions to whomever is taking the data, down there in Babar Mahal; he probably has never been in the Nepal Himalaya, let alone here in Rolwaling. For most Nepalis who weren't actually born in the mountains, the high Himalaya is almost like a foreign country. Residents of Kathmandu can rarely even see the Himalaya in the distance anymore because air pollution obscures what was once a magnificent view of the snowcapped mountains ranged to the north.

Down in Kathmandu it is late summer. Here, though the temperature dips below freezing at night, ice may still melt during the days and monsoon rains may fall. Summer is when the newly formed glacial lakes in the Himalaya like Tsho Rolpa are said to be most dangerous. It's the season when they could burst through the unstable natural dams that hold them, flooding river valleys, sweeping away farms and villages. Yet the house where I stayed for three days (and the staff inhabits year-round) was perched directly upon such a dam, a glacial moraine. I registered this unnerving fact, but no one around me seemed anxious. And I had no complaints. I had a mattress on a platform in a separate room to throw my sleeping bag on, and hot water for a bucket shower, thanks to a small hydropower plant Ram Bahadur and his colleagues managed. There was even an electric space heater for warmth in a part of the world where wood fires remain the major source of fuel.

Ram Bahadur was a pleasant host and a good cook. Three times a day—in an unappetizing-looking kitchen with one burner—he whipped up a tasty meal for me and my ad hoc crew of three young Nepali men. I had hired two of them in Kathmandu to be my guide and porter for a trek up the Tama Koshi River and the Rolwaling Valley to Tsho Rolpa. The third young man fortuitously joined us later.

This was not my first quest to a glacial lake in Nepal to see evidence of global warming. I had already explored the valleys to the east known as the Khumbu. Each year that area is visited by thousands of trekkers intent on reaching Everest Base Camp. It's also home to two of the largest glaciers in Nepal and to other glacial lakes that have breached their moraine dams, causing what hydrologists have dubbed a GLOF, or glacial lake outburst flood. On this trip to Tsho Rolpa I wanted to see for myself how engineers were trying to intervene in hopes of keeping some consequences of global warming in check.

The changes in these frozen reaches are not readily evident to the casual visitor; but longtime residents and hydrologists who track the melting of the ice see a shifting landscape and cause for alarm. The vast stretches of ice and rock at sixteen or seventeen thousand feet and higher—that are formidable challenges to reach for a trekker like me and seem very solid—are the essence of impermanence. Just turn up the heat as we are doing, and they can begin to disappear. What that ultimately will mean for a billion people in South Asia, no one as yet can say.

Though I am here to see evidence of the irrevocable changes in the planet wrought by my own commercial and industrial species, I'm often exhilarated as I huff and puff my way up to these obscure destinations: Tsho Rolpa, Dig Tsho, Imja, Ngozumpa, Gangapurna. None of them are easy to reach, but the trek to Tsho Rolpa was taxing in new and different ways.

I was uncomfortable and tired and worried a lot of the time—and on one occasion thought the trip was going to end abruptly. But I made it safely to my destination and back, bruised and covered in leech bites and other welts and bumps and itchy red spots, yet otherwise fine. I saw lush jungle and stark glacial lake. I stayed in pretty villages where people had little money, but bougainvillea and lovely flowers I couldn't name grew wild. Westerners in helicopters might pay thousands of dollars to see the vistas of river canyons and waterfall-studded hillsides I saw each day, slowly tramping up the trails.

I am going to tell you about the two-week journey, both to show how hard it was for an aging woman of relatively athletic temperament to get there and back, and also to convey the remoteness of these lakes. That remoteness partially explains why it is only very recently that people in Kathmandu, let alone other parts of the world, know what is happening up there.

I also want to show people who may never have a chance to visit Nepal, my home for several years, a glimpse of life in the remote foothills of the Himalaya where few Western tourists venture. In the years I've been spending time in this small nation, it has been in crisis—politically, socially, economically. Where it is headed is hard to guess. The country, its mountains, and its people have many fierce fans. But for most Westerners it's too challenging a place to visit.

Look on a map and you see a rectangular little loaf of land north of India and south of Tibet (which on most maps does not even show up as Tibet anymore— it's simply China). If you could iron out this vertical land of Nepal, home to half the highest mountains in the world and thousands of very steep hills, it would look a lot bigger. Several hundred years ago Prithvi Narayan Shah unified the various kingdoms and ethnicities scattered among the hills of Nepal into a single nation. On his deathbed he is said to have given advice about how to preserve the nation, characterizing it as a "yam between two boulders"—a vulnerable country sandwiched between its massive and powerful neighbors, India and China. It's a nation that has nonetheless managed to survive intact, somehow maneuvering between those two rivals. That's the irony now embedded in the yam metaphor; it's a tough and wily little yam.

Nepal's lack of good leadership has left it even more vulnerable these days and in the doldrums economically, but its mountains have always protected it from

Tama Koshi River in monsoon

being completely overrun. Its rivers hold the greatest wealth of water in all of South Asia. Some of that wealth is locked up in its glaciers.

The monsoon in Nepal typically tapers off by the end of August. It had not been a robust monsoon that summer, 2006, and I had no reason to doubt people who told me that a late-summer trek into the Rolwaling Valley would be fine. The night before I left Kathmandu the last week of August, it rained all night.

## Day 1

The morning was fresh and clear. Heading east from Kathmandu, the jeep climbed steadily from four thousand to about ten thousand feet, and I felt cool for the first time in two months. In the afternoon, about an hour short of the town where we intended to start walking, we encountered a landslide. There was a small bulldozer already trying to clear it, but Ale, our driver, seemed to think it might be hours before the road was open and he wanted to turn around and get the jeep back to

Kathmandu. I said okay, we'll start walking. My two companions and I picked our way over the landslide and headed down the two-lane road. My guide Purna Magar stopped to talk to some young men he knew and motioned me and Cholendra Karki, my young porter, to go ahead and take a shortcut down the slope to avoid the road's lengthy switchbacks.

Soon I had my first encounter with the slippery monsoon-saturated ground. I lost balance and came down on my right knee, sliding forward and sharply pulling back on my right thigh. I could almost feel the muscle pop as it stretched way beyond its normal range. I got up and kept walking, knowing I would need a pain-killer soon.

We continued a couple of miles, then motorcycles started to whiz past us. The roadblock had been cleared already. Soon busses began to roar past us with their South Asian horns rhythmically honking a short tune. Purna tried to flag one down, but the drivers ignored us. The busses were mostly full anyway. Finally one stopped and the conductor motioned me in. I started to step up, then saw there was no room: people were packed into the aisle, dizzying heat emanating from inside. I thought I might ride on top. Cholendra was already scaling the side of the bus with my red duffel bag. The conductor shook his head—getting on top and riding there was too difficult for me. Purna and I waved goodbye to Cholendra and kept walking.

A little farther down the road I looked up to see a police jeep speeding past, full of blue-uniformed young men. I opened my arms inquiringly. Purna used his own sign language. Something worked. The jeep screeched to a halt and backed up. There were already nine policemen in the small vehicle. We squeezed onto the back seats, perpendicular to the front ones, straddling the spare tire and breathing gas fumes. Soon it started to rain. As the kilometers whizzed by I was grateful there were no rules to prevent the Nepali police from picking up hitchhikers.

Cholendra was waiting for us in Charikot, which had been our destination and the planned beginning of the trek. It was already mid-afternoon, so we went to a hotel. I took a cold shower and some ibuprofen and the guys went to buy rope, plastic covering, and an umbrella.

It rained most of the night.

# Day 2

The hills were swathed in white mist in the morning, the sky overcast. As we began walking down the road it started to drizzle. We stopped for tea in the next town, Dolakha, as the rain began to pour down. Purna chatted with some local people he knew; I ate my Kathmandu croissant and fed some of it to a puppy with one blue and one brown eye.

An hour later the rain had not let up, but we decided we had to start walking. The rain slackened soon, but the path was very slippery. I fell a couple of times, walking more and more slowly to try to avoid another fall. The ground was rusty red, hard-packed earth, slick from the rain—the kind of clay villagers use to construct their homes, their walls, floors, fireplaces. I began to slip and slide uncontrollably as we neared the Tama Koshi River at the bottom of the long hill. My falls did not hurt, but I was covered in mud and frustrated from feeling at the mercy of the hillside. Purna took my pack, which wasn't heavy but perhaps contributed to my imbalance on the slick, steep ground.

We finally got to the bottom. Even on flat ground the rutted, rocky, muddy path was treacherous, so I picked my way slowly to a small village where we could get some lunch. I ate noodle soup while three teenage boys stared at me: a sweaty white woman, wet hair pulled back in combs, her mannish outfit covered in red mud. Even the restaurant's floor, made of the same red clay, was slippery for me. I decided my big leather boots weren't suited to a monsoon trek.

We were planning to reach a village called Singate, which Purna said would have a pleasant inn. By late afternoon we were still an hour and a half away, my sore right leg slowing me down, so we stopped instead in a village called Gumu Khola. Some villages are just a straggle of huts and houses along the trail, but this village had a small town square. Climbing up the steep ladder-like stairs in a building on the square, I found a room with two platforms covered by thin mattresses; the room was much like those in the lodges and teahouses in the major trekking areas, just not as clean as most of them, and very dark. I asked for a bucket of water and cleaned up a little. There was a row of outhouses in back, down by the river.

The restaurant and shop were on the opposite side of the little square. We ate dal bhaat and talked to several teachers from the local high school. They lived here in the village when school was in session, eating dinner each night at the inn. I asked them if they taught their students about global warming and the threat of a flood if the glacial lake Tsho Rolpa—twelve thousand feet above us, and our destination—should burst its natural dam. They said they had.

It rained lightly during the night.

# Day 3

The morning was clear. The long downhill walk had compounded the soreness from my fall two days before, so now both legs were stiff and tired. The path by the river was flat but just as slick and muddy as the hillside. My feet slipped out from under me soon after we started walking and I fell hard on a football-sized rock that jutted up from the side of the path. It jabbed into the side of my left thigh so hard the muscle in my leg went into spasms. I limped on, deciding that if the village an hour ahead had a lodge more suited to trekkers, we were going to take a rest day.

The village of Singate looked inviting. Between the houses and the river was a narrow expanse of cultivated land where greens, bananas, and squash grew. There was a little lodge run by a Sherpa family. Upstairs was a large, clean room with an electric light. I took a cold bucket shower, washed my muddied clothes, had a big lunch of tasty dal bhaat, and slept for two hours. I discovered some homeopathic Arnica in my bag and started taking it, alternating with ibuprofen. My legs were so sore by this point I had to creep down the ladder-like stairs to the ground floor.

Late in the afternoon I met an English-speaking Nepali employee of an Austrian aid organization called Eco-Himal. He and his colleagues had been working for some years to try to improve living conditions for the people in this region as well as to make it more attractive to trekkers. They began by helping villagers build toilets. Their office was right next door to our lodge, where indeed the squat toilet was clean and functional.

The Eco Himal worker gave me a tour of this bustling village, where—as a rare Westerner—I got stared at the whole time. At dinnertime I talked to some men who had come to eat at the lodge. They lived in Singate while they supervised work on the road that would someday extend from here up the river canyon, allowing construction of a dam and hydropower plant about eighteen miles up the valley. The road more than the additional electricity will change this district irrevocably.

The rain began again as I waited for dinner. I thought about Tsho Rolpa and imagined it filling up with monsoon rain.

# Day 4

Early in the morning we started walking deeper into village Nepal. As we pressed into the jungle it seemed to me we slipped out of time, back several centuries. It was warm down along the river, which the trail hugged for some miles. We had dropped thousands of feet from the cool hills at the start of the journey down to the bottom of a river canyon. We were still several thousand feet above sea level, but at this subtropical latitude the rich soil and monsoon rains create a lush terrain. The grass and fields and trees were an intense green against the clay-red earth. The sky was overcast, which kept the temperature in check, and often we were so close to the wide torrent of the Tama Koshi that a delicious breeze dried some of the sweat dripping down my back and face.

The trail rolled up and down along the river. It was easy and not too slippery: no falls—just waterfalls plunging toward the Tama Koshi. After the trail turned inland away from the river, we stopped at a farmhouse for tea. Purna knew the family. They came in from the field where they were working in the light rain and we sat on the red clay floor around an open fire in the middle of the room. He chatted with the mother while her pretty daughter-in-law nursed a fat, healthy-looking baby.

Farther inland we stopped in a hot, sleepy village for lunch. The smiling *sauni*, the owner, gave us each a handful of fresh little fish in a curry sauce. The fish had flattened heads and wide mouths; they looked like tiny sharks to me. Called *tite machha*, or "bitter fish," by the locals, they tasted a little too bitter for me, so I gave half of mine to Cholendra, who was happy to eat them with his mound of rice.

Farther down the trail and high on the opposite side of the river was a cluster of houses, the village where Purna was born. The din of the river drowned out voices. Purna whistled and got someone's attention. She recognized him despite the distance and motioned for him to come. He smiled and shook his head and we continued down the river. Purna was orphaned at five and taken in by a couple in a nearby village. His older disabled brother stayed on in a clay hut with a thatched roof that their father had built.

One day behind schedule we reached Jagat after crossing a sturdy steel bridge festooned with red Maoist banners. The little inn where we stayed was just feet from the rushing river, and though I loved its sound and the cool breeze, I started to wonder again what might be happening far above us. If Tsho Rolpa's moraine dam were to burst, a wall of water might come down the river, squeezed through the channel with tremendous force. If that happened it seemed people in this village would have no place to go: they're right on the edge of the river with a steep canyon wall behind them.

I mused on this as I sat on the floor in a little curtained room on the second story of the inn, reached by a ladder. By this time my legs were beginning to lose some of the crippling soreness from my falls. There was one window looking out over the river, its wooden shutters open. In the late-afternoon light the river was silver gray—fast and frothy. A chicken pecked its way into the room, looked up and saw me, and exited.

Downstairs Purna was sitting with his brother, who had heard of our arrival. He looked almost frail, though clearly he was strong enough to scoot up and down the steep hills. He seemed to be what Westerners would call developmentally disabled. He had a sweet though odd face, and beamed in happiness at seeing his brother. Purna bought him a pitcher of the local brew and gave him some new pants and a shirt, along with money.

I went to bed after dinner and slept well despite thoughts of the cold lake far above us.

# Day 5

We continued along the river, walking in tandem with two Nepalis who worked for Eco-Himal and had left Singate the same morning we did. One of them, a young Sherpa woman named Kandiki, worked at the health post in Simigaon, our destination on this fifth day. She carried a badminton set in her small backpack.

The trail eased up and down along the river under an overcast sky. A big black moth with iridescent green wings fluttered by me. Little waterfalls sprouted out of the emerald hillsides and we could see a snow-tipped mountaintop in the distance, probably Gauri Shankar.

We stopped for lunch in Gongar. While we settled in for the usual hour wait for our fresh dal bhaat to be cooked, Purna went to visit the couple that had taken him in when he was orphaned.

Purna returned for his food and two young men started talking to him. They were local Maoist insurgents, trying to find out who I was. I fished around in my little money bag and found the receipt from several days before when I had paid a small roadside "tariff" at a Maoist check post during our drive to Charikot. They were satisfied and moved on across the uneven pavement to another house. I learned that the government doctor posted to the district had been kidnapped by the Maoists for several months, forcibly recruited to tend to their health needs.

A light rain began to fall as we were getting ready to leave. Purna had disappeared, but he rushed up as I was shouldering my pack to say a friend of his would join us. There were some difficult stretches ahead, a river to cross, and the friend knew the terrain well. The friend, named Nabaraj Khadka, was taller than Purna and his open, smiling face loomed over Purna's shoulder as the latter told me about the new plan. Nabaraj would come with us all the way to Tsho Rolpa and then accompany us back to Kathmandu, where Purna would try to find him work as a porter. I was a little skeptical. I couldn't afford to pay wages, food, and lodging for three helpers in addition to my own expenses. But what was this difficult river we had to cross? I was a "boss" who was completely dependent on these young men. I had little choice: I shrugged OK and started walking.

Nabaraj was a fresh-faced, half Tamang, half Thakuri nineteen-year-old, jovial and energetic and very polite. Half an hour later I was glad he was with us. We came upon a place in the riverside trail that was narrow and steep. The original path had washed out. There was a ledge of loose rock, wide enough for a pair of feet. This strip of scree plunged at a forty-five-degree angle downward under a waterfall. The river rushed just below. I worried that even if I could safely get down and across, what about Cholendra and my big duffel bag? He walked to the edge and took a quick look down: "No problem," he said nonchalantly. Nabaraj took my large daypack and carried it for the duration of the trip. I scooted across and got soaked by the waterfall.

A short distance farther down the Tama Koshi we crossed the river on a relatively sturdy bridge and began our ascent from the three-thousand-foot elevation where we'd been walking for the past few days up toward Tsho Rolpa at fifteen thousand feet.

Stage one of that ascent was a very steep and rocky hill. The path was serviceable for sure-footed locals but not really intended for trekkers. A landslide had taken down the whole hillside in 2001. In some places the trail was washed out altogether and villagers had planted a series of wooden stakes at an angle in the hard, bulbous hillside and laid logs across them. I stepped across on the logs, leaning against the hillside. The whole way up I struggled to find footholds. I sometimes held on to Purna or Nabaraj to keep upright or to keep from slipping backwards. The ascent of the hill took more than two hours. It was raining but warm. I felt like a walking steam bath: my shirt and trek pants clung stickily to me; my glasses steamed up, streaked with drops of sweat and rain.

We reached a small Buddhist *stupa* at the edge of Simigaon in the late afternoon and continued to ascend on an easy path. In the space of our short climb we had moved from mainly Hindu villages to Buddhist ones. Simigaon was a handsome village, stretched out on the hillside, houses dotting the slopes. We climbed up through the village to the home of the health worker, Kandiki. Purna said he remembered her from his childhood, and she wouldn't say no if we asked to stay at her house. There were no lodges in Simigaon.

We arrived at the large farmhouse, which Kandiki had reached hours before us. We took off our muddy shoes and clambered up to a bright open room on the second floor. We sat on mats on the clay floor near the hearth; Kandiki gave us tea and plates of boiled potatoes. I gladly drank the hot tea, but sitting there in soaking clothes in a cool adobe house I wasn't very interested in eating yet.

It turned out there was one other *bideshi* (foreigner) in the region. I hadn't seen another white face since I left Kathmandu, but there was a Frenchman staying here at the house, working with Kandiki's father on a local bridge-building project. They were out working on the project when we arrived. I briefly felt anxious that I had no place to stay, but Purna said Kandiki knew of another house a ten-minute walk from here that would take a guest. My hungry crew of growing boys first had to finish their pile of potatoes, peeling them and dipping them in crushed chili.

Tsering Deki's hearth, Simigaon

We walked to the other house in the lingering light of late afternoon. The house was smaller than Kandiki's. At the top of the ladder-like stairs there was the typical open hearth of the village house, along with a charming hostess named Tsering Deki, who was Kandiki's aunt by marriage. Kandiki stayed and helped Tsering cook dinner for us.

Tsering wore a light brown *chuba*, an elegant kind of jumper that Sherpa and Tibetan women wear. Underneath it she wore a striped T-shirt, probably made in China, and around her waist a striped Sherpa apron. She had just finished washing her long hair as we arrived, wrapping it up in a scarf. Her outhouse was spotless and there was even a very small roll of toilet paper inside, a rarity in rural Nepal.

The evening was warm, so I got a bucket of cold water and washed in the little bathing room next to the toilet. Once clad in dry clothes, I felt fine again. I slept in a room off the main one, a kind of storage room with a platform bed and a solar electric light. It seemed palatial to me, the Best Western of the Rolwaling Valley. A gray cat curled up by me while we ate dinner and later slept on my platform.

# Day 6

The next morning's walk took us higher, up through dense woods, as we changed direction and headed northeast toward Tsho Rolpa. Little black leeches, the bane of all wet-season trekkers, plagued us here more than they had along the Tama Koshi. Hundreds stretched themselves out toward us, reaching blindly for warm prey. The underbrush along the narrow trail was alive with the hungry creatures. Tsering Deki had made a little wand for me with a small bag of salt on one end. If a leech took hold, pressing the salt on its head would encourage it to let go.

Before midday we reached a farmhouse, perched on a cleared hillside in the sunlight. Nabaraj knew the middle-aged couple who lived there, a Sherpa wife and Brahmin husband. He knew they would give us some lunch in exchange for whatever money we wanted to give them. We rested on the porch and took off our muddy boots and sneakers. Across the canyon, three monsoon-fed waterfalls burst from the densely wooded slope, like bright silver chains sparkling on malachite. The Sherpa woman welcomed us into their small main room, the walls and floor made of the familiar local clay. We sat on the floor on straw mats around the fire, sipping tea while she fussed over some curry she wanted to make for the young men. She also made the rural Nepali staple called *dhindo*, a porridge similar to polenta that can be made of corn or millet. The preparation took so long, two of my young companions and I dozed off.

The long break cut into the afternoon's walking. By late afternoon a downpour seemed imminent, so we stopped about an hour sooner than we had planned at a place known as Kyalche. It was a farm with a small lodge that had a row of dark little rooms. I took one and my three companions another.

I inquired where the toilet might be. There was none, said Purna, after asking the sauni, the proprietress. Gesturing around the muddy, slippery farm, he said, "You can go anywhere." I retired to a spot midway between the chicken roost and the goat's house, where I encountered *cisnu*—stinging nettle. I had been stung many times on the trail on my hands and arms, but this time my bare posterior encountered the poison.

The sting of the cisnu subsided to an itch as I sat on the floor in the smoky straw hut that served as kitchen and living room. The sauni was there alone with two small children; her husband was away working in India. She fed the toddler rice while she cooked dal and potato curry for us. I felt a sting behind my knee; on inspection I found two fat leeches. They must have jumped on while my guard was down near the chicken roost. They are tiny when they're hungry, arching their backs and moving along like inchworms; but they swell up many times their size after they've latched on and sucked for a while. The only clue is the mild sting. I ran for the salt poultice.

We ate our nondescript dinner as a spectacular mountain thunderstorm broke. Bursts of lightning flooded the dim room and thunder shook the walls. After dinner I went to bed, the only challenge on my mind at that point being the lack of a toilet. I had avoided liquids with dinner, thinking of the slick rocks outside. A fall and discomfort awaited me if I tried to pick my way across the slippery, uneven ground in the dark through the *cisnu* and leeches. I decided to turn my little washbasin into a chamber pot, but worried off and on all night how I would empty it without the sauni or her inquisitive kids or my crew of guys seeing what I was doing.

# Day 7

My water-resistant watch had gotten soaked and stopped working, so I was way behind schedule the next morning. Usually I was up half an hour to an hour before the boys. I managed to empty my impromptu chamber pot and wash it while my wet clothes dried in the bright morning sunlight.

The tiny shop at the end of the row of rooms had nothing inviting to take for snacks; there were just a couple of Cokes and Fantas, some cigarettes and noodles on the dusty shelves.

We drank tea but skipped breakfast, because there was another little inn known as Dong Kharka an hour up the trail, the last opportunity for food or drink until we reached our destination that evening. The pretty gap-toothed Sherpa woman who prepared our "brunch" was a much better cook than the one the night before. The guys sat inside the smoky room while I sat outside trying to dry out the contents of my daypack and chatted with a traveling Buddhist monk who was doing his laundry in a nearby stream.

My companions had opted for the usual time-consuming meal of dal bhaat or dhindo with a potato curry. The woman made a second curry using some of the meat that was curing, hung from a rack above the open fire. Purna said it was Himalayan Tahr, a species protected inside Sagarmatha National Park just east of here, but not here in the Rolwaling Valley. I tasted the meat; it was a little like venison. I preferred noodle soup with fresh greens and the red mountain mushrooms I had seen drying on mats outside. They were big and tender and full of flavor, a delicacy a Western chef might covet.

The isolated house where we were eating was a mere hundred kilometers from Kathmandu, but a conventional measurement like kilometers can cease to have a lot of meaning in much of rural Nepal. Our feet were the only vehicles available, there was little electricity, not much health care, limited work or educational opportunities, and agriculture was chiefly accomplished through human labor. The important measurement for me was that we were still a long half-day's walk from Beding, the next opportunity for shelter.

As we left this outpost of comfort, it started raining. We had followed the large Tama Koshi for the first few days of the trek; we left that formidable and monsoon-swollen river behind when we headed up to Simigaon. Now we were nearing the river that flowed from our destination, the glacial lake Tsho Rolpa. Though much smaller than the Tama Koshi, in the monsoon the Rolwaling Khola was also swollen and very fast and dangerous.

Within an hour we came to a stream that fed the Rolwaling Khola from a steep side canyon. I could see the water leaping and foaming behind the heads of my companions, who had come to a stop together. As I caught up with them I saw what had stopped them: a washed-out bridge. Pieces of splintered wood dangled from the rocks; the rest of the bridge had vanished down the river. There was no safe way across the roaring stream.

We would have to turn back, I thought, and my heart sank. Nabaraj shook his head to my desperate idea of climbing higher to see if we could get across. He started checking out the foothold below us on the muddy verge of the river as he pulled out his *khukri*, the long Nepali curved knife I had seen so often in tourist shops that I had forgotten it had some practical use. He walked around the wet woods eyeing small trees.

"What are you going to do, build a bridge?" I asked in disbelief. He grinned and started whacking at a sapling. The blade was sharp; in minutes he had hacked through the five- or six-inch trunk. The branches were tangled in adjacent trees. Cholendra set about pulling them from the trunk.

There was a smooth boulder in the middle of this small but powerful stream. Nabaraj planned to use it as the anchor for the two spans of his bridge. He found another small tree and set about cutting it down, then cut the two trees into lengths. Soon there were five slender logs. The young men positioned the lengths close together, lodging them on the riverbank and in the grooves of the boulder, weighing the ends down with large, flat stones. They leapt back and forth, repositioning the logs and stones and wedging everything securely.

In the meantime I had hatched my plan for the crossing. I would take off my boots and go down and straddle the "bridge," scooting across on my butt just above the foaming water instead of risking losing my balance and disappearing into the Rolwaling Khola.

They had other ideas. Boots off, fine. But I was going to step over on the bridge with them holding on to me. At first I protested. "What if I fall? Then I knock Nabaraj down and we're both in the river."

I lost that argument. I suggested they take all the packs over first, then come back for me. I don't know if I was just stalling or wanted to test the bridge. They said they would get me across first, which made sense. If they didn't get me over, there wasn't much point in taking my duffel bag.

With their help I eased down a couple of yards of dark, muddy bank and put my right foot onto the three tightly packed little tree trunks that constituted the first half of the bridge. Cholendra took my hand and went ahead of me. Nabaraj stayed behind with one foot anchored on the bank and one on the log bridge. With a few quick, slightly springy steps I was on the big, warm, bone-pale boulder.

Cholendra preceded me down onto the second span, Nabaraj came up behind to make sure I was steady. We crossed easily. Cholendra and I, barefoot, getting soaked in the rain, walked around on the rocks on the opposite bank, looking for a way back up to the trail. He went back for the duffel while I struggled to pull on my wet, leech-bloodied socks and muddy boots.

Once everyone was across we found a way through the underbrush up to the trail and continued walking, soaking wet but full of adrenaline. The fording of the furious stream didn't seem much more foolhardy to me than deciding to do this trek during the monsoon in the first place. And I felt I was in very good hands with Nabaraj and Cholendra. Purna was somewhat cautious, so his lack of objection to Nabaraj's initiative had been reassuring. I told him this had ceased being a trek and was becoming an expedition.

While we were still toiling up parallel to the Rolwaling Khola, we reached another side stream. Nabaraj said he'd carry me. I protested, then thought about the minutes that would be wasted dragging my wet footwear off and on, and agreed. He did not weigh any more than I do probably, but he's wiry and strong, so I hopped on piggyback for a short stretch to the first big rock. He wore rubber flip-flops.

In four days we ascended twelve thousand feet; so this day must have included about 3,000 of them. The ascent seemed gentle, but it was long and muddy and somewhat slippery. After the bridge-building adventure the walk became an increasingly wet, uphill struggle for me. The afternoon dragged on. Adrenaline ebbed away. The middle of the trail had become a small stream. My boots were soon saturated since I was too tired to try to avoid the water. We were still far from Beding.

Finally we came out into the open, away from the hillside forests of the past two days. The rain continued and the temperature steadily dropped as we went higher and the day waned. I pulled a windbreaker over two soaking wet shirts. Fatigue and the increasing altitude slowed me down.

Late in the afternoon we reached a cluster of houses, the vacated winter homes of the people who live higher up during the summer. They had glass panes in the windows, a sign of comfort and prosperity. I had hoped it was Beding, but it wasn't. Nabaraj went ahead to arrange our accommodations. It was almost an hour before we got to Beding and saw him again, waving us to come up by way of the village *gompa*. I scrambled over the rocks up the hill through the village.

We were now in the starker terrain of the high mountains. People here said they had seen yetis. Trees grew on the opposite side of the river, but those on our side had been cut or burned in a fire. High on the canyon wall, maybe fifteen hundred feet above us, were two small, whitewashed houses: retreats for Buddhist lamas like the one I had met earlier that day.

Tibetan Buddhists believe that the Rolwaling Valley is, like the Khumbu, one of the *beyul*, or hidden valleys blessed by Padmasambhava. The beyul are difficult to reach and were intended as safe places for spiritual practice in times of strife or religious persecution.

We stayed that night in a Sherpa house. Two girls in their late teens met us when we arrived. I kept waiting for the real proprietors, but none came. The girls, perhaps shy about talking to foreigners, spoke only to the boys.

The long, main room was typical of Sherpa homes, with glass windows all along the side facing the river. Benches stretched beneath the windows, covered with blue and red Tibetan rugs. Painted wooden tables with enclosed sides were lined up in front of the benches. On the opposite wall, shelves contained all the bedding for the family and guests, neatly folded. At the far end of the room was the family altar. Statues of Buddhist deities and photos of lamas and family members sat behind the glass windows of the cupboard, some wrapped with *khata* scarves, indicating the person had died.

I quickly drank a cup of Tibetan butter tea to warm up and tried to figure out where I could change clothes and later sleep. The Sherpa girls gestured at the room as if wondering why I would need more. I was chilled and had to get out of my wet clothes. Purna pointed to an enclosed area, shielded from the front door by a wall. It was a large sleeping platform. Cholendra put my duffel bag on one end of it and I climbed up to change, then took my wet clothes down to the ground floor to drip in the animal quarters. The animals were away for the summer, grazing in the high pastures.

My three companions and the two sisters along with a friend of theirs chatted and giggled for hours. They all sat together by the fire in a little windowless kitchen tucked away near the front door. I crawled into my sleeping bag with more tea to write in my journal and doze until food arrived.

In each village I had visited so far I had tried to get a look at what remained of a warning system built in 1998. It was intended to alert villagers if Tsho Rolpa were to burst its moraine dam. Nepali officials created the experimental early

Draining Tsho Rolpa's glacial waters

warning system with a loan from the World Bank. In seventeen villages strung out along the Tama Koshi and the Rolwaling Khola, they placed sirens mounted high on poles. If the moraine breached, an electronic warning would be generated up at the lake and would automatically set off the sirens. Villagers would know to head for higher ground, away from the river.

If Tsho Rolpa should burst, a third of its contents might spill down the Rolwaling Valley through the Rolwaling Khola, then into the Tama Koshi, inundating Na and Beding, perhaps missing the higher parts of Simigaon but wiping out the villages of Gongar, Jagat, and Singate down on the big river.

Even with the sirens installed, villagers along the Tama Koshi were in a panic the summer of 2000—as they had been during several earlier summers—because of all the talk of a sudden flood. Hydrologists had been passing through the villages for several years, working on installing siphons in the moraine in an effort to drain some of the water from Tsho Rolpa. Then they began to construct a trench to do the draining.

Residents of Gumu Khola and Singate, the first two villages where I stayed, packed up and headed for higher ground during the summer of 2000, coming back down to farm their land in the daytime but spending nights above the river. Two months later they returned to their homes. After 2000, villagers along the Tama Koshi felt safer because some of the water in the lake had been drained; in subsequent summers they remained in their homes.

Tsho Rolpa

The siren system functioned for five years, according to the Department of Hydrology and Meteorology; but by 2003 it was defunct. The solar batteries had been pilfered by Maoists or villagers who wanted them for other purposes. The towers still stood above the tile or thatched roofs of a dozen villages I passed through, their sirens silenced.

After dinner Purna interpreted as I asked the young Sherpa sisters what they knew about Tsho Rolpa and its danger. They giggled a lot and did not seem very worried. They were dubious about the sirens, one of which was just behind their house. They said it had behaved erratically, sometimes going off in the middle of the night for no reason.

I went to bed and dozed while the girls and boys continued to talk in the kitchen. Soon I heard the girls taking the bedding from the shelves and quiet fell on the house.

## Day 8

In the light of morning I saw that the three girls had made a bed for themselves together on the floor near the altar. Purna slept alone on the long bench by the windows, and Cholendra and Nabaraj were side by side on the narrow bench-staying warm together.

After tea and chapattis I gathered up my wet clothes and heavy soaked boots and eased into them for the walk up the valley. Purna was concerned about my going all the way to the lake that day, because it would involve ascending from about twelve to fifteen thousand feet with no acclimatization. We could see Na—the last and highest village in the Rolwaling Valley—for miles before we reached it. And far above the wide, treeless valley we could also see a little blue roof high on a ridge. That was our destination, right at the lip of the lake we couldn't see.

I wanted to go on to the lake. Having trekked to high altitudes before, I knew that the altitude would probably only slow me down, not make me sick, and also that there would be very little in the way of comfort in Na. Na's rustic cottages were just for summer. Most of the Sherpa families in the area have three homes, one in Na, one in Beding (presumably the nicest one) for fall and spring, and one lower for deep winter.

Nabaraj stopped at a small house in Na. This turned out to be the rustic summer house of the family we had lodged with the previous night. The father—formerly a climbing Sherpa, who set the ropes for mountaineers ascending Everest and other peaks—was here tending his animals and potatoes. He made us tea and put a pile of potatoes on to boil while we sat on a bench. He squatted on the dirt floor and crushed chilies and salt with a stone. We peeled our potatoes and dipped them in this mixture for a satisfying lunch. The man never smiled but was kind. He seemed a little distracted, worn out, even punch drunk, and I wondered about the long-term effects of multiple ascents to extreme altitudes. He was still able to do heavy labor; I guessed he looked much older than he was. We found out his wife had died seven years earlier, which meant he had raised the two giggling girls by himself.

After lunch Purna agreed we could make it to the lake. It was a pleasant walk up the valley until the end, when I had to ascend the steep hill that led up the final thousand feet. At the top of the hill was an unnaturally flattened area with clean contours and straight lines, clearly man-made. Piles of gray rubble, the debris from the glacier, had been sheared off and molded to create the flat surface where the blue-roofed building stood. To reach that house we had to cross a metal bridge about thirty feet across. Below was the drainage channel, its sides pitched at a forty-five-degree angle. Water rushed through the channel, steadily draining Tsho Rolpa. To the left I could see a corner of the stark, gray lake.

In satellite photos Tsho Rolpa looks curiously like a bent index finger, slightly beckoning, pointing toward the Rolwaling Valley. Now when I look at photos of the lake taken from a helicopter or satellite, I cannot imagine myself in a landscape that looks so cold, inaccessible, and forbidding. Yet while I was there I felt strangely safe.

The Nepal government first began studying such glacial lakes because it needed information for its potential hydropower investments. The cost of insurance for those ventures was pegged to how much warning could be provided in case of a flood.

Tsho Rolpa is the only glacial lake in Nepal—and one of only two in all the Himalaya—where engineers have tried to implement a solution to the threat of flooding. The other is in Bhutan. Such efforts are difficult and expensive at these remote locations, as well as dangerous. A single helicopter ride up to Tsho Rolpa cost nine thousand dollars.

Constructing the trapezoidal trench in the moraine dam required heavy equipment—bulldozers, tractors, excavators, concrete mixers, diesel generators—to be ferried in by helicopter. Other supplies and equipment had to be carried by porters traversing the same trail I had walked during the past week. The cost for this effort was almost three million dollars and the work took four years to complete.

At that point Ram Bahadur Khadka and the other two men who rotate duties at this outpost began their vigil at the lake. Ram Bahadur welcomed us warmly after we clambered across the bridge and up to the door of the blue-roofed house. The hydrologists down in Kathmandu who had briefed me about the lake had radioed up to let him know I was coming. He soon assured me he would prepare a big bucket of lake water, made hot with a large electric coil.

One of his colleagues had just returned the day before. He spoke little and we found out he wasn't feeling well, possibly from the altitude, which had apparently bothered him on previous returns.

I had a corner to myself in a cluttered room just off the main one. I guessed it was where the hydrologists from Kathmandu slept when they visited. There was a clean toilet outside; there was electricity, hot water for bathing, a space heater, and a bad radio with a little news about what the king had said that day down in Kathmandu, and good food—all the comforts of a teahouse trek.

That night after dinner as we sat by the electric space heater, we listened to the barely intelligible news on Ram's little radio. Purna and Cholendra translated: twenty-five thousand people in western Nepal had been made homeless by landslides during the late monsoon rains, and many died. A whole village in a district not far from us had disappeared in a mudslide one night.

# Day 9

We all slept late, bundled up in our bags or blankets until seven. Ram's colleague had a touch of altitude sickness or a cold or both. He drank some tea and headed back down to Na to rest at the lower altitude.

Ram's daily duties consisted of starting up the micro-hydro plant in the morning and turning it off at night. This involved scrambling down a steep hill to the river and back. He also checked the lake level several times a day and kept a log. I learned, through Purna's sketchy translations of what Ram told him in response to my questions, that it had not been raining up here. There had been no downpours like the ones along the Tama Koshi that had made me worry that the lake above us was swelling. The lake level was in fact falling every day. This gener-ally reduces the threat of a glacial lake outburst flood.

At that point I let go of what little anxiety remained, reassured by the spartan comforts of this outpost, perched directly on the natural moraine dam. The moraine is not that fragile, I reasoned, if during the construction of the drainage trench men were rattling over it in bulldozers, digging and trenching and reinforc-ing and installing the bridge and the floodgates we had crossed the day before. All that heavy equipment is still up at Tsho Rolpa, housed in large corrugated sheds on the moraine. I wondered if all that expensive equipment would ever be used again or would just sit corroding.

Later I learned that the trench and the building had been situated in that area of the moraine because it was actually the safest. Hydrologists had peered inside the moraine with a geo-radar device and found no ice. Thus there was no ice that could melt, making this section of the moraine more stable. If there were going to be a breach in the moraine it would more likely happen on the opposite side of the lake where there is still ice, called "dead ice" because it is isolated from the main glacier and tends to melt rapidly under the right conditions.

We did laundry with heated water in the morning sunshine. Sweaty clothes and bloody socks all got purified in the soap and sun. We ate Ram's plentiful dal bhaat, then went for a walk by the lake. I scrambled up in sandals while my wet boots baked in the sun.

From the top of the ridge above the house I could see all of the lake and the bottom of the Trakarding Glacier. The lake was a gray bowl, half full of opaque beige water. It looked like a crater on the moon that had somehow filled partially with water. The rim of the bowl was once the top of the glacier; the water is about a hundred feet below the rim. It is not picturesque like other lakes in the Nepal Himalaya such as Gosainkund or Gokyo—older mountain lakes formed by dif-ferent forces, where deep blue or turquoise water laps at the edges. Tsho Rolpa's beauty perhaps lies in its forbidding strangeness.

Winter freezes the lake's surface. The staff stays through most of the winter to record weather data, though sometimes in the deepest part of the winter they must leave to escape the cold.

In the six years Ram had been there, he said, the lake had grown about a hundred meters each year, extending farther up the valley as ice at the bottom of the Trakard-ing Glacier melted. He told Purna, who struggled to translate for me, that there had been more water in the lake in July because the glacier was melting from the summer warmth. But now, in late summer, the air was cooler so the ice was melting less than in June or July. Melting snow and ice swell the lake and put pressure on the moraine, not monsoon rains like those that had worried me as I trekked up the Tama Koshi.

I walked to the edge of the lake and peered down over the crumbling moraine into the cold water. Cholendra stuck right with me, making sure I didn't get too close to the edge of the unstable verge. A fall into those frigid waters would mean pretty much instant death. But the edge of the lake was less scary to me than the thought of walking down the steep hill above the Tama Koshi on wet, slippery rocks—the task a few days ahead of me on our return trip.

When Ram measured the lake level that evening he found it had dropped three centimeters from the previous day. Since July it had gone up only once; the rest of the time it steadily decreased.

# Day 10

We had to leave in the afternoon to keep to our planned two-week schedule. I could happily have stayed another day, watching the clouds come and go, the colors of the shadows and the sky change, the sheep graze in the little valley below. But I had gotten a good look at the strange lake and its well-built drainage channel, and I had extracted as much information as possible in the absence of an expert who could explain in English the lake's hydrology.

After lunch our little troop strung itself out along the valley, Cholendra reaching Beding an hour before the rest of us while I took up the rear, gazing around, watching the yaks watch me, savoring a walk without worrying about getting drenched or falling down.

This was to be a festive evening in Beding, a monthly community gathering at the gompa, the local Buddhist temple. The girls' father had come down from Na. I soon discovered that Kandiki's father and the Frenchman had come up from Simigaon. I imagined that many of these people were going to stay in the same room with us, so I claimed my cubicle again and wrapped up in my sleeping bag with some hot tea, watching the comings and goings. As he put on his maroon robes, papa joked that he was a "baby lama" and would chant the prayers with the full lamas.

The festivities had begun down the hill in the gompa by the time we arrived. Men and women sat on the floor or on benches along the wall. Some drank tea as they talked, while many drank *chang*, a kind of thick beer. Pretty girls circulated around the room, keeping everyone's cup full. "Shay shay, shay shay," they said, smiling as they poured, encouraging us to drink more. Children ran in and out or crawled on the floor. I had been escorted to the far wall to a space on a bench next to the one other foreigner, the French bridge builder. We sipped the chang and talked with an English-speaking Sherpa while the lamas chanted prayers. Later I met Kandiki's father, who was also a trekking guide. They and their crew had made good use of Nabaraj's bridge of saplings two days before on their way up to start construction of a small bridge that would provide an alternate route between Beding and Simigaon.

Seated on a dais in the center of the room, the lamas played drums and one blew a conch as they chanted, taking time out for sips of chang. When the chanting ended, small offerings—potatoes, cookies, candy—that had been sitting on the altar were distributed. We all took one or two items. A little later the women brought in enormous kettles of sherpa stew, a delicious potato soup.

Outside the temple the younger generation—including Cholendra and Nabaraj—
were dancing. Some sang while one beat a Nepali folk rhythm on a drum. They
illuminated themselves periodically with little LED flashlights, pulsing in the
darkness like fireflies.

Purna and the girls escorted me up to the house, then went back down to the
dance. As I was starting to doze I heard papa and Kandiki's father come in noisily.
From behind my closed eyelids I could see their flashlights shining on me and I
fancied I was in papa bear's bed and he had forgotten through the haze of chang
that he had a guest. Kandiki's father settled down on the bench perpendicular to
my platform. Papa went to the other end of the room. Much later the guys and
girls came up and made their beds.

# Day 11

At half past five our host and his daughters started talking loudly and moving
around as if no one were asleep, which indeed soon no one was. Kandiki's father
seemed to groan at the rude awakening, but soon he was up and on his way. Purna
stayed under the covers longest, as he always did. Cholendra and he had slept together on the floor this time; Nabaraj had stayed with some other friends. My
host up at Tsho Rolpa, Ram Bahadur, came by on his way up to the lake. Papa soon
headed back to Na to tend to his animals, though he came back into the house to
tell me goodbye.

The girls slowly prepared tea, then boiled potatoes and made chapattis for us to
take for the long day's walk down. As we walked into the trees the day grew warm
and sunny, transforming the landscape I remembered as cold and gray into a
pretty green park. We reached Nabaraj's bridge soon after our picnic lunch and
found it just as sturdy as the day he had built it.

I started walking faster, knowing we were less than an hour from the house
where I had eaten the wonderful mushrooms, and less than two hours from the
rustic place with no toilet. I hoped to override Purna's plan to stay at the latter
again. He thought that going all the way to Simigaon today would be too difficult
for me, but I wanted to spend a comfortable night with Tsering Deki, my favorite
hostess of the trip. Dong Kharka, the mushroom place, was closed now, as Purna
knew it would be. After a short rest I walked even faster and soon approached the
sans-toilet motel. It too was deserted, which we had not expected. I didn't have to
convince Purna I could make it to Simigaon, because we had no choice now. He
was hungry again, so I gave them a chocolate bar to share. It started to rain.

The next couple of hours I walked fast downhill on an easy trail. The guys were
looking forward to tea at the little farmhouse where we had stopped on the way
up. The day was still bright when we reached the house, Cholendra having gone
ahead again to speed up preparations of whatever they might offer us to eat. There
was no milk for tea, but there was plenty of fresh, tangy homemade curd. Cholendra

had put on a pot of potatoes. The Sherpa woman wasn't there, only her husband—who was keeping an eye on the two kids from the no-toilet place while their mother walked down to the nearest phone to try to call her husband, who was working in India.

The potatoes took a very long time to cook. It was four o'clock before we left, and I was worried about the remaining distance. There was a lot of up-and-down between the farm and Simigaon. The path became slippery again. We started seeing people on the trail for the first time in days: schoolkids going back to their village, groups of men and women from the valley. Cholendra went ahead to Tsering Deki's with a request for some hot water for bathing and sherpa potato pancakes for dinner. Cholendra had hoped to go to Kandiki's father's house, ostensibly because the latter had told us the night before he would make us potato pancakes. But the other guys were teasing Cholendra about being sweet on Kandiki. I said they could all go where they wanted, I was going to Tsering's.

Cholendra skipped on down the path with my duffel bag. The rain grew heavier and darkness began to fall on the woods. My legs were tired, my glasses fogged and wet; I could barely see my feet on the rocky trail in the dim light. If I put the umbrella down to rest my upraised arm and sore shoulder, my glasses got wet and I couldn't see through them at all.

The rocks were increasingly slippery, the path steeper. I picked my way carefully. I was irritable, almost crying. I kept thinking we must be close to the final curve to the left that would lead down to Simigaon, but the trail kept turning to the right and left, then right again, down and down. Finally I saw a faint light. The ever-vigilant Cholendra was coming back with a flashlight and a big umbrella. Fifteen minutes later I dared to hope the house looming up was Tsering's. It was.

We were back in Simigaon, whose name means "place where beans grow." Inside the house Tsering gave me a hug and offered us hot lemonade. Then Cholendra took me down the slippery incline to the bathroom, carrying a candle and a bucket of hot water for me to wash my hair and scrub off the blood from the day's leech bites.

Back in the main room Cholendra grated potatoes for the pancakes and Tsering Deki prepared a pitcher of her own home brew. Her chang was far superior to the drink I had tried for the first time the night before at the gompa in Beding. Hers was a light, delicately flavored nectar instead of the heavy, soupy chang of the night before. The effect of several cups was a gentle, relaxed high.

The pancakes took a long time, so I sipped a few cups of chang and petted the cat, slipping it little bites of butter. Tsering was a widow; her husband—a "climbing Sherpa" like our host in Beding—had been killed in a climbing accident. Her young son was in boarding school in Kathmandu. She told me the walk from Beding always took her two days and I had done it in only one. I didn't really believe her, but warmed again to her kindness. Her potato pancakes, smeared with butter, were delicious.

Gentle rain fell all night.

# Day 12

In the morning there was no rush to leave Simigaon. Our destination was only Gongar, the village where Nabaraj lived with his family. He and Purna had been saying for days they wanted to spend the night there on our return. I worried that stopping in Gongar would make our final day's walk to Singate too long for me, but when Nabaraj said he would be very happy if I stayed at his home, I knew I had to say yes.

The walk to Gongar would be short, but I was dreading the descent of the steep hill that had been so difficult for me to climb. I wore sandals, hoping to gain better traction; but even so I had to hold on to Nabaraj half the time because the rocks were so slippery. In a mere hour and a half the descent was over. We were at the Tama Koshi again and the temperature had jumped twenty degrees. My feet were covered in leech bites.

At the bottom of the hill were neat piles of cut stone, ready for use in building a new trail—one more congenial to people like me. We stopped at a house on the other side of the river for some noodle soup. My feet were still dripping with blood, so I sat outside on the porch of the second story, dangling them over the edge. The anticoagulant in the leech's saliva makes the small wounds bleed for a long time. Cholendra came out, saw the blood dripping from my feet, and started pulling the tassles from ears of corn that were drying in the eaves. He used them to wipe the blood from my feet.

Reaching Gongar before mid-afternoon, I went in search of a place to swim. There was a deliciously chilly pool in a side river that joined the Tama Koshi. It was in open view and I had no bathing suit, but I hoped no one would be around. I was sitting in my underwear and shirt on a rock, washing the leech bites and getting used to the cold water in preparation for my plunge into it, when I looked up to see the entire population of the village school ranged on the bridge watching me. A group of them came down and perched on a big boulder just above me. All that white skin was just too curious. I tried to make them go away in English and ineffective Nepali, then tried ignoring them. Nothing worked, so I gave up and scooted into the rocks and got my trek pants back on just before a handful of little boys came up from the other direction to watch at closer range. I stomped back to Nabaraj's house.

His father, Lal Bahadur Khadka, owns some land above Simigaon, but the farm could not support a family. He and his pretty wife Cham Maya brought their growing family down to Gongar and established a small shop with a restaurant. They sold tea and homemade chang and cooked meals. The children all went to school.

The afternoon we arrived, Cham Maya was working with other villagers to remortar the little brick schoolhouse on the other side of the bridge. A little farther down the path a handsome new three-story building fitted with several bathrooms stood empty. Engineers will live in it while they're building the new hydropower plant and dam upriver. That work can't start until the road is completed in about six years. In the meantime, the large new house, built to Kathmandu standards, will sit empty while the little schoolhouse of stone and clay will fill with Nepal's noisy and nosy next generation.

The Khadkas were in debt for their eldest son's wedding. They had hoped to pay the debts and start a new business after selling their land to the government for a good profit. The original plans for the new road placed it directly on the path we had traversed, the one that goes right through town in front of the Khadkas' house. Later, government officials decided to put the road down by the river, bypassing Gongar, the disappointed Khadkas told me.

The Khadkas had taken in several people, a couple of kids and an older disabled woman from Simigaon, apparently deaf and mute but cheerfully daffy. They fed us well that night. Nabaraj and his sister had picked a basket of the stinging nettles that were such a bane to bare skin. Once cooked they become benign and nutritious in a bright green soup. There was rice and dal and potatoes and *dhindo*, and they offered me some dark chang made from millet, which I liked far less than Tsering Deki's smooth brew.

The old woman babbled happily as she cleaned up the rice and dhindo we couldn't finish. The family and guests had devoured most of the soup and curry by the time our hostess served her, at the end of the meal, but she scraped the bowls to add flavor to her starch.

After dinner I tried to find out how Nabaraj's family felt about the possibility of a flood from the lake, close as they were to the river. Cham Maya told us she had been very worried in 1999 about a sudden flood wiping out her village. In a musical but agitated voice she explained to Purna that she thought all of them might die. That was at the height of the construction of the channel when workers were passing through Gongar and there was constant talk all along the Tama Koshi about the big lake high above. But the sirens had been installed the year before and soon the draining of the lake began.

Cham Maya said she felt better when the Department of Hydrology and Meteorology staff reassured the villagers that the siren would warn them in time. She said the staff told them they would have an hour after the lake burst to go up to a higher place; that's how long it would take the water to reach Gongar. Then the Maoists came and took the siren batteries and she was very unhappy again. She said she saw them carry the equipment away one night. They took the sirens from the nearby village of Jagat too, and now she worries again about a sudden flood because her family is so close to the river. She says she prays to God to keep them safe.

The entire family along with the old woman trooped upstairs to sleep. My three sidekicks and I went to the boys' room outside, which doubles as the guest room when travelers come through the village. Nabaraj had straightened it up and swept the floor. The room was dark but spacious and had four beds. I spread my sleeping bag on the one closest to the door, then opened the wooden shutters that

Nabaraj had already closed. I wanted to feel the fresh breeze off the Tama Koshi and see the bougainvillea when I awoke in the morning.

The cowshed was directly below our room and the toilet just beyond it, but to reach it I would have to step carefully over the cow muck. In the middle of the night I slid the heavy wooden bolt from the door as quietly as I could, then went out onto the path at the edge of the village to squat under the stars.

## Day 13

In the morning Nabaraj gathered up the damp jeans and shirts he had washed in preparation for his trip to the big city. He packed up some local dried fish to take to Kathmandu for making curry. The indispensable *khukri*—the knife that had saved the day and helped him build a bridge—stayed behind as we made our way out of the village and back down the Tama Koshi.

By afternoon the heat was so intense I felt faint. Every time we came to a stream I splashed water on my head and soaked my kerchief, tying it loosely around my neck. At times we would round a bend and the river breeze might take the ambient temperature down five or ten degrees. After many hours the sun fell into the west and the intense heat abated.

By evening we were back in bustling Singate. The boys went to buy our bus tickets for morning, and I headed to my bucket shower, tired and stiff. I went to bed soon after dinner.

## Day 14

The bus was supposed to leave at a quarter past six. I woke before four and never went back to sleep. I lay in the dark listening to the river and the quiet. I would soon regain all the comforts of Kathmandu but lose moments like this: cradled by the peace of the night and the hills, safe yet aware I was very far from the American life I had known and from all my old notions of safety.

And now I knew the lake I had feared just a few days before was stable, certainly for this year. The lake may have grown another hundred meters in length as ice melted during the past summer, but the channel was doing its job, draining the extra water and relieving the pressure on the moraine. As we walked to the bus, the Tama Koshi was a dull silver ribbon in the dawn. Mists drifted up the hillsides. Tangerine, gold, and dark peach hibiscus burst from the lush green. The bus left at half past six, almost right on schedule. We could see a tall pile of freshly cut logs on the other side of the ravine as the bus labored up the steep hill. They would be used to extend this road past Gongar and on to the new hydropower plant. Scattered up the hillsides on both sides of the road were solid, two-story clay brick farmhouses, the first stories painted white, the second a burnt orange.

Our bus would take us to where we could catch an "express" bus to Kathmandu. Soon it made its first stop. I looked over the shoulder of the man in front of me to see goats in the road ahead. I assumed we were waiting for the goatherd to get them out of the way. Some men from the bus went out to help. Then I began to suspect the goats might be coming with us, because I had been on local Nepali buses where dozens of chickens were loaded onto the roof.

In a few minutes I saw the well-fattened goats sliding up past the window, pulled by the scruff of their necks or their horns. One man stood on the ground, one clung to the ladder on the side of the bus, one braced himself on the edge of the roof as they passed the goats up quickly, hand over hand. The next time the bus stopped I got out for tea and saw the goats sitting on the roof apparently content, blissfully unaware they would soon serve a double purpose, a sacrifice to a Hindu god and food for a family.

The driver skillfully managed the hairpin curves on an impossibly narrow road above the steep canyon, guiding the heavily loaded bus ever upward, crossing small rivers that spilled over the road as the water made its way down to join the Tama Koshi.

# 5

# The Shrinking Third Pole

Glaciers in the Himalaya are receding faster than in any other part of the world and, if the present rate continues, the likelihood of them disappearing by the year 2035 and perhaps sooner is very high if the Earth keeps warming at the current rate.
—IPCC Fourth Assessment Report, 2007

Dig Tsho is another glacial lake high in the Himalaya of Nepal. On a summer afternoon in 1985, the lake's waters burst from their bowl of ice and rock. An inland tsunami flooded the valleys below, sweeping away potato fields, yaks, and a hydropower plant.

It was a Buddhist festival day in the Sherpa village of Thamo. Thamo's residents are descendants of families that five hundred years ago came over the mountains from nearby Tibet to settle the region known as the Khumbu, below what Westerners call Mt. Everest.

People were drinking *chang*, laughing and having fun. At four o'clock in the afternoon one woman, standing on a ridge above the Bhote Koshi, heard a sound like the roar of an airplane, then felt the ground begin to shake. The woman yelled to the other villagers, who came down to see a wall of water approaching from upriver. Those who lived on the slope closest to the river ran into their houses, grabbed religious items—portraits of monks, statues from family chapels, and Buddhist texts—along with leather trunks holding money and family jewelry. Some ran uphill to neighbors' houses and waited, while others carried images of Buddhist deities down to the riverbank and pointed them at the advancing flood, pleading for the river to change its course.

Elderly men and women in Thamo and nearby villages believe they know what caused the flood. They say a Sherpa man was tending his yaks in the high, sparse pastures near Dig Tsho that August. The morning of the flood, a stray dog ate his bowl of curd. The herder was so angry he grabbed the dog, tied its legs so it couldn't swim, and threw it into the lake. The act of cruelty angered a local deity, who caused a big chunk of the glacier to break off and fall into the lake. The water surged out.

There were no human casualties in the Sherpa villages high in the Khumbu, but lower down the channel, along the Dudh Koshi, people drowned in the churning river.

I heard this story in the spring of 2005, just a few days before I joined a pair of Nepali hydrologists who were trying to gather data about how much the vast glaciers in the Khumbu are shrinking. We planned to hike to Dig Tsho as well as to Imja, a glacial lake that was beginning to look as threatening as Rolwaling's Tsho Rolpa.

A hydrologist's explanation of the Dig Tsho catastrophe would differ somewhat from that of the Sherpa farmers. No dog, no yak herder, no deity caused an enormous piece of the steep, hanging glacier above Dig Tsho to break off and plunge into the lake. We all had a hand in this, along with our forebears, especially those of us who drive cars, use electricity generated from coal or gas, enjoy products made in factories that emit carbon dioxide, or eat a lot of burgers and drink milk from cows that belch methane.

The Dig Tsho flood was the worst of the fifteen glacial lake outburst floods that have occurred in Nepal in recent decades. The Khumbu's Dig Tsho and Imja, along with Tsho Rolpa—the lake I visited in the summer of 2006—are all in the Eastern Himalaya. So are several other glacial lakes that have burst in Nepal and Tibet, including Tam Pokhari and Zhangzhangbo. Glacier specialists see this region of the Himalaya as a glacial lake and GLOF "hotspot." It's wetter, there are more glaciers, and the lakes the glaciers are leaving behind as they melt grow larger each year.

The seventy large glaciers throughout the Himalaya and the Hindu Kush in Pakistan, along with thousands of small ones, together create a reservoir of water that sits at the source of a triad of perennial rivers in South Asia: the Indus, the Ganga, and the Brahmaputra. With more than a hundred thousand square kilometers, this is the largest body of ice outside the polar caps.

Global warming started to affect glaciers half a century ago, most scientists say. In the early years of this century, while the administration in Washington, DC, muzzled its own scientists as they tried to warn about the increasingly scary consequences of greenhouse gases, scientists in Nepal were trying to figure out a way to track and document the changes they saw in the vast masses of ice in their mountains. They knew glaciers were receding, and faster in recent decades, but the only reliable eyes that could track the process every year were miles high in the sky: satellite cameras. These were helpful, but did not yield information that would allow for good predictions about what the consequences of such melting might be.

Until recently, the Himalaya was one of the least studied glacial regions. The researchers trying to track climate change in Nepal have been few and underfunded, and the object of study very hard to reach if you have a desk job in Kathmandu. Like my trip to Tsho Rolpa, visits to glaciers in the Nepal Himalaya may require weeks of walking at very high altitudes. There are few roads and they don't go anywhere near the glaciers.

But in the spring of 2005 two hydrologists who then worked for Nepal's Department of Hydrology and Meteorology headed out on an expedition to check the status of glacial lakes in the Khumbu. I had interviewed one of them, Arun Shrestha, two years before when I was working on a story about hydropower in the Himalaya. I got in touch with him again, fortuitously enough, about the time he was planning a trip to Dig Tsho and Imja. Inspired by a recent report, I asked if I might come along. I would be welcome, he told me.

That same spring, the World Wildlife Fund (WWF) released a report summarizing some of the data that scientists had gathered about Himalayan glacial lakes that had already burst or might do so in the future. The report sought to draw attention to what was, at the time, a largely ignored phenomenon, remote from the consciousness of people in Kathmandu, let alone New York or London.

The WWF report ventured an estimate that Himalayan glaciers might be substantially melted within forty years and suggested disasters of biblical proportion should this come to pass. First floods, then searing droughts would come as the glaciers shrank away to nothing, traumatizing the Gangetic plain, where half a billion farmers and city dwellers now live, including some of the poorest people in the world. A WWF researcher told me there was no hard data about when those floods and droughts might arrive. At this point a few small glaciers had already melted.

Some news stories picked up the direst predictions in the report because they made good headlines and could be more easily summarized than the data about how many meters the glaciers were receding each year. Advocacy organizations like WWF are in somewhat of a bind. If they hope to affect public policy, they have to highlight the more alarming predictions about an issue, even if they are tentative.

In 2005 my mind balked at the idea that (should I have the longevity of my forebears) I might live to see the climax of this creeping cataclysm. These glaciers are indeed mere ice, but they are so massive that their disappearing in a generation seemed unthinkable. On hot summer days in California's Sierra Nevada I had seen pieces of the winter snowpack still sitting in a patch of shade, refusing to melt. How do we imagine a world with naked Alps and Andes, a planet without its poles or snow-covered Himalaya? Some call the Himalaya the third pole, so much water is locked up here.

As I walked up to Imja Lake in the spring in 2005 with Arun Shrestha, it was reassuringly cold and there was plenty of frozen water. Arun and a few colleagues from the Department of Hydrology and Meteorology were planning to do some reconnaissance of the glaciers. I went along to chronicle their efforts and try to understand how glaciers melt.

I was to join the hydrologists in Dengboche, a large tourist village where a couple of dozen inns for trekkers spread out on a windswept, treeless plain. With my guide Dambar Thapa Magar and porter Cholendra Karki, I reached the village in the afternoon. I asked Cholendra to look around the village for the hydrologists.

They came, already bundled up against the chill wind, and we made our plans in the warmth of the waning afternoon sun inside a glassed-in porch.

In the morning we hiked up the valley to Chukhung, a small village frequented chiefly by foreign climbers who want to tackle a modest Himalayan mountain called Island Peak. It's the last village before Imja. We reached it before lunch, but needed a night to acclimatize to the higher altitude before going farther.

The following morning before seven we started walking up the valley to Imja Lake. The hydrologists, much younger than I, strode purposefully. We had a lot of ground to cover that day because we were returning to Dengboche that afternoon. Dengboche itself was high above the river and perhaps somewhat safer in the event of a GLOF from Imja Lake than villages lower down the valley and closer to the river.

I walked as fast as I could, and fortunately the trail was flat by Himalayan standards. The gradual incline did not slow me, only the mere fact of the altitude: we were already above sixteen thousand feet. I lagged behind but managed to keep my companions in sight. The landscape was dun-colored except for the white mountains looming in the distance. We passed big mounds of loose rock, the remnants of the glacier. Acres of rubble and loose rock made the land look like the site of a cataclysm. The rubble was punctuated by ponds of opaque icy water. Inside some of the mounds of rock and rubble we could see ice—jagged and greenish white, hanging under the rubble in abstractly sculpted shapes.

Well before eleven and ahead of schedule, we reached the bottom of the terminal moraine. It was a hill of loose rock, requiring of me a breathless scramble to get to the top. Cholendra grabbed my hand to steady me as I teetered on a precarious chunk of the moraine, trying to catch my breath. At the top we found a flat place to sit and look out over the lake. I'd already written some radio stories in which I mentioned moraine dams as if I understood what they were, though I really did not. Now I was sitting on one. They look and feel like a big pile of big rocks. The surprise is that the rocks are often sitting atop and held together by ice. Turn up the heat a little, the ice melts, the rocks slide, there goes the moraine dam, here comes the flood.

It was cold and uncomfortable, not a place for the picnic of hot tea and fried Tibetan bread Cholendra was carrying for us, so I rushed to start asking questions. My brain went suddenly blank, as if the cold thin air I had been breathing had anesthetized it. "Is anybody's brain working?" I asked, smiling and sniffling from the cold. "Not mine," replied Arun, smiling back weakly behind his dark sunglasses. So we sat there for a while in silence. Arun pulled a big blue and black Gore-Tex jacket over his emerald green fleece with matching cap. I pulled a down vest over my thick fleece jacket with multiple layers underneath.

Arun Shrestha is taller than most Nepalis. He is strong and fit, so I was a little surprised to find he too was affected by the altitude, especially since I had been struggling to keep up with him. I had met him several times since I became interested in Nepal's mountains, and he had always been easy to interview because of his clear, precise English. His round face showed the beginnings of a mustache and beard after a week in the mountains. He is friendly and courteous, but not inclined to a lot of small talk. We talked mostly about the glaciers, the reason we both had made this demanding trip. The trip was work for him, but I sensed he enjoyed being up here. I had heard that many South Asian scientists were city people and did not relish research like this. But Arun had tramped through the hills often while he was growing up; his father, a government employee, had been posted throughout Nepal and his grandparents lived in the hills of eastern Nepal. He was named for the Arun, one of the powerful rivers that flow out of the eastern Himalaya.

I had seen Imja Lake once before, in 2003. Just out of curiosity, I had struggled partway up the steep hill behind Chukhung village in a stinging November wind to catch a glimpse of the gray-green lozenge of a lake in the far distance. Now the silt-laden water was frozen; the lake surface twenty-five feet below us was nearly white. Arun said things looked different from the last time he had been here because this was a colder spring and less ice was melting into the river. But there are big seasonal variations, he hastened to add; it was the overall trend that they were trying to understand.

The hill of loose rock on which we were sitting is called a terminal moraine because it's at the bottom of the glacier. It stretched two thousand feet across, creating the lower shore of the almost rectangular lake. The rubble was deposited before the glacier started melting. If the moraine were to break under the sheer pressure of the water or during an earthquake, a wall of water along with many tons of boulders and debris would inundate the floodplain for miles down the valley. That could destroy part of the most developed tourist region in Nepal. The trekkers bound for Everest Base Camp may tarry for days in the river valleys of the Khumbu as they work their way upward. There are villages, farms, and dozens of lodges. It's the only trekking area that continued to produce good revenues during the height of the Maoist insurgency. In a peaceful Nepal it is even more productive. In essence, it's the country's cash cow.

Like most of these large valley glacial lakes, Imja started growing half a century ago. It started out very small, but in recent years has been growing rapidly. Imja holds about thirty-six million cubic meters of water, making it half the size of Tsho Rolpa, the lake with the drainage channel.

Imja Lake is fed by a complex of glaciers: Imja, Lhotse Shar, and Amphu Lap-cha. The place where they meet, called the terminus, has been retreating about 250 feet a year. We could see the three glaciers curving toward each other in the distance, looking somewhat like long, sweeping ski runs. Behind us the lake drained into a cloudy river, the Imja Khola, which joins the Dudh Koshi.

The ice inside the moraine was gradually melting because it was in contact with water. Thus the wall of ice and rock that embraces the lake was slowly thinning. "Depending on the rate, that can be dangerous," said Arun. "It can create a GLOF. But the end moraine of Imja is quite big, so that lowers the threat for now," said Arun in his usual calm and measured tone.

My guide Dambar scrambled back up the side of the moraine, breathless. To help the hydrologists, he had volunteered to go down closer to the surface of the water with a portable GPS device to get a reading on the level of the frozen water.

"We don't understand the system," said Arun. "It might change in such a way, such an unpredictable way, that we might find ourselves helpless to do anything at that point. So that's why it's wise to do regular monitoring now." At that early date Arun thought Imja Lake might be a candidate for the kind of draining system installed at Tsho Rolpa, and for some kind of warning system.

Dipak Gyawali was Nepal's water minister for about a year between upheavals in Nepal's political landscape. He's a polyglot bundle of energy who shows no signs of slowing down in his late fifties. He has the distinguished looks of a minister, with his salt-and-pepper hair and slightly thickening waistline, but his manner is jovial and frank. His face tends to have a perpetual look of slight surprise, though at this point in his life he's not surprised by much, especially South Asian political shenanigans. He was born in Varanasi, India, because his mother had retreated there for safety at a time when there was political turmoil in Nepal. His father was later Nepal's attorney general for many years.

He took a degree in physics in India, then went to Moscow to study hydropower engineering. After several years of working on hydropower for the Nepal government, he became interested in economics and was awarded a Fulbright grant to study at UC Berkeley's Energy and Resources Group under John Holdren, later President Obama's science advisor. By then, Dipak says, he had begun "to think engineering was more part of the problem than part of the solution."

Dipak speaks English just as fast as he does his native language. I sometimes think he talks fast because he has so much information in his head about so many subjects and he's always trying to get the ideas out while he has a few minutes in the midst of a busy schedule of meetings and travel abroad. He cofounded Nepal Water Conservation Foundation, where he still works, writing, lecturing, and directing research. He's a sharp critic of what he sees as flawed ideas and development projects based on them. I met him in 2003 when I was trying to learn about hydropower, and have sought his advice and ideas about South Asian water issues ever since. I sought him out in 2005 after reading the World Wildlife Fund report.

He dismissed its predictions as alarmist. "We all have to be worried about glaciers receding and what long-term impacts there will be, but to claim there's going to be floods in Bangladesh because of all this is bad science, pure and simple."

Glaciers melt in the hot season from March to June, he pointed out, before the monsoon begins. This is the season that has the lowest flow in the river. In a normal Himalayan snow-fed river, the difference between the dry season flow and the monsoon flow is dramatic. In the Karnali River of western Nepal (one of the Ganga's many tributaries) "there are about 250 cubic meters per second, or cumecs, in the driest months. During the monsoon, in July and August, there can be 15 to 20 thousand cumecs. And that's what causes floods in the lowlands," Dipak said. "No amount of melting glaciers and melting snow can create the kind of flood you have in the summer."

But people in the mountains are still in danger, Dipak acknowledges. They will have to cope with the threat of deluges from glacial lakes. "The entire volume of a glacial lake can empty out in a few hours, and that could cause mayhem in Nepal, China, and the mountains of India. These floods carry not only water but boulders and trees and whatever else they can jar loose. This will be the effect from the source of the flood for about 40 kilometers downstream, but then it will dissipate."

Estimating how much of the water in the Ganga and its tributaries comes from melting snow and glaciers during the hot, dry months before the monsoon has been a central concern of researchers. If a substantial amount is from glaciers, and glaciers were to shrink drastically, there could be a great deal less water in the rivers in the dry season. That would mean the farms, villages, and cities that rely on that vast network of streams and rivers flowing downhill from the Himalaya might be in deep trouble. Should the glaciers disappear, as various experts were speculating, the hydrology of the Indian subcontinent could be profoundly altered.

A few years ago, estimates of the amount of meltwater in the rivers from March to June went as high as 70 percent; recent studies say 10 percent is more likely for the Ganga. "The role of groundwater soaked up in the Himalayan hills seeping into springs and streams during the dry season has been completely missing in all the hype about glaciers," says Dipak. The middle hills of the Himalaya are like sponges that soak up the monsoon and release it gradually through the rest of the year. The streams and rivers that carry this water may not have any snowmelt in them, let alone glacier melt. Only the large rivers that have their source in the sort of places I was walking with Arun can bring snow and glacier melt to the Gangetic plain.

Dipak thinks there's a far more serious threat from global warming for people downstream of the Himalaya. "The extremes are going to change and we have no idea what that will mean. Cloudbursts seem to be more frequent." Massive amounts of rain may fall in just a few hours. That's happened in Bombay, Calcutta, and near Kathmandu in recent decades, and most recently in Pakistan in the summer of 2010. These cloudbursts drop far more water than even the heaviest monsoon rains typically bring. People and animals die; homes and livelihoods are destroyed. "Hot and cold fronts collide, creating intense cloudbursts. The atmosphere is carrying a lot more energy in terms of heat because of global warming and that energy has to dissipate.

"If I were a minister now I'd spend my scarce resources on studying bishyaris, not GLOFs," Dipak told me in 2010. *Bishyaris* are floods from lakes created by monsoon landslides. Part of a hill can slide into a river and block it temporarily. The pressure behind this landslide dam builds as the water level rises. Within hours, or at most a few days, the river bursts through the landslide and can cause havoc downstream. "The phenomenon is much like a GLOF, but bishyaris are far more dangerous because they happen in densely populated areas and are almost impossible to predict. Satellite photos can tell me a lot about potential GLOFs but no one as yet can tell where a hill slope failure will occur." If monsoon rains become more intense due to global warming, as many predict, "With our steep slopes here in Nepal, just a little increase in volume of precipitation means the devastation of landslides is going to increase."

At the other extreme, there could be drought and delayed monsoons. "If seasonal springs dry up, or if the dry season is two weeks longer, you're talking about wiping out your livestock," says Dipak. "This is the kind of impact we don't know and needs to be studied." He would like to see South Asian countries get together and put up a satellite exclusively to monitor the monsoon, "the lifeblood of this part of the world. It's not much—less than three fighter planes in the Indian Air Force! And," he added with characteristic conviction, "it would provide much greater security to a lot more people.

"Climate change will manifest itself through water. It will affect every sector of life through precipitation, snow, rain, whatever," says Dipak. "Livestock, forestry, soil, sanitation, disease, everything. But we have no idea how it's going to happen."

Having seen foreign aid money mismanaged from time to time, Dipak Gyawali is leery of what he calls a "racket" developing in the monitoring of glacial lakes. He believes that panic was created unnecessarily by government officials and foreign consultants to justify the Tsho Rolpa project and that this could happen again. He felt that too much money was lavished on the project.

Dipak also thought the warning system for Tsho Rolpa—the sirens I saw mounted on poles in the villages along the Tama Koshi and the Rolwaling Khola—was mishandled. He said dedicated warning systems never work for droughts or floods. FM radio stations would be much more effective. They would function all the time, be maintained, and double as a flood warning system. "So when a flood does come, people will get the warning."

Arun Shrestha agreed that any warning system for Imja Lake should be very different from the one set up for Tsho Rolpa. That kind of system works only if villagers take ownership of it and keep it working. But Arun believes building the wide trench I saw at Tsho Rolpa to lower the level of the lake by three meters was money well spent.

Arun is also an internationally educated, polyglot professional. After an undergraduate degree in engineering in Nepal, he obtained a degree in hydraulic engineering in the Soviet Union before going to the United States for a PhD from the University of New Hampshire. He arrived there in the early 1990s, in time to be on the cutting edge of research on climate change in the Himalaya. A PhD program in glacier research, later broadened to climate change, was getting underway. "From the day I started the program I could discuss the connection between climate change and the Himalaya with the professors there. They had been aware of the issue for a long time."

In 1998 Arun returned from studies abroad to his job at Nepal's Department of Hydrology and Meteorology. The warning system for Tsho Rolpa had just been installed, and work on the trench was beginning. Tsho Rolpa is twelve times the size of Dig Tsho, the lake that burst in 1985. The decision to build a trench to drain some of Tsho Rolpa's eighty million cubic meters of water does not look like a frivolous one. Yet there continues to be debate about how it was done and how much it cost. Arun says the trench was the right thing to do and wonders how critics—who say that the job could have been done more cheaply, without the expensive heavy equipment brought in by helicopter—would tackle such a large moraine without it.

Cheaper alternatives like manually digging trenches to drain some of the water, as is currently being done in Bhutan (but only in the summers), would have been time-consuming and not as effective, Arun says. A series of siphons would not have worked very well either. Too many pipes would have been needed to drain the lake in any significant way, and in the winter they would freeze, then have to be repaired. The wide trench at Tsho Rolpa stays open and drains all year. Arun thinks another benefit of that project was how much Nepali hydrologists learned about glacial moraines while the work was being done—knowledge they can apply to other Himalayan lakes.

Arun's main goal is to accumulate good data about the glaciers. He says the sheer lack of data makes predicting the consequences of melting glaciers in South Asia a risky business. "In the Alps some glaciers have been monitored in a very detailed way for a long time, so scientists have a very good record of how glaciers have been changing, even in past centuries." In the Himalaya monitoring has been both limited and intermittent.

Arun says the way to start to correct this inadequacy is to establish several benchmark glaciers throughout the region and find out what they are doing, whether they're retreating or advancing and by what amount. He and his colleagues are doing this work in Nepal; another glacier specialist, D. P. Dobhal, is doing the same in India.

With this data it will be possible to translate changes in ice mass into amounts of water. Then glacier experts can extrapolate what amounts of water will be available at different times of the year as the ice mass changes in the future. They can use modeling techniques for their predictions, but only with years of field data will the predictions be sound. "It's not a one- or two-year project; it has to be steady and long-term. Models don't work without field data," Arun says. And the models developed in Europe or the United States don't work well in South Asia anyway because the Himalayan environment is so different.

Himalayan communities will have to adapt to GLOFs, Arun says, and his job is to alert those communities. First he has to determine whether there is a possibility of a GLOF, then quantify the potential discharge of water, how far it might travel, and what will be the peak discharge. To prepare for floods in the future, mountain communities will have to build bridges that allow flood surges to pass under them. And the design for new hydropower plants can be adjusted to handle glacial outburst floods. Builders can spend a little more money now to keep their investments safe in the future. The powerhouse can be placed underground, for example, spillways can be made large enough to accommodate floodwaters, and intakes can be devised so that they would receive minimal damage.

Arun agrees that floods will dissipate as they travel downstream and at some point become harmless. But, he says, "we don't know where that boundary is." He too is cautious about predicting disastrous drought from melting glaciers. "You have to distinguish water from glacier meltwater. The farther you go from the Himalaya the more important rain is compared to snowmelt. Even in the dry season there is base flow that is not snowmelt. Water comes through the ground." He says predictions of future shortages are probably accurate, "but we don't really know what would be the impact as you go further and further from the glaciers. You can safely say there will be an impact in the greater Ganges system because the contribution from the Himalaya is so large. But we don't have concrete examples of what could happen."

Like Dipak, Arun notes that faster melting of glaciers might mean more water for irrigation and other uses in the dry season. "But we don't know how long that increasing trend will last. As the ice reserve decreases, of course, then the flow has to decrease; but we don't know when that point is between increasing flow and decreasing flow."

The more immediate danger is erratic weather because of global warming. "Extreme events are much more important than the average yearly flow. The most important thing, for agriculture and other water users, is not the average amount of water flowing in the river but the timing of the peaks and lows. Droughts and floods might be more severe even though the annual amount of precipitation is the same.

"Alarm should be backed by science," Arun says. "Many people are saying there will be disaster in the future. That may be true, but the claim should be backed by science. We have put forward the possible impact of glacier retreat. Science says if the retreat continues there will be a change in the flow of rivers and streams. This hasn't happened yet, but it might happen abruptly in the future. For now, the glacial lakes are the problem we can see clearly. There could be a dramatic increase in the number of floods in the future. We have confidence in this prediction. We don't have so much confidence in predicting stream flow changes."

I've made half a dozen trips into the high Himalaya to see glaciers and I've lost count of the number of hours I've spent talking to glacier specialists. But I am just beginning to grasp very basic information about these ice rivers and how they are knit into the entire meteorological and hydrological landscape of South Asia. There are so many facets to the implications of global warming in this region, there's no wonder Arun and other scientists are not ready to make specific predictions. Arun told me the tools he and his colleagues have remain too primitive to assess those implications in a scientific way for the combined Ganges, Brahmaputra, and Indus basins.

Early on I tried to elicit from Arun a comment that would make a nice sound bite, something that would convey alarm similar to that in the WWF report. He stuck to scientific caution. But over the years that I spoke with him, Arun's understanding of the threats did shift. When I went to the Khumbu with him he emphasized glacier scientists' certainty that GLOFs would continue to be a threat, even though they could not predict what would happen on the Gangetic plain as a result of global warming.

Later he said GLOFs are certainly an issue, but not the big issue. His scientist's intuition told him the danger lies in how and where the water comes. The monsoon already concentrates rain in South Asia into a few weeks of the year, and the trend seems to show that the monsoon is becoming more compressed. It's all too likely that there will be too much or too little water, at the wrong time and in the wrong place. "When the consequences of our altering the climate and the behavior of the glaciers do become clear in South Asia," Arun says, "the news is not going to be good."

D. P. Dobhal works at the Wadia Institute of Himalayan Geology in Dehradun, Uttarakhand, India. He's been there since 1989 and has been recognized for his innovative approach to measuring glacier retreat, most recently on the Churabari Glacier. The Churabari is just south of the Gangotri, the glacier that gives rise to the Ganga. I visited him in 2007 after I had traveled up to the foot of the Gangotri. I hired a driver to take me over from Rishikesh one sunny morning. It was a brief but terrifying ride at the hands of a young Indian speed demon. At one point I asked him to slow down. Instead he turned off the blaring radio. That was somewhat an improvement; I felt calmer, though I didn't feel much safer.

Sitting in his office in Dehradun as a spectacular late-September monsoon downpour began to darken the skies and chill the hot day, Dobhal reeled off the number of glaciers in each South Asian nation. He showed me how all the rivers in Uttarakhand descended from glaciers then merged and eventually found their way to the Ganga. The ice rivers in the Himalaya vary from half a kilometer to seventy-two kilometers in length, he said. Fifty years ago the Gangotri was forty-five kilometers long; now it is less than thirty kilometers.

Dobhal, who prefers to use his last name, says the glaciers in the eastern Himalaya used to get good snow in winter as well as in the monsoon, what's called "the summer accumulation period." But now snow is falling less in October, November, and December and more in January and February, just before the spring thaw starts. That means the late snow does not have time to solidify, so it melts faster, because it's snow, not ice. A good winter snowfall protects the older ice, which then melts slowly. The thaw doesn't begin until June. But now the melting tends to start early, in April.

This lack of snowfall is the chief problem Dobhal has seen in the high mountains in recent years. As a consequence, the glaciers are depleting, not developing. With good winter snowfall they stay in balance and the melting rate is not a cause for alarm. Melting glacier ice accounts for 30 percent of the water in the rivers, he says. The rest is from snow and from the monsoon. Now that there is less snow, the spring flow in the rivers comes directly from the older ice.

When glaciers lose their volume, the rate of melting increases. It's the difference between a melting block of ice and a melting ice cube, says Dobhal. "Big glaciers create their own climate. They make cold weather. Big glaciers can be more powerful than the sunlight that reaches them. But as glaciers shrink, the power of the sun to melt them grows."

Dobhal is measuring the depth of glaciers to see how much they are shrinking vertically. He places sticks at different elevations, creating a network. "We are getting only fifty centimeters accumulation per year. And we lose 1.5 to two meters."

Dobhal chose to study the Churabari glacier because it faces north and he wanted to compare its behavior to other glaciers, most of which, like the Gangotri, face south. The Churabari is near Kedarnath, which along with Gangotri is one of four Hindu holy sites in northern Uttarakhand. He says the heat from the villages where pilgrims stay, as well as heat from construction and from people walking up and down the trail to the glacier, are all taking a toll on it. Shoes on the upper surface of the ice disturb the glaciers and accelerate melting. "Compared to global warming the effect of people visiting is small, but it contributes." I think of the barefoot sadhus I had seen during my brief visit to Gangotri. Perhaps centuries of worshipping sadhus have done no harm. But the hordes of middle-class Hindus, shod in sneakers as they climb up to do puja, have melted more ice than the centuries of itinerant sadhus.

The decrease in snowfall near his birthplace was what first got Dobhal interested in studying glaciers. He was born in a village near Badrinath, another of the four holy sites in Uttarakhand. Dobhal believes a complete retreat of the South Asian glaciers would take many decades, maybe a century, because glaciers have their own environment and create their own temperatures. "Snow and ice cover only 10 percent of the earth now. Long ago it was 30 percent. It's a long process," he says.

Dobhal answered my questions in rapid-fire but heavily accented English, intent on his subject. He agrees with Arun Shrestha: glacier scientists need many more years of data before they can tell what is really happening. India started gathering data late, in the 1970s, he says, and it's still too soon to say what the trends are.

As we were talking, the rainfall outside became more fierce, pounding on the roof and the courtyard. "You can't play with the climate; you can't play with natural things," he said finally.

Dobhal showed me around his lab, with photos of the work camps on the glacier, then escorted me to the entrance of the office building, from which we could see that my driver had fallen asleep in the car at the far end of the parking lot. The windows of the small Maruti Suzuki were steamed up, but my driver wasn't watching for me in any case. We saw him slide over from the front driver's seat to the back in order to nap more comfortably. I yelled and gesticulated, hoping to get his attention so he would drive over to pick me up, but to no avail. Finally I made a run for the car with no umbrella, splashing across the puddles in the parking lot, getting soaked in the downpour.

My trip to Imja Lake with Arun was my second trip to the Khumbu. The first time I had gone to that region, back in 2003, I was still a working journalist doing interviews for a story about how electricity came to Khumbu villages nestled 12,000 feet in the Himalaya, changing life and tourism there. That's when I first heard about the Dig Tsho flood that wiped out a hydroelectric plant, and about the then little-known phenomenon of glacial lake outburst floods. It was also on that trip that I fell a little bit in love with a river called the Dudh Koshi.

Like many people, I am drawn to the magic of rivers. I felt the tug of them in California and the Pacific Northwest. Then the mountains and rivers of Nepal kept drawing me back, halfway around the world. The cloudy aquamarine waters of the Dudh Koshi, plunging down from the Ngozumpa glacier, awed me. And I liked its name: *dudh koshi*, milk river. My guide said it was named for a lake, the Dudh Pokhari by Gokyo village. But I liked to think it's because of how the water looks: cloudy and nourishing. The river carries light-colored sediment from high in the Himalaya all the way to the Ganga; that, technically, is why it's cloudy. From a distance it's a pale blue ribbon; up close it's an awe-inspiring tumble of rock and water and noise. I've followed it up the valley from both sides on my trips to Gokyo, getting closer and closer to its tumble as the trails take us higher and higher up to the bottom of the glacier from which the river is born.

Gokyo village, perched on the slopes of the moraine about an hour's walk up from the bottom of the glacier from which the river springs, is a popular tourist destination. Most tourists head for Everest Base Camp, but others go instead to Gokyo to see its string of turquoise blue lakes. Intrepid ones may cross the rocky and difficult Cho La, a pass that connects the two valleys.

The Ngozumpa is the largest glacier in Nepal, about six miles long—a third the length of the Gangotri in India. It winds down from Cho Oyu, one of the eight-thousand-meter peaks that lure climbers to Nepal. I find the valley of the Ngozumpa far more serene and beautiful than the glacier valley to the east that leads to Everest, known as Chomolungma to the Sherpas and Tibetans, for whom it is sacred. The blue lakes strung out along the valley, Gokyo village perched next to the largest of them, soften the stark gray glacier.

I met Arun and one of his colleagues, Raju Aryal, in Gokyo five days after our visit to Imja. We had returned to Dengboche by late afternoon the day we visited Imja. The next day we went up the valley toward Chomolungma to take a look at the Khumbu glacier, which winds down from the world's highest mountain. We went our separate ways until our rendezvous in Gokyo. There the hydrologists planned to look for any changes in the Ngozumpa glacier that might be visible to the naked eye. They also sought a good location near the village for a device that would record temperatures. They were hoping to find a villager who would keep an eye out to prevent what happened to the solar panels and batteries on the warning sirens in Rolwaling.

They talked at length over tea with the *sauni* at the Gokyo Resort—a rustic inn by most standards, but the height of luxury in a tiny tourist village that is vacated by all but a handful of people in the frigid winters. Every drop of cooking kero-sene, every bite of food, every piece of wood, not to mention porcelain toilets, thin foam mattresses, and colorful Tibetan rugs, must be carried up to Gokyo la-boriously, by a man or a yak.

The Sherpa sauni and her Brahman husband agreed to take responsibility for overseeing the automatic weather station. People in the Khumbu were beginning to understand the need for such data gathering and the threat of glaciers melting. Arun or one of his colleagues would have to return in a few months to position the weather station and give the proprietors of the inn some basic training on how to tend to it. The only way the hydrologists could get the recorded temperatures from the electronic gauge would be to hike up at regular intervals with a laptop and download the data. Each journey would take a minimum of five days' hard walking from the nearest airfield in Lukla where small planes could land.

We left the resort and climbed up on the moraine. It was a short walk, but the sudden exertion made us all a little breathless. Whitewashed chortens were ranged along the top of the ridge and faded prayer flags fluttered. In mid-April the Dudh Pokhari below was still frozen; only near the shore at one end had the ice melted, revealing the lake's otherworldly shade of turquoise. The Dudh Pokhari and two smaller lakes lower down the valley were formed by glacial meltwater that came to rest in the valley's depressions during some earlier eon. Their water flowed through the rocky ground, which cleansed it of glacial silt so the clear water could reflect this stunning blue-green. Water from these lakes then joined the silty torrent emerging from the bottom of the glacier to become the Dudh Koshi.

We sat for a while on the ridge above Gokyo village, 16,000-odd feet above sea level. From time to time we could hear the ice crack, like an ice cube popping in an enormous summer drink. The Ngozumpa glacier was easing its vast bulk down the wide valley it has carved over the millennia. Below us stretched the gray surface of the glacier, which is not white and pretty like those in the Alps. The young, steep mountains of the Himalaya shed masses of sediment, which fall on the tops of the glaciers. Some small glaciers in the Himalaya are clean and white, but long valley glaciers like the Ngozumpa are covered with rocks and gray grit. This can contrib-ute to melting: when the debris cover is thin, the glacier absorbs a lot of heat from the sun, just as people do when they wear dark clothes.

The sediment covering the glaciers is one of the reasons the models used in the West to predict runoff from glaciers don't work here. They have to be adjusted to account for the effect of the debris on rates of melting and water flow. A thick cover of debris may slow the rate, but a thin cover will—depending on the thick-ness of the ice underneath—speed it up. In 2004 the government of Pakistan, worried about a six-year drought, made an ill-advised attempt to increase the melt rate and provide more runoff from the glaciers by putting charcoal on them, a dangerous short-term solution to a water shortage.

In some ways the surface of the Ngozumpa glacier here near Gokyo looks like a demolition zone. Some would call it a moonscape: desolate, cold, and forbidding.

All those descriptions seem apt enough, but I find the glacier strangely beautiful and very much alive. The magnificent sweep of the Ngozumpa stretches as far as we can see up and down the long valley.

"How would anyone know that it's melting?" I ask Arun. "Pools of melted ice are one clue," he tells me. He says there seem to be more of those than he had seen in previous years, both here on the surface of the Ngozumpa and on the Khumbu glacier we had visited just a few days before.

Satellite photos also showed the number of ponds to be increasing. The pools are like enormous pockmarks in the rumpled skin of the glacier, often rimmed by exposed ice. The pools are deep, and the surface of the water is many feet below the surface of the glacier. The ice is blue, which means it's old. The hydrologists planned to keep a watch on these pools to track evidence that the glacier is shrinking. Some of the pools are separated by fairly narrow slices of debris-covered glacier, which can erode, fall into the ponds, and melt away, thus joining two ponds. Eventually a series of such ponds could become a large glacial lake. None of the ponds we saw posed any immediate threat.

Glaciers have their own complex plumbing system. They are not completely solid but have channels running throughout under the surface. Water can flow through them or be stored in them. These ice caves are completely natural, Arun says. Some Westerners explored ice caves in the Khumbu in 2009 and assumed

Gokyo Lakes and debris-covered Ngozumpa glacier seen from Gokyo Ri, 18,000 feet

they were the result of global warming. Some may be; others may not be. Some of the pools on the surface of the glaciers will drain out through such channels before they can expand to become large glacial lakes.

We were sitting above what's called the "ablation area" of the Ngozumpa, where snow and ice are melting or evaporating. When there is more depletion of ice than addition, the balance is negative and the glacier is receding. Up above, where we could see a clean white sweep of glacier coming down between the mountains, is the accumulation zone. It's usually clean, even here in the Himalaya, while the ablation area is often so covered with debris the casual observer doesn't know there is ice underneath. There has to be a very large accumulation area above to sustain such a large ablation area, Arun tells me, acres and acres.

"A glacier is not a static block of ice," he says. "It's actually moving, flowing down the valley from a higher to a lower place. In the higher places it is colder and precipitation comes as snow. If it's cold enough it doesn't melt. It accumulates over time, and with pressure from more falling snow it changes into ice. Then it flows downstream to a warmer place where the melting starts. Ice from higher up continues to feed downward." It would take decades for snow that fell in the accumulation area of this long glacier to flow down to the ablation area as ice.

If the snowfall and melting are in balance, the glacier is stable. If there is more snowfall than what melts, the glacier advances. "But the opposite is happening now," Arun continues. "The melting is greater than the replenishment of ice from

Melting pond inside Khumbu glacier

above, and so the glacier retreats. And it's not just the lower part that is going back, it's not a linear thing; it's three-dimensional. The whole glacier is shrinking."

Precipitation isn't necessarily a good thing under these conditions. Rising temperatures could mean rain instead of snow. Arun says that "is twice as bad as simply getting less snow, because rain melts the existing ice on the glacier," thus doubling the loss of ice. Arun thinks the amount of snowfall here probably isn't changing so much yet. Warming temperatures may be shrinking this glacier. But the relationship between rainfall and glacier size can add complexity to the puzzle: "The decreasing glacier may indirectly cause less rainfall, since the amount of summer rain is partly dependent on the cooling effect from the glaciers."

For unknown reasons temperatures in the high mountain regions of the world are rising faster than they are at lower elevations. To Arun this is one of the chief causes for alarm: "Studies I have done on temperature trends in Nepal in the last few decades show that temperatures are increasing at a very high rate compared to the global average." But scientists are baffled as to why temperatures are rising up here faster, says Arun. "It's generally observed that trends are amplified in mountains. This is seen in other parts of the world, like the Andes and Alps. We don't understand why."

Strangely enough, down below in Kathmandu and farther down on the Gangetic plain—where summer heat is already beyond anything I can bear—temperatures show only a small increase above past norms.

It's still so cold up here relative to the rest of the planet I have trouble grasping how massive chunks of ice five miles long, a mile wide, and a thousand feet thick can melt. But these glaciers are like polar bears; they thrive best in extremely cold weather. And especially in the intense South Asian sunshine, they need low temperatures. I think of the yaks and yak-cow crossbreeds that thrive up here and can't go below 11,000 feet or they will die. Their shaggy bodies are supremely suited to the thin air that sometimes leaves me struggling for breath. It's a shaky analogy, but somehow it helps me to grasp what's happening more than do the statistics on the number of meters the Ngozumpa has retreated.

The Ngozumpa is so long and thick that changes here will be slow and detectable only through very accurate measurements, not just the hydrologists' quick reconnaissance. "A decade would be a minimum time to say with confidence what is happening here," Arun said. "If this trend continues, and we think it will," he continued, "the glacier retreat will be even faster in the future. So I think this is a pretty alarming situation."

Arun and his colleague Raju finished looking around the moraine above Gokyo for a good spot to place the weather station. I told them goodbye, laughing that we had walked for three days to get together for half an hour, here on the moraine above Gokyo.

We were going to meet again in three days and walk up to Dig Tsho, the lake that burst in 1985. I would walk to our meeting place in the village of Thame at my own pace. Arun and Raju headed down the valley with long strides; they were going to download data from a monitoring device they had already established above Namche Bazaar, then they too would proceed up to Thame. They might sometimes walk for several days for a half hour's downloading, only to find that solar panels had been stolen or the device had malfunctioned. Those were just the typical hazards of gathering data in the Himalaya.

As I dropped down the trail at the bottom of the glacier, a blisteringly cold spring wind ripped around the boulders. Just a dozen yards from me amid the tumble of freezing water was a boulder the size of a house, carried down by the ice river perhaps hundreds of years ago.

By now I had been tagging along with the two Nepali hydrologists for more than two weeks of hard walking in this rugged, thin-aired landscape. Dig Tsho—our final destination and the scene of the most devastating GLOF in Nepal's history— was for me the most mysterious and alluring of our goals.

But first I wanted to talk to villagers who had witnessed the 1985 flood—the flood that some Sherpa people attributed to an angry goddess, others to global warming. I asked Ang Danu Sherpa, who was then the managing director of Khumbu Bijuli Company, the Sherpa-run electricity cooperative, to help. I had met him in 2003 when I came to learn about the small hydropower plant—among the highest in the world at 13,000 feet—that produced power for Thame, Thamo, Namche, and other nearby villages. The Dig Tsho flood had destroyed an almost completed hydropower plant on the Bhote Koshi, a tributary of the Dudh Koshi that comes over from Tibet. After the flood, the Thame Hydropower Plant was rebuilt on a small side stream where it would never again be in danger of a glacial lake flood.

Ang Danu met my companions and me at his office in Thamo. He gave us tea and a lunch of pasta, then took us to meet some of the older villagers who remembered the day of the flood. Mingma Gyaltsen Sherpa, the son of the owners of Panorama Lodge in Namche, where I often stayed, came with me to help translate.

Darfuti Sherpa, who was seventy-seven when I spoke with her in 2005, ran to warn the villagers at the festival about the approaching flood. She had lived in Thamo all her life. Her family farmed and kept livestock, then built a rustic lodge in 1990 after tourism developed in the Khumbu. I ate lunch there whenever I passed through Thamo. Her daughter was a mountaineer. Big plastic-rimmed glasses perched a little precariously on Darfuti's well-lined face, anchored in the loose gray hair above her ears. She wore a warm red jacket under her *chuba* and a little green cap was pushed high on her head.

She heard a noise that day in 1985, then ran up and told her fellow villagers there must be some strange airplane coming. I couldn't understand her words as Darfuti talked to my interpreters, Ang Danu and Mingma, in the Sherpa language, which is a dialect of Tibetan. But I could understand as she imitated the sound she had heard, first with a high whine, then a buzz. She said first the sound was in the air, then there was a shaking, and women started crying.

Pasang Namgyel Sherpa, also seventy-seven, and his wife, Dal Lamu, were friends and neighbors of Darfuti. He had been a monk and she a nun before they married. One of their sons was killed in an expedition to Dhaulagiri. Their sitting room, overlooking the river, was warm with sunlight. Wooden cabinets painted red and gold, along with pictures, hangings, and multihued prayer flags filled the small room.

When the flood came, Pasang Namgyel told us, they were celebrating the Buddhist festival called *fengi*, in honor of farmers. During fengi, villagers imitated a wedding, sending a mock groom from one group, a mock bride from another. They were all laughing and drinking as they waited for the wedding group to arrive. That's when Darfuti ran up to tell them about the noise and the shaking. Then they saw the loose stone wall of a potato field tumble over and fall into the river from the vibrations in the earth.

They went down to the ridge so they could see upriver. More than half a mile upstream a wall of water advanced slowly toward them. It had to squeeze and bludgeon its way through the narrow river gorge, creating the broad floodplain we see today. The riverbanks slowed the flood's progress and gave the villagers time to go back to their houses, get their valuables, go down to the river, and plead to the god to change course and spare their village. Pasang Namgyel said it took almost an hour for the brunt of the flood to reach them. A huge boulder moved downstream with the flood.

The villagers really weren't afraid for their lives after they saw the water coming downstream so slowly, Pasang told us. The adults had been drinking *chang* and were very relaxed, and the children were too young to know what was happening. But they were very afraid for their fields and houses. They watched a bull on the other side of the river. He could have gone up the slope, away from the flood, and escaped, but in his confusion walked down toward the river. He was carried away in the flood. So were government buildings and two bridges, along with the hydropower plant.

Pasang Namgyel, with his wife Dal Lamu chiming in loudly from across the room to embellish the story, talked about the yak herder from the tiny village of Dig who became so angry that he drowned the hungry stray dog. That happened before midday, they said. Within a few hours, Dal Lamu told us, the ice broke and by late afternoon the water was nearing Thamo. Pasang Namgyel said the yak herder kept his actions secret for months because he felt sure the deity of the lake had caused the flood.

Later I stood with Ang Danu Sherpa on a ridge above the river, near the offices of Khumbu Bijuli. The land belonged to Pasang Namgyel; this was the spot where the villagers had stood, wondering whether the swollen river would angle its force in such a way as to take this field as it swung back and forth, bouncing down the river canyon.

I asked Ang Danu why he was convinced the flood was the result of global warming. He said it could not have been caused by an avalanche that time of the year, but only by a block of ice falling from up high into the lake. There had not been heavy snow the previous winter in any case.

Ang Danu pointed to the wide, scoured-out channel, the new floodplain of the Bhote Koshi near Thamo. Before the flood it was narrow, with riverbank forests on both sides. This part of the Khumbu is a gentler, more sheltered landscape than the high valleys at the bottom of the glaciers. Walking over the hill from Namche Bazaar to Thamo one passes through firs and pines and rhododendrons, and in the spring wild blue mountain iris flank the paths. Junipers and willows and silver firs used to grow on the riverbank before the flood; now there is a sand bed where there once was forest. Prosperous inn builders in Namche use it, mining sand to make concrete for construction.

After the fury of the flood abated, the valley all the way down from Dig Tsho was dotted with boulders. In some places the river channel was ten times wider than it had been. The worst damage was along the channel of the Bhote Koshi, the river that flows from Tibet; but below its confluence with the Dudh Koshi a few miles downstream, there were also signs of scouring.

Namche Bazaar, the tourist hub of the Khumbu, is high on a hill above that confluence and was safe. But Mingma's great-uncle, Kancha Sherpa, told me people watched the flooded river channel below them in fear, as boulders bobbed and big trees were torn from the hillside. They could not be entirely sure the flood would abate. By the next morning all was quiet, but the bridges on the well-traveled route from Namche to Lukla had been ripped out by the flood.

At our inn in Thame, Arun and Raju quizzed the proprietress about the best route to reach Dig Tsho. She showed the way on a map but recommended we follow two women she knew who were going up to Dig, the last little village before the lake. The women had a *goth* up near Dig, which is a shed people use for themselves and their animals during the summer grazing season. They were on their way up to plant potatoes and tend to their yaks.

We followed the women for over an hour the next morning, winding up the trail above Thame, passing through villages, seeing schoolchildren in uniforms hurrying down to school. The young ones were only going as far as Thame, while the older ones were going all the way to Khumjung to the high school Edmund Hillary established. For me that would be a half day's walk. The young Sherpa girls and boys could do it in a little over an hour.

Arun continued to ask villagers if we were on the right path. I was keeping pace with the men but working hard at it. The two women on their way to Dig with a small child were walking still faster and we lost sight of them.

We came to a steep hill, a bare knob of earth in the middle of the gently sloping land. There were two routes from here, a long one around it, another up over it. An elderly Sherpa man and woman had just started up. A yak herder said we should take the hill. At this point we were already above fourteen thousand feet, and walking uphill was tiring. I looked up the slope and groaned, knowing I could do it but not knowing how much more lay beyond. I voted for the long route. I was overruled, gently. My companions pointed to the elderly pair already far above us.

It took me about half an hour to reach the top. The Sherpa couple sat down to rest and fell behind me briefly, but by the time I reached the top they were already out of sight, as were all my companions. I caught my breath and walked up the ridge to where Arun and the others were standing, looking toward the far end of the valley that now spread below us.

Arun pointed to the rocky hill far away at the end of the valley: that was the terminal moraine of Dig Tsho. He said the light-colored ribbon of earth on the right side of the dark mound was the breach through which the floodwaters had flowed. We continued walking in the direction of the lake, staying high on the ridge. Two trails, one lower down on our side and one on the other side of the river, had been washed out in the flood. That was why the women had told the hydrologists to take this high route.

A short while later we came directly opposite the village that had received the brunt of the flood, Mingmo. Three houses and many fields had been washed away. The village looked deserted. Its handful of residents were away, perhaps tending their yaks. There was one tiny figure in the distance, a short way down from the top of the steep, sandy riverbank. An all but invisible path angled sharply down from the village toward the river channel below, and he was sweeping it. He appeared to be sweeping sand from a hill of sand.

The tiny laboring figure made me think of Sisyphus, toiling endlessly at a hopeless task. There was only silence and sand, slipping down, seeping everywhere. Above the slope of sand, the village's potato fields angled down, leaning toward the slender trickle of river. The massive boulders strewn in the river channel, thousands of them, must have boiled out of the lake when the water surged over the moraine. That moraine is seventy-five meters high, and from two to three hundred meters wide. But only a swath of it had exploded, creating havoc all the way down to Thame and beyond. "When it's in a neat pile it doesn't look like so much. Spread it out and it's a stunning sight," Arun said.

The sand deposits at Mingmo suggested to Arun that the Langmoche Glacier (a piece of which had broken off and caused the Dig Tsho flood) once extended to where we were standing and all the way down to the confluence with the Bhote Koshi, the river we had crossed the day before on our way up to Thame. Probably there had been some ponds on top of the glacier. Sediments tend to settle in them, causing the kind of sandy deposits we could see in the river channel now. After the ponds dry out, the deposits are exposed. "We think that such ponds occurred in the region around Mingmo and that's why we see such thick deposits of sediment. Probably one large pond extended across what is now the present streambed," Arun told me.

The floodplain was now a kilometer wide. The original channel would have been much narrower. Mountain rivers don't have wide floodplains, and people don't expect the torrential monsoon floods that are common in the plains, so they live very close to the rivers. This makes them all the more vulnerable to GLOFs.

We walked for two more hours, until about eleven. The hydrologists continued clambering over rocks near the top of the ridge. I went down lower, ignoring the Sherpa women's advice, thinking I might find the walking there less difficult. I didn't. There were boulders to climb over, and no trail, only intermittent flat places near the river channel. My guide Dambar accompanied me but was not much better than I at picking a trail through the rough underbrush.

We reached the goth and stopped in a sheltered spot next to it for lunch, out of the wind. Cholendra had carried an oversized Chinese thermos of hot black tea, sweetened with sugar. In his pack he had a mound of hardboiled eggs, a pile of round fried Tibetan bread, and a jar of plum jam. We were all polite and took only our share, but we ate fast, hungry from the uphill walk. The bells around the yaks' necks tinkled. Dambar chatted and laughed with the Sherpinis.

After lunch the men set about trying to find a way up the steep moraine—much higher and steeper than the one at Imja. First we traversed the wide, rocky floodplain, working our way across to the opposite side of the channel the flood had blasted open. We climbed onto hillocks covered with sparse mountain grasses. The men walked fast. They explored a dead end over a hill to the right and returned before I could catch up with them. They said there was no way up to the moraine from this side. They would have to go back across the field of boulders and ascend the moraine from the bottom.

I was breathless and tired already. I knew I could not begin to keep up with them, even though given enough time I could have gotten up the moraine. It was only 250 feet above the river channel, but much steeper and more treacherous than the hill we had climbed that morning. We had a long walk back, and a cold wind was coming up; I would need my energy to get back to Thame. This was really a two-day excursion that we had tried to pack into one, lacking both time and camping gear. I was disappointed but had no choice but to stay behind.

Cholendra stayed with me, though I knew he would have liked to clamber up the moraine with the hydrologists. He was still a new acquaintance at that point, but that day he started becoming my good friend. He watched to be sure I was all right, and noticed that my guide had gotten far more interested in impressing the hydrologists than in looking after me. We watched as the three men became small figures, scrambling down across the boulder-strewn floodplain, then up the steep moraine. Low-hanging clouds obscured the tops of the surrounding mountains. The men soon became flyspecks in the rocks. Cholendra could see better than I as they emerged from the boulders near the top and slipped over the lip of the hill to what lay beyond.

We watched for a long time. It was about an hour but it seemed like more before Cholendra and I started walking back to the goth. Eventually the threesome came over the lip of the ridge again and felt their way down the moraine. Back below we gathered at the goth to drink the last of the warm tea. "It's a nice little glacial lake," Arun announced. He said most of these lakes originate on debris-covered glaciers and therefore are dirty, like Imja. But this one is not in contact with the glacier, so it's blue and picturesque.

I was even more sorry I hadn't been able to ascend the moraine. Later I saw the hydrologists' photos of the pretty lake. It looked like a deep bowl. The steep, white curving glacier at the far end looked like a giant ice slide. Rocky slopes hugged the round lake, its waters bluer than that of most of the glacial lakes formed since the planet started getting warmer. The hydrologists were only able to walk immediately to the left and right of the outlet because the moraine embracing the lake was so steep. The level of the lake is about fifty feet lower than it was before the GLOF. All that water went down in the flood.

I was depleted and worried about the long return to Thame. Arun said he too was exhausted. We agreed that doing this in one day was not such a good idea. But he had been pressed for time and had hurried in the morning because he didn't know how long the walk would take. Once we had reached Mingmo and he could see the moraine, he knew they could do the round-trip in a day. "It looked very near then," he said. "But not as near as we thought."

This was the first time Arun had been up to Dig Tsho. The erosion and bank cutting was far more dramatic than what he had seen in photos. Arun hazarded a guess that Dig Tsho had been there about 250 years, calling it a well-established lake. He was not sure whether there was ice or rock and soil inside the moraine because the surface features didn't give clues. But there was no melting evident, so he thought it was unlikely there was ice buried inside, which made the moraine more stable. The breach, the lighter-colored V-shaped channel made by the fast-moving water, showed what was inside the moraine itself: mostly dusty rubble. The breach goes all the way to the bottom of the end moraine. The bottom of the lake is deeper still.

"Why are you confident that what happened in 1985 was caused by global warming?" I asked Arun.

"Hard question," he replied. "We are confident, but we don't have strong evidence to prove it. The increasing frequency of GLOFs and the size of lakes in this area and also the increase in surface air temperature all point to global warming."

The outlet seems to be draining the lake and keeping it from growing. The lake has not grown since 1992 and has now been removed from the list of threatening lakes in this region.

Many people think that after a glacial lake's moraine bursts, the lake is safe. In the case of Dig Tsho this is probably true; the lake holds less water now and it's draining steadily through the breach. But breaches can fill up with debris again, allowing a lake to rise. When water pressure is sufficient, or when a big chunk of ice falls, there could be another flood. A lake just over the border in Tibet called Zhangzhangbo has burst twice. The first time, in 1964, there was little damage. But the second time, in 1981, the GLOF destroyed the "Friendship Bridge" between

China and Nepal, ripped out parts of the new road from the border, and badly damaged the Sun Koshi hydropower plant downstream in Nepal.

Pradeep Mool works at the International Centre for Integrated Mountain Development, known as ICIMOD, where Arun Shrestha now also works. Like Arun, Mool worked for the Nepal government before coming to ICIMOD, a mountain think tank for scientists from around the world. They focus on glaciers and the high mountain species, people and ecosystems of South Asia. Mool specializes in what is called "remote sensing" of glaciers. His training was in geology before he specialized in glaciers. Remote sensing uses data and photos from satellites and other electronic devices to ascertain what's happening up in this forbidding terrain. That data often gives glaciologists the information they need to decide which lakes they must study at closer range, and it's much easier and cheaper to acquire such data now than when glacier specialists first started their inquiries.

Mool says glacial lake bursts have happened about every five years in the Himalaya, some in remote and unpopulated areas of Tibet. "You can see army trucks in the Arun basin brought down from China, hanging on the walls of riverbanks," says Mool, referring to a flood from a lake in Tibet called Gelhaipco that burst in 1964. In the same watershed there were three GLOFs from a Tibetan lake called Ayaco. Mool says there are many small floods from GLOFs that are never reported in the news, or only sketchily reported. The only way to study such floods is to have a project and a budget. Even experts in an organization like ICIMOD can't just pick up and go investigate such lakes without a lot of planning. Thus hydrologists must continue to make guesses about trends. Even so, the pattern Mool sees alarms him.

Satellite imagery, according to Mool, shows that the end moraine at Zhangzhangbo has closed and water has risen again to the brim. An earthquake or increase in temperature could lead to another breach, and now there is even more development downstream. Other lakes in Tibet are growing on tributaries of the Yarlung Tsangpo River, threatening India and Bangladesh too. In one case China spent millions trying to create a safe outlet but wasn't able to do so; the only alternative was a warning system.

Soon after he began specializing in glacier research, Pradeep Mool and some colleagues flew over Dig Tsho in a helicopter and mapped the lake. They estimated it held six to ten million cubic meters of water, which makes it not very large by Himalayan standards. The only way to be sure about volume is to get out on these lakes in little inflatable boats to do depth soundings. But the size of a lake is only one factor in determining how much damage a GLOF could do. With a hanging glacier above it, a small lake like Dig Tsho can do tremendous damage, emptying all its water in a short time. A big lake connected to a flatter, valley glacier could empty slowly, as Imja seems to be doing.

There are various ways of estimating the threat posed by glacial floods, Mool told me. A lot of it has to do with what is downstream, in the path of the flood. If there are villages, bridges, trails—or roads and hydropower—a lake that might potentially burst could become a candidate for draining or a warning system. Some less tangible issues have to be taken into account to assess damage in rural areas. How much productive agricultural land might be taken out of commission? What would be the impact of a lost bridge? How many villagers depend on it? How far would they have to walk to get across a river if a bridge is wiped out for months or years? Many of the bridges in Nepal's hills and mountains have only been built in recent years: they may spell the difference between children getting to school, agricultural products getting to market, and sick people getting to a hospital on the one hand and complete isolation on the other.

Mool and his colleagues were completing a new inventory of the glacial lakes throughout the Himalaya, the first in almost ten years. In those ten years both Imja and Tsho Rolpa had become a little less worrisome, though they were still on the short list of six lakes in eastern Nepal that had to be tracked. A lake I had not heard of named Thulagi was on the list, a lake at such a low altitude there are trees and grass growing around it.

By 2010 regular monitoring had become the watchword for these lakes, not engineering interventions. Over the ten years researchers had tracked glacial ponds, like those I had seen on the tops of the Khumbu and Ngozumpa glaciers. The scientists discovered that some of them vanished, drained perhaps through their glacier's complex plumbing system. That has reduced the count of potentially worrisome glacial lakes.

In 2009 Pradeep Mool went up to Imja for further reconnaissance of the lake with some colleagues from ICIMOD. They paddled out into the cold waters in inflatable boats to do depth soundings so they could measure the amount of water in the lake. They had acclimated on the way up to the lake, and had been working there for several days before one of the men complained about symptoms of altitude sickness. By the time they sought help, it was too late. Before a rescue helicopter could reach him, the scientist died.

Late in 2009 I saw a small article on the web, suggesting that a date cited in the 2007 Intergovernmental Panel on Climate Change report might have been a typographical error. The estimate 2035 was cited as the point at which Himalayan glaciers might disappear. The web article said 2035 was perhaps intended to be 2350. The latter date had been suggested in the late 1990s by a Russian scientist, who speculated that by then the entire ice-covered area of the world would have shrunk to 20 percent of what it is now.

The part of my brain that had been puzzled for five years by the idea that the ice rivers I had walked on could be gone in a few decades relaxed; 2350 seemed to me much more plausible. But it still would not be good news for the planet in general, and the later date might encourage people to shrug off global warming and go back to the business of spewing hydrocarbons into the atmosphere with less guilt.

A short while later, a flurry of articles showed that reporters had gotten on the story. By January 2010 a whole new scenario emerged, one even less flattering for the IPCC than sloppy proofreading. The specialists who compiled the glacier section of the three-thousand-page IPCC report had apparently picked up the dire estimates of glacier disappearance from the same World Wildlife Fund report that had inspired me to tramp to glacial lakes with Arun Shrestha and had been roundly criticized by Dipak Gyawali. "Glaciers in the Himalaya are receding faster than in any other part of the world and, if the present rate continues, the likelihood of them disappearing by the year 2035 and perhaps sooner is very high if the Earth keeps warming at the current rate."

The IPCC report also quoted information from a 1996 report by Russian hydrologist V. M. Kotlyakov. But the IPCC report associated his conclusions with the date 2035, not with the 2350 date he had used in his prediction that the total glaciated area in the world outside the poles would likely shrink from five hundred thousand square kilometers to a hundred thousand. The original report had noted that the Himalaya would be one of several high-altitude regions where glaciers would persist, though reduced in size.

A reporter tracked down the source of the WWF Fund's forty-year estimate, which turned out to be an article in the highly respected, London-based magazine *New Scientist* quoting Indian glacier specialist Syed Hasnain in 1999. Interviewed in 2010, Hasnain said he had merely been speculating at that time.

As would be expected, vocal global warming doubters had a field day over this flap, one which lasted for a few weeks. Combined with observations that some glaciers in the Himalaya and Hindu Kush region had actually advanced while others were retreating, they claimed all the worry over Himalayan glaciers melting was alarmism.

Arun Shrestha sighed in his office at ICIMOD as we talked about these issues in the spring of 2010. He said this was what could happen when other than peer-reviewed material is used in reports like the IPCC's, which will become gospel for laypeople and will be used by scientists as well. "That's the problem of review," he said. "This is all review. People don't go to the original paper. They just review the review, and then review the review of the review and on and on."

I winced a little. This is how journalists can perpetuate errors. Arun's own work had been reviewed in the 2005 WWF report, but he confessed he had never read that report's summary, which was the source of the infamous "forty-year" estimate. But he had seen the date in the IPCC report. He said he and other scientists talked about whether to take it seriously or not, then dismissed it as something that was unfortunate but not worth trying to dispute with their research, and got back to their own work.

A different discussion was emerging about the same time as the flap over 2035, which seemed to me to have much more significance. Some scientists were talking about what has been called the "Asian brown cloud," which is composed of black soot from wood fires and burning garbage, diesel exhaust, and dust from cities in India and China and other parts of Asia, as well as from desertification. All this pollution floats around in the atmosphere, absorbing heat and getting deposited on the land through rain. Some of it is deposited on snow or ice, accelerating their melting.

Little work has been done to ascertain how much this heat-trapping has contributed to warming temperatures in the mountains and to glaciers melting. Scientists at ICIMOD were talking about the issue but had not yet done serious work on it. But it was important to get a handle on this, because if the soot were a major cause of glaciers melting, then Himalayan glacier retreat became a much more regional issue, not simply the result of Westerners churning out greenhouse gases. If a major cause was in Asia, then a partial cure was under the control of its people and governments. Perhaps the Himalayan glaciers were not doomed after all. The West needed to get off oil and coal, certainly, as did Asia. But only in Asia was garbage being burned for lack of good waste policies, and wood being burned for lack of good, cheap alternatives for poor people to cook.

By 2009 melting glaciers were finally being widely discussed. The prime minister of Nepal and his cabinet flew by helicopter to a hill near Everest Base Camp to hold a meeting, with much fanfare and plenty of oxygen tanks. The goal was to get the topic in the newspapers and to call attention to global warming, highlighting the fact that the people in the high mountains were the potential victims of GLOFs and shrinking glaciers though not the villains in the global warming scenario.

A few months earlier, some well-meaning young Nepalis had staged a run from Imja Lake downhill to the Khumbu villages in order to highlight the path of a potential GLOF from Imja Lake. "Beat the GLOF Action Run" its promoters called it. About a hundred runners participated. The website promoting the 2010 repetition of the original event dramatized the issue: "The runners will run on the paths and bridges that will one day be washed away. They will see the forest and the farmland that will become desolate landscapes. They will cross bridges that will cease to exist. Villagers and their homes will all vanish."

Some Khumbu villagers were unhappy about the run. It created bad press for them, perhaps drove some tourists away, and upset local people unnecessarily. They wanted to be better informed about these issues, and they wanted to be involved in whatever solutions were devised—not passive recipients of what the government or donors dreamt up. Khumbu native Mingma Gyaltsen Sherpa told me Sherpa people felt exploited by event organizers who came and made money and just left. Now some of them are even suspicious of researchers and want them to ask permission from Khumbu residents before pursuing their investigations.

By the time you might read this, things will have changed. There will be more data, and perhaps better predictions. Recent research suggests that a receding terminus does not necessarily reflect what is going on with large glaciers, because so much of the glacier resides at very high altitudes in very cold temperatures. Lakes may form and GLOFs may occur lower down, but possibly the overall mass of a glacier could remain about the same—for now.

Still, if someone reads this forty years hence, the Khumbu glacier may have receded all the way up to Everest base camp. Older guides like Mingma's great-uncle Kancha say it used to be much easier to get up to the base camp. They could walk all the way up on the glacier from a small village called Thugla. As I write this, the glacier has receded as high as the village of Lobuche, and there are so many ponds and soft places on the glacier, everyone has to take a longer route around it now.

Another day up in the Khumbu in 2005, near the village of Thame, I talked with a weather-beaten sixty-two-year-old man named Ang Phurba Sherpa who was planting potatoes with his wife. His creased, sunburned face was framed by a well-worn blue knit cap bearing a white Nike swoosh. He said he had lost his house and potato fields in the 1985 Dig Tsho flood. They went down the river with the dead yaks and presumably one drowned dog.

Mingma accompanied me to translate. All the older villagers speak mostly Sherpa; younger ones speak Nepali as well. We sat outside on the bare chocolate-brown earth where Ang Phurba and his wife Ang Lakpa were taking potatoes from a big basket to plant in roughly plowed ground. "Planting potatoes," he laughed when I asked him what he was doing. I invited his wife to come over and sit with us so I could interview them both. She declined, but just like Pasang Namgyel's wife Dal Lamu, she chimed in on every question. Ang Lakpa started talking even before Mingma finished translating what I was asking.

Ang Phurba said his main house, the one with the family chapel, had been swept away in the flood, leaving an eroded riverbank. The lost home was in Hor-sho, the next village up the valley, just over the ridge behind us but exposed to the Langmoche Khola, the river that flows from Dig Tsho. His was the only house lost that day in Horsho. After the flood his family moved down here to Thame and built a new house on some land he owned. Thame is on a ridge some distance above the confluence of the Langmoche Khola and the Bhote Koshi. It was untouched in the flood.

Ang Lakpa decided to come closer. She told Mingma her husband was away that day and didn't see the flood. He was up high on the mountain looking after his yaks. She and his parents saw the flood. They were celebrating *fengi* just like the villagers down in Thamo when people saw the water coming down the valley and started crying. By the time Ang Phurba came home, everything was gone. Ang Lakpa told Mingma she had heard that the lumber from her house washed down to Namche and someone had used it to build a house there.

Ang Phurba said he had heard the story about the yak herder and the dog, but he and his wife believed that something else had caused the flood. Mingma translated: "The river was going like a snake, going up and swinging. They saw a black object in the river bobbing up and down. He said he heard there was some huge black thing in the river. It wasn't a boulder. They just heard it was something black. They believe that the water had some kind of animal in it, and the animal can't find the goddess in the river." Mingma saw my puzzlement but wasn't sure how to make clear the belief system beneath the bare translation. "The river is breaking loose and looking for something," he said, which only puzzled me more.

My guide Dambar, who speaks English quite well even though he was not educated in an American college like Mingma, tried to help. The couple spoke in a combination of Sherpa and Nepali. "It seemed like this thing was leading the flood, something supernatural was leading," Dambar said. As the river swung in one direction it seemed to take all the land with it, then it would swing to the other side of the river and take the land, like it was leading the flood.

Ang Phurba and Ang Lakpa said they feel safe now here in Thame, but during the monsoon sometimes they get scared there will be another flood. Ang Phurba sees that the glaciers are melting, even though he doesn't connect this melting to the Dig Tsho flood that took his home. He says everyone here knows that less snow is falling, and the winters are warmer.

Ang Phurba, whose son is a Buddhist monk, said Tibetan scriptures prophesy the melting of the glaciers on Nangpa La, the pass between Nepal and Tibet that he crossed many times, taking potatoes to trade for tsampa, buckwheat flour. When he was young, Ang Phurba walked on ice as he crossed those passes. Now, so he hears, there's only earth and rock.

"This will be bad," he said, "for the universe."

"Not for him, or for us, but for the whole universe," said Mingma, echoing Ang Phurba in English. "It is in the book. It is meant to happen."

# 6

# In the Valley of Dhunge Dhara

How did they build something so durable? How could they have
envisioned such a system and built it? Our ancestors were no
ordinary humans, they were incarnations of Bishwakarma and
Manjushree.
—Naresh Shakya Bansa, elderly resident of Patan

The Kathmandu Valley was once a lake. Ancient stories tell us the valley was created when the Boddhisattva Manjushree came to worship a divine lotus planted in
the lake long before by a messenger of the as yet unborn Buddha. Manjushree
could not reach the lotus because of the deep waters, so with a sword he smote the
rocks in a narrow gorge and drained the lake.

Geological evidence supports the mythic lake that Manjushree is said to have
emptied. The Kathmandu Valley is a basin at an altitude of approximately 4,000
feet between the lower and the middle hills of the Himalaya. As the Himalaya
were shoved north into the Tibetan plateau, many valleys were created between
the folds of the hills. If a landslide were to block the main exit from such a valley,
it might begin to fill up with water from rivers and springs. Around two million
years ago, it seems a large lake formed in this fashion in the Kathmandu Valley's
bowl of wooded hillsides.

Long after, perhaps because of a big earthquake, or a series of jolts over many
years, a channel opened a gorge at the west end of the valley. What would later
be called the Bagmati River spilled out, finding its way down to what is now the
Ganga and leaving the valley dry by around 10,000 years ago. There were, as far as
we know, no people living in the path of any such Bagmati flood, so none were
harmed. Instead, the draining of the valley led to the superb conditions the ear-
liest settlers would eventually exploit: terraces and knolls, rich soil, springs,
rivers, and shallow aquifers.

It is enticing to imagine that the myth captures some distant human memory
of the events that helped to create this perfect valley. We know these hills and
mountains have been a crossroads for restless mankind since before any recorded
history. Perhaps even for thousands of years before the oldest inscriptions give us
hints about settlements and rulers in the valley, people were peacefully going

*Map 4* The Kathmandu Valley. Created by Kanchan Burathoki

about their business here. Eventually their agrarian life in the fortuitous valley cradled by a ring of rough, protective hills might have caught the eye of passing strangers. The travelers might have carried tidings of the valley's climate, water, and soils. And so came the invaders. Many times over the centuries, the Kathmandu Valley has been invaded, sometimes conquered, by outsiders.

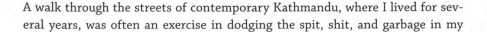

A walk through the streets of contemporary Kathmandu, where I lived for several years, was often an exercise in dodging the spit, shit, and garbage in my path. But in spite of diesel exhaust, acrid smoke from burning plastic, and smelly canals, there is still magic in the city. There are still tiny quiet streets, musty and mysterious, and big white Buddhist stupas crowned with gold, their eyes gazing in the four cardinal directions, seeing all. By late October hibiscus, poinsettia, and bougainvillea bloom in concert, while huge chartreuse pomelos and bright orange persimmons hang on trees. It is the season when neglected street dogs and cows are honored, fed sweets, and for a day wear red tika on their foreheads and garlands of orange marigolds around their necks. During this Hindu festival of Tihar, or Diwali, Lakshmi the goddess of abundance is invited into homes on her special night. Candles beckon at doorsteps and small footprints painted in red and white—symbolizing blood and milk—lead the way from hand-painted *mandalas* in front of each house. The footsteps and candles lead upstairs to private altars, where fruits and flowers mingle with offerings of money.

Among the ugly concrete structures built in the architectural style currently favored in Kathmandu, there are still many attractive red brick homes with tile roofs, as well as red brick palaces and tiered-roof temples. These dwellings, shrines, and temples are the remains of the valley's traditional architecture, often rich with stone and wood carvings and metalwork. Most of them are from the era referred to as the Malla period, roughly coinciding with the European Middle Ages and extending up to the 18th century. But Kathmandu was inhabited by highly skilled artisans for centuries before the ascendance of the Mallas. What we can see now is often the reconstruction of earlier, perhaps virtually identical, houses and temples. Inscriptions on stone water spouts document the history of the valley from 550 CE; older inscriptions capture bits of information about life in the valley from several centuries earlier.

Throughout the city there are also sunken structures of the same red brick. Inside them people can be seen stooping to take a drink from an elegant water spout carved in the shape of an animal head, washing their faces and feet, or doing laundry. These are the *hiti*. Hiti, in the Tibeto-Burman language of the Newar people (who held sway in the valley for at least a millennium until the end of the Malla dynasty), refers to a square or rectangular sunken structure made of brick, a water source. Stone spouts, or *dhunge dhara* as they are called in the Nepali language, related to Sanskrit, protrude from the brick walls. From their carved serpent mouths there may still flow a strong stream of water.

Long before there was any talk of diverting water from a snow-fed Himalayan river to supply Kathmandu's burgeoning population, the hiti of the old Newar towns were the visible manifestations of a superb valley-wide water system. That system began more than two thousand years ago, flourished until very recently, and in some pockets of the valley remains a mainstay for residents. It is even being revived in places because the municipal system is so bad.

Kathmandu missed out on water from rivers flowing down from the high Himalaya because they all bypass the valley. They descend toward the Ganga to the east and west, leaving the Kathmandu Valley high above them but not dry. The Bagmati—the Kathmandu Valley's holy river—and its tributary, the Bishnumati, flow down from the hills. Springs throughout the surrounding hills feed streams and underground aquifers in the flat land below. Having to manage with monsoon-fed springs and streams was not a problem when the population was only a few hundred thousand. Now that the valley's population is between two and three million, more water than these sources can provide seems to be needed. How to get it has been a topic of hot dispute for years. I will delve into those disputes in the next chapter. For now, I will celebrate the genius of the people who established the ancient water system in the Kathmandu Valley.

The loss of much of this elegant water system is tragic, as I see it. It was beautiful as well as functional. It fulfilled all the water needs of the valley's residents and was central to their religious life. Its partial survival in the face of urban encroachment and neglect is testimony to the skill of its builders and to centuries of communal upkeep. And its destruction is part of what has led to the current water crisis in Kathmandu.

The beginnings of this water system arose sometime in the first millennium BCE. The people who started it may have descended from an even more ancient culture. Sudarshan Raj Tiwari is an architect and former dean of the state university's well-respected engineering college. When he was a young professor, after studying architecture in Delhi and at the University of Hawaii, he became interested in the valley's ancient history. He began deciphering that history from inscriptions, chronicles, legends, and rituals. In the course of his work, Tiwari studied the ponds, drainage systems, and other artifacts of the ancient Indus Valley culture that produced Mohenjodaro and Harappa, located in what is now Pakistan. This is the culture that also used toilets and a sewage system 3,500 to 4,000 years ago. Tiwari believes the Indus culture and the ancient Kathmandu Valley are connected.

The Indus Valley culture—which extended far beyond the Indus when it was flourishing and whose beginnings may stretch as far back as eight thousand years—declined after the invasion of nomadic Aryans. That invasion began approximately 2000 BCE and was complete, archaeologists believe, by approximately 1200 BCE. Tiwari speculates that some of the people forced from their land by the Aryans carried their technology for burnt brick construction and ponds for storing water into exile with them. Some of their descendants may have moved east into the Gangetic plains and later up into the hills of what became Nepal. Finding an alluvial plain similar to the Indus Valley where their ancestors flourished, they revived the technology for burnt bricks and established the first elements of the Kathmandu Valley's water architecture.

These early settlers were known as the Kirats, though there were already people living in the Kathmandu Valley when they arrived. Wherever they came from, the Kirats are the ones who created the small hilltop settlements that dot the valley, Tiwari believes. Remnants of ponds dug to capture rainwater and contemporary place-names from their language support his conjectures. The Kirat dominance in the valley extended roughly through the first millennium BCE, ending around 100 CE, by which time Hindu settlers under the rule of a dynasty known as the Lichchhavis had moved up from the Gangetic plains of what is now Bihar, in north India.

Tiwari has a spacious office at the engineering campus of Tribhuvan University, Nepal's state university. The campus, a collection of buildings from various eras, is a quiet oasis of green in a noisy city. The building that houses architecture and urban planning, built about twenty years ago, is red brick but not particularly aesthetic in its design. As I walk toward its front steps I see a broken window that looks like it might have been that way for a while.

Tiwari has the stereotypical looks of a scholar. Prominent eyes loom from a pale face. He is balding and graying, and has the look of someone who spends most of his time indoors thinking, not in physical exertion. He speaks slowly and deliberately, but the tremendous energy he has put into his study becomes clear as he relates story after story about ancient times, digressing from one to another but always coming back to his central thread. When his early papers on the valley's history attracted the ire of historians who thought he was poaching on other scholars' territory, he went back to the university for a doctorate in culture. His advisor told him to submit material he had already written about the valley's ancient culture for his dissertation.

A cool autumn breeze blew in through a bay window as Tiwari talked about the valley's early inhabitants, making them seem like near relatives who had merely forgotten to leave much information for their descendants. But to the sharp eye of Tiwari, they left an abundance. Tiwari is rather like a detective, finding clues and links, weaving them into a fabric that can hold water. He enjoys speculating in order to make a continuous story out of the clues he finds, but his speculations are based on a web of facts.

Tiwari says the Lichchhavis were an urban society who found an already thriving urban center when they moved from the north Indian plains to the Kathmandu Valley. This is part of why he believes the earlier Kirat settlers in the valley may have been descendants of the ancient Harappan urban culture of the Indus Valley. He has a hunch that the people the Lichchhavis encountered in the valley were already using not only ponds but a stone conduit system of some sort, which the Lichchhavis exploited and greatly expanded. The first verifiably Lichchhavi conduits were built by 550 CE. That date is engraved on a spout in Handigaon, the early capital of the Lichchhavis and perhaps a central place for the Kirats as well.

What happened between 100 and 550 CE, Tiwari believes, is that the two groups merged to some extent. Settlements grew larger, and the centrally located ponds the Kirats used were no longer adequate for all the people at the periphery of the expanding towns. The Lichchhavis had relied on wells in the Gangetic plains; but wells on knolls and hilltops often have to be very deep to be useful.

Thus, a system of sunken pits that could receive water channeled from shallow aquifers began to evolve. The brick hiti could be located farther and farther from the city center if they were at a slightly lower level. The underground aquifers were often recharged by ponds, both natural depressions and man-made ones. Eventually canals bringing water from the hills were also used to recharge the ponds. Gravity carried the water downhill whether from pond or aquifer.

Tiwari thinks an ancient story of the reputed sacrifice of a Lichchhavi king may capture a time when the population was experimenting with its new water system. According to the story, which relates events that supposedly occurred around the middle of the fifth century, some stone water conduits inside the royal palace had dried up. Astrologers told the king that sacrificing a near-perfect person to the goddess in front of the conduits would restore the water. The king knew he and his son were the only ones with all the qualities of perfection required. Wanting to sacrifice himself, he devised a plan for his son to execute him as he sat shrouded and unrecognizable near the water source. The son beheaded the disguised man, as his father had ordered.

After the sacrifice was carried out, according to the chronicles, the water began to flow again. Other details in the story hint to Tiwari that there was at that time some tension between Hindus and Buddhists, to whom the old king was partial. He thinks the murder, or "sacrifice," might have been motivated by a desire on the part of the Hindu priests to capture the state religion.

But one thing seems clear, says Tiwari: the story shows that water conduits were in the valley by around 450 CE, during the rule of Dharmadev, the father. Tiwari believes the stone spouts and the system that delivered water to them were probably in an experimental phase during the fifth century. By a century later, inscriptions cut in stone indicate the system had proliferated successfully. Thus, anxiety over the fledgling water supply system became embedded in a story that chronicled a regicide and a parricide, and perhaps priestly struggles for dominance. The story also shows that Kathmandu's water system was from the beginning imbued with mystery—its survival as much a matter of the spirit as of engineering.

Over the centuries the water system became very sophisticated. There were filtering chambers made of gravel, sand, and burnt brick. Conduits made of stone or clay or wood were protected by a covering of bricks and a layer of local impenetrable clay. These kept the water that reached the filtering chambers clean; the filtering chambers just made it even better for drinking. Still today, residents who enjoy this water say it tastes far better than water distributed through municipal pipes.

Tiwari relates an account from a Chinese traveler about a palace with water that reached up to the seventh story, about one hundred feet, then cascaded down in a series of waterfalls. He thinks that the Lichchhavis might have been using copper piping to transport water from a reservoir, under pressure, to the top of the tall palace. If so, this would have been an extremely sophisticated development, rare even today in the subcontinent. Tiwari found records from the time showing a water engineer for the palace who received a high salary.

"These people studied nature very well," says Tiwari, referring to the manner in which the early water engineers in the valley tapped water from its shallow aquifers and constructed drainage from the hiti. Sometimes, he says, they used a series of conduits, one below the other. Drainage from one hiti was filtered through the ground and then fresh water came out at the next hiti. "Kind of a stepped thing," he says. "I can't imagine how they could make a drainage system like that."

The various towns in the Kathmandu Valley each had its own underground octopus with a multitude of arms or conduits delivering water to the dhunge dhara. Many of those arms were severed in recent decades as builders dug deep into the ground for the foundations of big hotels and shopping malls. The worst disruption came from constructing something called the "ring road," a thoroughfare of sorts that circles Kathmandu, built in the 1970s with help from the Chinese. This disorderly ring of honking, spewing vehicles lies roughly at the foot of the hills that embrace the valley and from which for centuries flowed enough water for the towns of Kathmandu, Patan, and Bhaktapur. The ring road has blocked the old canals from the hills.

The Lichchhavis from the Gangetic plain were largely Hindu and brought with them the idea of flowing water as pure water. They prized water that was pure in multiple ways—ritually and in ways that would satisfy current standards of hygiene. Though they had left the actual Ganga behind, symbolically they brought her with them. Since the Ganga is pure, any pure water may become Ganga. So flowing water or ritually protected water becomes Ganga. The valley's ancient water system shows great care taken to ensure purity on both the practical and the symbolic levels.

On the practical level, they used the effective filtering system I mentioned. Tiwari cites an old inscription that mandates that the water for visitors to a dhunge dhara must be "cool, tasty and clear." On the symbolic level, the equating of pure water and Ganga can be seen in temple architecture. Aquatic creatures decorating the corners of roofs or placed under pillars essentially consecrated the temples through symbolism that evokes Ganga. Most of the dhunge dhara similarly convey that their water is pure and is of the Ganga. Their water typically spurts from the mouth of a *makara*, a mythical reptile sometimes equated with the crocodile or the gharial and always associated with the Ganga. And beneath the spouts there is often a carved figure of Bhagirathi, the king who persuaded Ganga to come down from heaven.

From early Chinese travelers' tales we know that the people of the valley were given to bathing three times a day, a habit Tiwari thinks they must have brought with them from a warmer climate where daily purifications of mind, body, and soul would not be uncomfortable. They would, in fact, be very desirable on the warm, dusty plains.

Of the valley's three large, ancient "royal" cities—where you can still find spectacular religious and royal architecture—Patan is the most interesting. Bhaktapur is essentially one big museum, beautiful but somewhat removed from the stream of life. In Kathmandu's sprawl of concrete there is still a handsome central temple and palace area with a crush of tourists. But Patan is a living ancient city. Once you get away from the choked main thoroughfares that bring traffic from the Kathmandu side of the Bagmati River, the crazy pace slows down. You can walk down quiet, narrow streets and lanes lined with red brick dwellings, some of them very old. There is a magnificent durbar (court) square that still makes my jaw drop when I come upon it through one of the narrow backstreets. I try to visit it later in the day when most of the tourist touts leave and I can relax and enjoy it.

In the warren of lanes all around this durbar square, small, multistory dwellings line the streets; there are shops on the ground floor, living space above. Gaps in the rows of houses, sometimes with archways, lead from the street to shared community courtyards or *bahals*. And down many of these streets in Patan you will also come upon hiti of all sizes and shapes, in all states of use and disuse.

The oldest functioning hiti is right in Patan Durbar Square, next to a palace that now houses the Patan Museum. Called Manga Hiti, an inscription indicates it was built during the early Lichchhavi era, 570 CE. The deep hiti is built like an amphitheater with two wide stair-stepped shelves—terraces where people can rest in the sun or spread out their laundry to dry. Two worn stone lions flank the stairs at street level. The bottom of the hiti is twenty steps below the square, some twelve or fourteen feet lower than the plaza. From its south wall protrude three stone spouts.

The typical dhunge dhara is a single serpent mouth, a makara with a trunk-like snout resembling an elephant's, curled up to create the opening for the water. Manga Hiti's central spout is the most elaborate of the three, combining four animals, each born from the mouth of another. The water spills from what appears to be the mouth of a fish, which is emerging from the carved head of a mammalian creature with many evenly sized small teeth. This head in turn has emerged from the larger head of the crocodile-like makara. The makara's snout-like lip curls forward as usual, but in this elegant and richly detailed carving the snout forms the head and beak of a bird. The bird seems to be surveying the birth of the water from its proudly arched neck.

As I sat one afternoon on the eastern terrace, marveling at the quadruply mythical creature, ten or twelve people clustered around the spouts. They were waiting in line to fill copper or tin urns or plastic jugs. A woman placed her hand on a makara to steady herself as she waited for the water to fill her tall copper pot, almost as if she were petting the stone creature. The drainage system was not working well. Murky water had gathered in the shallow pit below the spouts, but worn stones positioned directly under the spouts allowed the women to place their jugs and buckets above the water while they filled them.

People in Patan take water from Manga Hiti's spouts

A little girl crawled into a big stone vessel in the middle of the paved floor of the hiti. She played there while her older sister filled the family jugs. Her pink pants and quilted jacket were dusty and faded. Heading back up the stairs, she pouted until her sister let her carry one of the plastic jugs; then she struggled up the stairs on her own for a few steps before her older sister came back to carry the heavy jug.

Once when I visited Manga Hiti I talked to a young man named Raju Kapali who said he had used the water from this hiti since he was a small child. "This is pure water," he said. "It comes from tantric. Very powerful goddess. Very precious water," he said, pointing to the spout on the far right that must be used in the Newar festival called Kartik because it helps kill evil. He said he came every morning and evening to fill a water jar, especially for *puja*, for worship at home. Water from some of the spouts is also believed to have medicinal value.

Back up on the plaza, middle-aged men in traditional dress sat in the sun below the elaborately carved Krishna Mandir, a temple dedicated to the dark-skinned, flute-playing god of the Hindu pantheon. A slender boy in short pants and flip-flops carried two yellow buckets with red lids away from Manga Hiti. He walked quickly with small, shuffling steps, the heavy burdens hanging from his arms, and disappeared down a dark, narrow lane.

South Asian society is far more festive than Western society, and the residents of the Kathmandu Valley are particularly good at celebrating all aspects of life. There seems to be a festival happening every week; and water is often central to these

Man drinking water at Manga Hiti, Kathmandu

festivals. During one spring event, a chariot—laden with a goddess and her mythical jewelry—is pulled around Handigaon, the oldest capital of Kathmandu. After the procession the chariot is submerged in an ancient and famous pond. Other festivals require going to the local river for a ritual bath at dawn, though now people often have to bring a pot of water from home to avoid the filthy water in the rivers.

Sudarshan Tiwari has studied many of the Kathmandu Valley's festivals, especially those of the Newars. These are descendants of the people who dominated the valley before the eighteenth century, and whose Tibeto-Burman language is related to that of the Kirats, rather than to Sanskrit, the Indo-European language of the Lichchhavis. Years of observing the rituals performed during these festivals have led him to believe that the festivals were ways of making sure the water system was working. A particular festival might ostensibly be dedicated to Vishnu or Shiva, "but it was for the people's mundane needs. The people would die without water. You don't live for superstitions," says Tiwari. "Man does things for his own good, his own use. Not to please Narayan," a manifestation of Vishnu.

The stupendous festival of Rato Machhendranath, the greatest of the valley's water festivals, is an annual event in Patan. Rato Machhendranath, known as the red (*rato*) god and the fish (*machha*) god, is essentially a rain god; he is also seen as an incarnation of Shiva by Hindus and of Avalokiteshvara by Buddhists. His image is carried around the streets of Patan for weeks each spring, housed in a chariot with a towering superstructure. The chariot's tower is so tall it probably could have been seen moving from a long distance away. Its roughly pyramidal shape rises more than sixty feet above the little temple where the five-foot image of the god sits.

The chariot is first constructed of wooden scaffolding, which is then wound about with long vines before being decorated with green vegetation and red silk streamers. The skill of the men who wrap the vines determines the strength of the structure. Even when it is well built, it tilts precariously as it moves, pushed from behind and pulled from the front by dozens of strong men down dusty, bumpy roads. The only way to stop it is for several men with trapezoidal cuts of heavy wood to throw these portable brakes in front of the enormous moving wheels of wood. Sometimes they are injured, even killed, as they push, pull, and brake the chariot or ride its long curved prow. Thousands of people may walk along with it as it moves from stop to stop, maybe only a few hundred yards in a day. Just touching the moving chariot is believed to bestow blessings.

If the chariot falls over, it is believed to presage ill for the kingdom: droughts, the death of kings, the fall of governments. In 2000 it fell over. In 2001 the crown prince allegedly murdered his father. Most people in the valley believe the chariot's mishap presaged the evil deed to come.

The god housed at the bottom of the chariot is made of red clay and painted red. He is virtually shapeless and androgynous, truly primitive, not sculpted like the highly wrought Buddhas and Shivas from the same era. The god is just a lump of earth, more like the aniconic stone gods and goddesses that spring from the earth throughout the valley. He is a farmer's god, though wrapped in rich gold cloth during the festival. A simple face is painted at the top of the chunky red body. A tiny mouth and big wide eyes give him a look that is strangely childlike and wizened at the same time.

According to the stories, a terrible twelve-year drought afflicted the valley in the reign of one of the Lichchhavi kings of the seventh century. The king's priest said he would have to go fetch Lord Machhendra and somehow persuade him to come to the valley to relieve the people's suffering. The god came back with the king in the form of a bee. As they stopped to rest ten miles south of Patan, the heavens opened and rain poured down. To celebrate the god's gift, the king founded the town of Bungamati, which has become known as Machhendra's birthplace.

Every twelve years a more elaborate festival pays homage to Machhendra's arrival in Bungamati. A completely new vehicle is built. The effort takes months. The cart's massive wheels, long prow, and towering mast must be constructed not with modern tools but more or less in the same way they were constructed in the first millennium. And the work has to be done by representatives of ancient families. Multiple animal sacrifices are required, even though many of the priests are Newar Buddhists. When the construction and rituals are complete, the chariot must travel all the way from Bungamati to Patan and back, a round-trip journey that can take more than two months. To Tiwari the reason for all the work and ritual is very clear: "Every twelve years you have to repair your canal and reservoirs."

Every twelve years, for example, the Raj Kulo or king's canal that carried water down from the hills into Patan had to be cleaned out. Though this Raj Kulo is now blocked by the ring road, in recent years Newars have been working to restore it so that it can be a benefit to the Newar towns that lie just outside the ring road. It can still be useful for irrigation, and even bring water down to recharge Patan's derelict ponds and aquifers.

Ponds and canals have the least water in them in late spring, the time of the Rato Machhendranath festival. By this time, at the end of the dry season, almost a full year has passed since the last monsoon. People can walk into the ponds and canals, clean out leaves and dirt and get the channels ready to carry and store the next monsoon's flood. I met some Newar men who were promoting the effort to rescue the Raj Kulo. They were not old, most in their forties, and could still remember being brought as children to help with this work. The flourishing of the canals and hiti was not ancient history.

During the Rato Machhendranath procession, women brought water from the dhunge dhara in their neighborhoods to throw over the chariot. "If they don't have water they can't come. So if the water system in your *bahal* has no water, you can't participate in the festival. And if you can't participate, you are not a citizen," Tiwari says. "This means everyone must participate in maintaining the water system." To a great extent, the breakdown of this ownership has led to the current water crisis in the Kathmandu Valley.

Yet still, each spring, the idol made of earth housed in a chariot wrought from vegetation careens through the city much as it has for centuries. Lurching and swaying down the roads, the chariot resembles the top of a rough-hewn stupa, a floating pagoda. Together the idol and the chariot form a prayer to the fickle monsoon to come to earth and bring forth another year's vegetation so that the people will have grain and food. Not so very many years back, when the Raj Kulo was unbroken and the dhunge dhara flowing with water, the Rato Machhendranath festival exhorted the Newar people to the hard, communal work of water management as much as it begged the god for rain.

As I leave Tiwari's office and plunge into the screeching herd of cars and motorbikes careening across town, I catch a rare glimpse of Ganesh Himal—a Himalayan range northwest of the valley—crisply caped in snow. Looming high above us, it seems near, but at the same time part of another realm. For a moment I try to feel my way back to the quiet of the ancient valley, full of birdsong, intense sun, soft air. I wonder if the hard work of the farmers might have been lightened by the sight of those mountains, so bright in the clear air. Perhaps, or perhaps I am romanticizing. In any case, the Kathmandu Valley had rich natural beauty and an abundance of resources sufficient for a population of a million or so, perhaps more—if well managed. The population is now approaching three million and management has been pretty bad.

Kathmandu's water problems could not, at this point, be solved simply by restoring all its hiti, even if this were possible, given the many that have been irretrievably ruined by urban development. But some people are trying to correct decades of mismanagement by aiming for sustainability through a combination of restoration, conservation, and rainwater harvesting.

Of the valley's 389 dhunge dhara, more than two hundred still have some water. Sadly, their filtering systems seem to have been damaged—or simply cannot cope with the current level of groundwater pollution—because low levels of coliform bacteria show up in all tests, along with ammonia and various minerals. Nonetheless, the city supply often has worse, and in any case some people receive it only a couple of times a week if at all. Restoration of some of the hiti, especially in Patan, has brought a regular supply that people can choose to treat by boiling or other methods.

Alko Hiti is situated near the northern edge of Patan, at the bottom of a slight hill where the land begins to flatten out toward the floodplain of the Bagmati River. The nearby neighborhood is densely populated; typical Newar homes cluster around courtyards. Alko Hiti was neglected after city pipes brought water into those homes; it became dirty and overgrown like so many hiti throughout the valley. But a few years ago, as the piped water dwindled, locals went back to the hiti, first in desperation, then in a more entrepreneurial spirit. In 2004 they cleaned it up, formed a user's committee, and started storing the water for later distribution.

Alko Hiti is one of the most prolific hiti in the valley, producing 173,000 liters a day in the rainy season, half that in dry season. There's a legend about the genesis of Alko Hiti, which dates from the Malla dynasty. A famous tantric priest had received five pebbles from a serpent god. He kept them locked up in a pot in his storeroom. His wife, curious, went and looked into the pot. When she found only stones, she threw them out the window in disgust. Where they landed, water spurted up. These became the five spouts of Alko Hiti.

The hiti is approximately thirty feet square at its bottom level. Because it is at the bottom of a sloping hillside, it is shallower than Manga Hiti, located at the center of Patan. Only ten steps lead from the street level to the bottom of the pit. A stagnant pond behind the hiti indicates its drainage system has been disrupted, but it is a popular hiti and with good reason. Its five spouts are always flowing with water. In twenty minutes as I watched early one afternoon, half a dozen people came to fill tall aluminum jars or plastic jugs, some of them more than once.

I visited Alko Hiti with a young man named Keshab Shrestha, a program officer with a local group that helps families install rainwater-harvesting systems. We spoke with Asakaji Maharjan, who had come to fetch water from the spouts. He told us in a reedy voice that he collected water for various households throughout the day to make money. A slight man with a big smile and one slightly unfocused eye, Maharjan wore blue striped pants with the cuffs rolled up. He said the pipes are dry in the homes of the local people for whom he fetches water. Using an array of plastic buckets and rectangular jugs, he often takes each of them seventy-five liters of water a day.

Maharjan quickly filled his buckets from one of three large makara-headed spouts that protrude from the north wall of this Malla-era hiti. One small, cat-faced spout delivers water from the east wall, while a small, worn serpent spout does so from low on the south wall. The people who came for water seemed to choose a favorite spout. Two of the large spouts are said to come from the same source, while the other three each proceed from different conduits, though no one I asked seemed to know where the sources are.

The mystery surrounding the sources of the water flowing from the dhunge dhara captures the spiritual and ritual nature of the old water system. Perhaps the diagrams of the plans were lost over the centuries, or perhaps priests have always known the water sources and continue to protect them from potential enemies by keeping their knowledge secret. Today that protection might include being vague with curious Western journalists.

Many people in the valley retain a sense of magic about the water sources, a belief that propitiating the *naga* or water serpents and practicing the proper rituals could keep the water flowing. Temples and shrines were often located at water sources to protect them. The water that sprang from the crocodile mouths was a gift from the gods, not to be taken for granted, whose desecration must not be risked. Thus what has happened to the hiti seems all the more tragic in the face of the water scarcity in the Kathmandu Valley today, caused partially by a burgeoning population, but perhaps even more by a loss of unity of the sacred with daily needs.

The water here at Alko Hiti is so abundant and reliable that a group of nearby residents have devised a way to store some of the flow and distribute it directly to households. Rising high above the hiti to the south is a four-story building with an oversized version of the hard black plastic water tanks that sit atop most houses in Kathmandu. This one holds ten thousand liters.

Kedar Shakya, a member of the Alko Hiti Conservation and Drinking Water Supply User Committee, came down the steps of the hiti and greeted me and Keshab Shrestha. He and Keshab explained how the system works. The committee was started five years ago. In the evening, after people stop coming to get water, operators attach a portable gutter to the dhunge dhara to siphon water into a small tank. The water is then pumped up to the storage tank atop the building next door, which houses community offices.

Out on the street Kedar showed us a row of points at ground level behind a metal grille. The water from the big tank above flows down to these spigots. Each day, according to a prearranged schedule, operators in the community center open the taps so that a resident can attach a long hose to one of the points and run water back to his or her home for an hour each day. Five or six houses at a time can take the water to fill up their respective household tanks. Gravity is sufficient to carry the water from the community tank into a home, even to an upper story.

More than two hundred households use the system. Each household pays 145 rupees a month for the service, a little more than two dollars. The sum sounds small, but it has been enough to maintain the system and even buy the community a generator, which helps pump the water up to the roof in the evening during the dry season, when electric power is being rationed throughout Kathmandu.

Public services here are unreliable at best, so communities like this one are taking the initiative instead of waiting for the services to improve. The system was set up with an initial investment of $110 per household, combined with some donations of equipment and technical expertise from the municipality and from UN Habitat. The community itself supplied the labor. Roshan Raj Shrestha, the chief technical advisor to UN Habitat's Water for Asian Cities program—whose eco-home with eco-san toilet is a model of water conservation for Kathmandu— says his organization has provided only a little money and advice to support efforts like this. People are eager to do what is necessary to regain the use of their neighborhood dhunge dhara. "So many stone spouts have dried up that people are starting rainwater harvesting to recharge the aquifer. In many places people just put rainwater in dug wells. Through this small intervention some spouts begin to discharge again," Roshan says.

Restoration efforts like Alko Hiti have been more widespread in Patan than in other parts of Kathmandu, because it is still dominated by longtime family residents, mostly Newar. In the central areas of Kathmandu, most of the residents are renters.

"Either stone spout water can be recharged or rainwater can be stored," says Roshan. "People are so frustrated now. They are waiting and nothing improves. And it will be worse in the future. So people are doing very innovative things. People need just a little intervention and then they will do a lot of things on their own."

A short walk up the hill from Alko Hiti is the Chyasal community, whose four thousand residents used to rely exclusively on their dhunge dhara. The two spouts of the Gaja Laxmi hiti had always had abundant water until 2007; then they suddenly dried up, perhaps because of weaker monsoon rains, perhaps because of construction. Now there is only a trickle of water coming from one of them. A steel cage protects an ancient statue of Gaja Laxmi, or "Elephant Laxmi," sitting at ground level to the right of the spouts. The stone carving is so worn I can barely make out the two elephants that are almost floating above Laxmi's shoulders, pouring water for her. The inscription on a stone opposite the spouts has been worn away too. An elderly man from the community who sits on its managing committee said the hiti is 1,500 years old.

To preserve this ancient water source, the municipality helped the community dig a well inside the hiti to store rainwater. The community helps any willing nearby resident install a drainage system to capture rainwater for recharging the community well. Unlike the clean, clear water from the old spouts, the well water is turbid and full of iron. The community set up a system to pump the water up to a filtering system in a nearby building. There the water percolates slowly through sand, similar to the filtering system the Lichchhavis used in the filtering chambers they placed behind the dhunge dhara. The sand removes iron and usually absorbs bacteria. If tests show the water is not fully decontaminated, then some chlorine is added before the water is bottled in twenty-liter jars and sold to residents for about fifteen cents each.

Sangeeta Awale, a woman in her thirties with a shy smile, has worked here for two years, overseeing the filtering and sale of water. She says she likes her job. She lives near Gaja Laxmi but used to have to go to Manga Hiti or another hiti half a mile away, then carry big jars of water back home. She had to do this four to five times a day to gather enough water, so she started using less water for drinking and cooking. For laundry and bathing she sometimes used the inferior water from wells or taps in her own or a friend's house. Now she and her neighbors have plenty of safe water for drinking and cooking.

North of Kathmandu, outside the "ring road," the land slopes gently upward toward forested hills. Halfway between the smoggy valley and the crest of the Shivapuri hills is a reclining god, a six-meter statue of the Hindu god Vishnu dating from the early Lichchhavi era, around 640 CE. The statue is made of granite and polished to a dark metallic sheen with techniques used at that time. The god rests inside a rectangular and somewhat stagnant pond protected by a fence.

This pilgrimage and festival site is called Budhanilkantha. The statue is known as a "sleeping Vishnu." Sudarshan Tiwari says the statue was called the white Vishnu in Sanskrit, or "white sleeping in the water." He is black, yet he is called the "white" Vishnu. This puzzled Tiwari. Vishnus don't normally sleep on water in the first place, he says, and wonders why a black stone statue is referred to as white.

Tiwari believes the answer once again shows the vigilant water management of the valley's former denizens. He thinks the Vishnu was placed at the bottom of a waterfall in the Lichchhavi times, thus he was sleeping on white water. As long as the Vishnu was bathed in white water, there would be water in the cities below. People were forbidden to cut trees or in any way disturb the watershed.

The hills to the north are still the source of some of the valley's water, giving rise to the springs that feed the Bagmati and Bishnumati rivers. But the sleeping Vishnu is now surrounded by increasingly large and elaborate hillside homes built by wealthy Nepalis for themselves or their foreign tenants. Up above the Budhanilkantha neighborhood, in Shivapuri Park, there has been some effort to curb deforestation, but not enough to bring back a waterfall. In ancient times, according to Tiwari, absolutely no cutting would have been allowed because that might have deprived the Vishnu of white water. "The Lichchhavis knew it was their watershed. They wanted to protect the source of their water."

# 7

## Melamchi River Blues

> We started an organization called Save Bagmati Campaign.
> It ran well for two years. Unfortunately corruption
> set in after that. Green insects infected it.
> Do you know what a green insect is? It is the dollar.
> —Huta Ram Baidya, founder, Nepalese Society of
> Agricultural Engineers

While I lived in Kathmandu, I regularly visited the American Mission Association. Members call it Phora, while some Nepalis call it "mini America." It's a club, and expatriates with the right kind of visa can apply to become members. It has a pool and tennis courts, a small gym, a field for baseball and soccer, a children's playground, movie rentals, manicures and massages, a commissary and wifi café, and very polite Nepali staff. It has a certain colonial feel to it, which bothered me at times: yet it was also a haven where on a weekday afternoon I could exercise, read the papers, and eat lunch.

Phora refers to *phohara durbar*, which in Nepali means "fountain palace." The extensive, well-tended grounds where dozens of expats and their children gather for hours on weekends was once the site of a Rana palace, a place for parties and dances, performances and cinema. It got its name because there were fountains throughout the gardens as well as inside the building. The ornate, neoclassical palace is long gone. In serious disrepair by 1960, the palace was demolished and the land sold to the American government.

But *phohara durbar* has other claims to fame. It was also the site of the first piped water in the Kathmandu Valley. To explain how this came about, I'll tell you a little more about the valley's history and culture.

The Lichchhavis and Mallas kept the city from growing beyond certain limits. They prohibited building outside a ring of shrines to various mother goddesses, like Kali. They knew that disturbing the land beyond that ring would be "killing your own food, your economic base," says Sudarshan Tiwari, the architect and cultural historian who has reconstructed aspects of ancient life in the valley. There is still some agriculture in the Kathmandu Valley, because a few of the old landowners stubbornly hold on to their fields even as a sea of "wedding cake," multistory, pastel houses engulfs them. But daily the green plots of rice and vegetables shrink as the valley succumbs, like the ancient water channels, to unplanned urban development.

Bringing the gift of water to the people was once a way to gain merit, or expiate sins. Early in the Lichchhavi period the stone spout was called "kirti," which means merit. Merit accrued to anyone who built a stone spout for public use. Giving the gift of water to people was thus sanctioned as a public duty, a responsibility that fell to those who possessed the means to deliver it.

The history of modern Nepal began in 1768 when Prithvi Narayan Shah and his army conquered the Mallas, who reigned in all the royal cities of the Kathmandu Valley, having taken over from the Lichchhavi dynasty. Shah was the head of a small Hindu kingdom to the west of the valley called Gorkha. He coveted the riches of the Kathmandu valley and its strategic importance for creating a nation. He consolidated the many territorial gains made by the Mallas into a single kingdom. The Shahs ruled until very recently, except for a long hiatus during which some uppity courtiers of a family known as the Ranas managed to usurp their power.

The Ranas were hereditary prime ministers to the Shahs. The first of them wrested power from a Shah king in 1846. For the next century the Ranas kept the hereditary rulers confined to their palaces and trotted them out only for ceremonial purposes. The Ranas brought the changes that, combined with Western influences, Western development ideas, foreign aid, and later the rapid growth of Kathmandu's population, brought the metropolis to the place it finds itself in today. Sudarshan Tiwari says the Shah kings didn't change all that much about life in the valley; they followed the Newar culture that evolved under the Lichchhavis and Mallas. But the Ranas did make changes. For one, they built big palaces outside of the main towns, making agricultural land urban.

The Ranas were a contentious bunch, given to internal squabbling and intrigues. The stunning regicide-parricide of 2001 is not a complete anomaly to students of Nepal's history. When crown prince Dipendra Shah allegedly gunned down his mother, father, brother, and other members of his family, the deed shocked the world and remains the fact most people manage to summon up about Nepal when I mention I lived there. Yet from the Lichchhavis to the Ranas and Shahs, palace murders have been common, if not so tragically spectacular.

The story of a prince executing his father in the mid–fifth century, which Sudarshan Tiwari believes captures a period of transition in the water history of the valley, was only the beginning of a long string of palace murders. Another in the late nineteenth century is connected to *phohara durbar*. To gain merit—and expiate his sins for having been involved in the murder of one of his cousins—Prime Minister Bir Shumsher Rana set about building a new water system in Kathmandu. In 1891 the Bir Dhara Works began piping water from the hills north of Kathmandu, near the source of the Bagmati, to a holding tank not far from the current American Embassy, just inside the Ring Road. Pipes carried the water to the houses of the Ranas and other members of the elite. Public taps allowed the rest of the population, who also took water from *dhunge dhara*, to enjoy some of the new supply.

For a while two separate water systems, the Newar and the Rana, functioned side by side, serving different purposes for different populations in the city. But the coming of the Rana system was a turning point. The well-engineered and graceful water system of the ancient Kathmandu Valley, whose careful mainte-nance persisted through much of the twentieth century, would eventually take a back seat to the kinds of changes the Rana dynasty, with its taste for European luxury and Western modes, had introduced.

Until 1951, though the Ranas themselves were fond of foreign influences, Nepal was closed off to foreign visitors. It opened to outsiders only two years before Tenzing Norgay Sherpa and Edmund Hillary reached the summit of Ever-est. After the opening of the country's borders, rapid change and westernization began, including foreign aid. The Shah dynasty, restored in 1951—when with India's help the Ranas were overthrown—would eventually set about expanding the piped water system with some of that foreign aid.

By 2000 about two-thirds of the valley's population supposedly had access to the piped system. That optimistic estimate is relatively meaningless, since so much of the system works only minimally or not at all. Most people manage these days, depending on their income levels, with some combination of wells, public taps, water tankers, and erratic piped water, whose quality is generally too poor for drinking and must be treated. Some still use the dhunge dhara, and a few have instituted water harvesting to augment their supply and recharge their wells.

Things have gotten so bad that poorer people in the country's capital are no better off than those in rural areas where citizens have to carry water from stand pipes or streams. In the dry season I saw long lines of women waiting for the brief period of time when water would spurt from the old tap at the *chowk* down the lane from my apartment building. Some rural denizens may even be better off, because their public taps continue to yield water year-round; and though the water may not be entirely pure, it's often safer than what's in Kathmandu's pipes.

Meanwhile groundwater in Kathmandu for those who use wells is frequently contaminated with iron, zinc, chlorine, and ammonia (not to mention dangerous bacteria) as polluted surface water percolates into shallow groundwater. Hotels and other big users continue to drill their own deep tube wells, further depleting aquifers already overused by the municipal system. And as the water level in aqui-fers falls, land threatens to subside.

As early as the 1970s, government engineers said the valley's water would not be sufficient for much longer and set about looking for a new supply. They were right, since more and more of the Newar system was encroached upon as the years went by, and the Rana system was not only inadequate but aging and leaking.

A few years after the search for a new supply began, the first sketches of the proposal that has come to be known as the Melamchi project were put forth. The project requires an inter-basin transfer of water from a Himalayan river that flows through a valley east of Kathmandu. According to this plan, engineers would siphon the waters of the Melamchi Khola through a twenty-six-kilometer (sixteen-mile) tunnel blasted through the hills northeast of the city.

During a span of ten years, as I made trips to Nepal and traveled into its mountains, I heard about the fits and starts of the Melamchi water supply scheme. I knew I would eventually have to try to write about it for this book, but I delayed. I started, then dropped it to try to figure out other confusing issues, chiefly India's water imbroglios. I rather dreaded having to take up Melamchi again. The history of the project involved forty years of ups and downs. There were too many angles, too many people involved, too many different interests—even the suicide of a former head of the Melamchi administration attributed to the ill-fated project, and a former prime minister prosecuted for graft in connection with it.

Still, having lived in Kathmandu during most of the time I spent in South Asia, I perhaps learned a little more about this city's problems than about those in other South Asian metropolises. And while Kathmandu's problems are unique to this country, its economy, and its government, they may also offer insights into the South Asian continent as a whole. It's a messy story, and seems far from finished.

It's ironic that Kathmandu missed out on the glacier- and snow-fed abundance of Nepal's big rivers; Kathmandu—long known as Nepal Mandala because of its centrality—has snatched Nepal's resources and limelight for centuries. Thus, grabbing some of the Himal's abundance—as the project known simply as "Melamchi" has been trying to do for the past several decades—looks completely justifiable to most of the valley's people.

Yet the loans—now totaling several hundred million dollars—which the Nepal government has accepted to build this project are on the heads of all citizens of Nepal, not just the Kathmandu Valley's residents. Spending part of that sum in rural areas where people are dying from waterborne diseases might have gone a long way toward solving a continuing and serious problem in Nepal. Even with all the money lavished on the project, work on the tunnel that is supposed to carry the water has been delayed again and again.

Ajaya Dixit, cofounder and director of Nepal Water Conservation Foundation (NWCF), says he's been talking about the Melamchi project since he was in his twenties. Now in his fifties, he's still talking about it and says he would not be surprised if that is the case fifteen years from now. Educated in Nepal, India, and Scotland, Ajaya (like his colleague, former water minister Dipak Gyawali, with whom I discussed glaciers in 2003) is something of a recovering engineer.

Ajaya is a serious man with a sense of humor. At an exuberant inauguration ball organized in Kathmandu to celebrate Barack Obama's swearing-in, he wore a costume that combined clothing from all the countries of South Asia, including a *lungi* from Bangladesh and a Pakistani *topi*. Lettering on one side of the hat read O Bama, on the other side O Rama.

Respected as a water resources expert throughout South Asia, Ajaya says the heady days after Nepal's first venture into democracy in 1990 energized him and provided freedom to do the kind of research and writing he had always wanted to do. Good at math, and once fascinated by computer modeling, he leans more toward work that weds engineering and social science, seeking to solve human problems.

"The Melamchi project represents nothing more than a child's grasping for a lollipop held out by an indulgent international banker," says Ajaya, a "quick fix" for Kathmandu's perceived water shortage. He questioned the justifications for Melamchi from the outset and has continued to do so as circumstances changed during the Maoist insurgency. "We have had massive migration into the valley in recent years and still people are drinking water so this whole notion of scarcity needs to be deconstructed."

Throughout the ups and downs of the Melamchi plan, Ajaya and many other water specialists in Kathmandu maintained that proper plumbing, conservation, and strategies like rainwater harvesting could get the job done for a fraction of the cost. But the government and foreign donors, along with many informed Nepalis, remained fixated on Melamchi. Myopically so, according to Ajaya, for they glossed over fundamental questions about how water was being managed.

In the fifty square kilometers of the Kathmandu Valley (about 20 square miles) there are 100,000 taps, public and private, Ajaya points out. In the driest month of the year, they yield sixty million liters a day from surface and groundwater sources. He says if you figure that 40 percent of this water is wasted through leakage, there are still 36 million liters per day. Each tap should get 360 liters, which Ajaya says is a lot of water—not scarcity worthy of ransacking a neighboring valley's river.

But not each tap in Kathmandu actually receives that much water. So either there is less than sixty million liters available in the valley, or there is a much higher rate of loss, say 70 percent. And then probably some get a lot of water while others get little. Quite possibly all of these conditions prevail. Such a situation calls for careful monitoring and analysis, says Ajaya. It calls for meters, for regulations, and for plugging all the holes in the system. Those measures have not been taken. All the money and attention have been going to Melamchi.

NWCF cofounder Dipak Gyawali, with whom Ajaya has collaborated for the past two decades, has plenty to say about Melamchi too. He recommends implementing all the less expensive alternatives first. Each such improvement might ameliorate about 10 percent of the apparent shortage. Then, if necessary, proceed with Melamchi.

Critics of the Melamchi scheme said the project would just make a few people rich, chiefly the contractors and politicians who favor it. Without better management, along with conservation and rainwater harvesting, Melamchi by itself won't succeed in solving Kathmandu's inequitable water supply anyway. But government officials and foreign lenders have maintained there is no other solution. The valley has a water shortage; Melamchi must come.

In 1992 Ajaya wrote an article for the popular South Asian monthly *Himal*, outlining the failures of Kathmandu's water system and the misguided thinking that led to the belief that Melamchi is the panacea. Much of the article could just as easily be published twenty years later. "It's not for want of spending that Kathmandu's water supply remains poor and erratic," he asserted then. At that point the World Bank was heavily involved in costly schemes to improve the valley's water system. None of the programs under these loans succeeded in improving anything, thanks to what Ajaya calls "extraordinary ineptness in their implementation," implying that the blame fell fairly equally on bad management from the bank and within Nepali institutions.

Yet the World Bank's approach did indicate some agreement with Ajaya's skepticism about a genuine scarcity. Any loans for an inter-basin transfer like the Melamchi scheme were to be contingent upon fixing the valley's water management system and infrastructure.

The World Bank had gotten involved in Kathmandu's water in 1974; loans were arranged after the coronation in 1972 of King Birendra, victim of the 2001 palace murder. The aim of those loans was to beautify the capital, whose sewage management was already failing, polluting the Bagmati River.

Information I gleaned from a source at the World Bank in Kathmandu (who wished to remain anonymous) indicated a lack of common objectives from the outset. As he saw it, the Nepal government simply wanted more water, while World Bank consultants wanted to find out whether the current supply was being managed as well as possible. Sealing leaks would have been a lot cheaper than diverting a river. The Nepal government estimated that 40 percent of the valley's water was lost to leakage; outside estimates set that figure almost twice as high. The World Bank program aimed to reduce leakage to 35 percent, a leakage rate seen in some western cities.

Unfortunately, thirty-five years later the leaks are just as bad, so it seems a blessing that the water from Melamchi still has not arrived. What a waste that diversion would be if half of the water leaked out of the system. It's hard for an outsider to fathom the causes for such spectacular lack of progress. How can it be so difficult to impose some order on the system when the technology involved in water and sewage management is relatively basic?

There may be no single explanation. Certainly there is corruption within Nepal's institutions; but on the other hand foreigners continue to blunder as they push Western money and Western modes on South Asian structures. There is also the chaotic development of Kathmandu, which has grown with what seems a complete lack of city planning, in turn related to weak government as well as to an influx of refugees from rural areas during the Maoist insurgency.

Ajaya Dixit sees a "deep structural contradiction, a malaise" at the heart of this conundrum. "Luxury guided the introduction of technology" in Nepal, he points out. "Technological intervention in Nepal of whatever kind—hydropower, transportation, water supply—it was for luxury, not for general use." The Ranas brought piped water to *phohara durbar* and other parts of Kathmandu "for their own use, and made sure it worked and that it continued to work for them."

When the Rana regime was overthrown, no sense of public responsibility filled the gap. The Rana's class interest did not translate into a larger societal interest because their technological innovations had nothing to do with the smooth functioning of a modern, inclusive city. When the original consumers of a particular luxury were no longer in a position to maintain it, it was likely to deteriorate. In Kathmandu, the Rana's water system was simply extended, not fundamentally changed.

Financing for attempted improvements has come from the World Bank or the Asian Development Bank. But when the World Bank and ADB try to implement methods that are logical in the West, says Ajaya, it doesn't work. "Our social and political systems and technological ones just don't match. That's the dilemma.

"A century of modern technology in this country and we still haven't figured out a way of using those systems to deliver basic services for the population as a whole," he says. "Technology in the West was part of the capitalist transformation. In Britain pumps and water came together. People knew they had to pay for water. But here the water in the *dhunge dhara* was free.

"Nepal is still trying to respond to this shift. We have had hydropower for 100 years yet we still have 18 hours a day of load shedding," Ajaya noted, referring to the more or less orderly rotation of blackouts in Kathmandu's quadrants and other regions of the country each day, spreading around what little electricity there is in the country. More than a century since the pipes of the Bir Dhara Works distributed spring water from the hills to Rana palaces, there is still no efficient water delivery in the Kathmandu Valley.

Back in the 1970s when the World Bank got involved in improvement schemes for Kathmandu's water delivery, its staff concluded that Nepali institutions needed reform because they were bureaucratic and inefficient. The government created a new board to oversee the work of the Nepal Water Supply Corporation. This seemed like a good move at the time; but according to my World Bank source, it broke down. Accountability was "cloudy." Managers saw themselves as more accountable to the politicians who appointed them than to the board. Officials in the corporation gave their relatives jobs and bought expensive foreign vehicles for personal use.

The World Bank insider said the bank's objectives were good but they misunderstood the culture they were dealing with. He told me that expatriates did the work while Nepali professionals often ended up as their gofers and did not gain experience to do the jobs themselves. And they were paid vastly less, which provided little incentive to try to excel.

Ajaya Dixit sees the influx of foreign money as corrosive, saying that the Nepal Water Supply Corporation's ability to manage the valley's water seemed to decline significantly after large infusions of money in the form of loans in 1974, 1977, and the early 1980s.

A report requested by the Nepal government in 1987 recommended decentralization of management, control of leaks, and the institution of a public utilities commission to set fees. Dipak Gyawali, who then worked in the water resources ministry, was on the committee that prepared the unvarnished report. It listed a host of problems with the government agency's use of the bank's loan, while also faulting the inadequacy of the bank's management of the project.

According to the report, the water supply corporation's problems did not stem from lack of money but from "a lack of responsibility towards the money it handles." The company was good at disbursing the big foreign loans but was unresponsive to the public and unable to provide safe water and sewerage. Because of its "profligate administrative culture" officials succumbed to political pressures by hiring unnecessary staff; they made both staff and loan-purchased vehicles available for uses that had nothing to do with the work of the water supply company.

The corporation's "attitude to controlling leakage and wastage is indifferent at best," the report went on, noting resistance to installing bulk meters. Meters would have allowed Kathmandu's water officials to assess how much water was actually flowing through the system and eventually to charge users appropriately for it. Dipak says both the government and the World Bank ignored the findings in the report.

The bank insider maintained that the bank team made efforts to rehabilitate both the institutions and the distribution system. Leak detection was a major project. Yet detecting leaks in a system that does not supply water twenty-four hours a day is difficult. Around 1996 they devised a method using tankers to flush isolated parts of the system to detect leaks. They discovered that most leaks were coming from the small pipes that delivered water from the main system to individual houses. These were not buried very deeply and were easily damaged.

Repairs on these pipes would be cheap compared to the size of the loans coming from the World Bank, but this was not necessarily good news in the water bureaucracy. There was little glory and no kickbacks involved in making repairs to these pipes, even though such repairs are crucial both for controlling leaks and for keeping the water clean. Water supply and sewage are inevitably linked in the valley's water system. The pipes that supply homes are located perilously close to the ones that carry sewage away. And empty, damaged pipes create suction, thus drawing contaminated water from leaking sewage pipes into the distribution system. These are some of the very problems that continue to plague the system today, combined with illegal connections to the main distribution pipes.

Even the work done to patch up the old system, according to Ajaya Dixit, was guided more by the need to disburse foreign donor funds than by what the system really required. A minister might say a project had been achieved, even when there was no improvement in the water system, because contracts had been assigned, materials procured, and budgets spent. "You set a target, you buy so many kilometers of pipe. Your achievement is 100 percent. Managing the contract, disbursing the money—that is the incentive, the task, and the goal. Maybe all that happened. But there wasn't water in the taps. Add a corrupt political system with no account-ability and what do you expect to happen?"

How much water was moving into the system and being distributed, how much to charge for that water, or whether the organization that collects money is finan-cially viable—"those kinds of questions were never, never, never asked," Ajaya said, his voice rising. "Why would an institution that's basically designed to pro-cure and disburse funds be interested in conserving water? You have to have an institution that has an incentive to conserve, that recognizes the value of water for what it is. What we had was basically an outfit for doling out contracts."

According to my World Bank source, the bank team was in the process of training Nepal Water Supply Corporation's staff to detect leaks and repair the city's pipes. Then the government dropped the ball on that effort, failing to use the trained staff to continue the program. The World Bank saw the problems as intractable, at least for the near future, and withdrew from trying to fix Nepal's urban drinking water problems. The bank's investment, even according to bank officials, had been wasted: far too little of the money spent actually went to im-proving the leaking system. Further, an opportunity to improve the system with-out resorting to the expensive Melamchi tunnel and diversion project had been lost. Nepal still pays regular installments on the outstanding debt from "soft," low-interest loans that accomplished far less than they should have.

After twenty years of having primary responsibility for Kathmandu's water, the World Bank "realized they were getting into quicksand," as Ajaya Dixit describes it. The World Bank bowed out finally in 2002; by then the Asian Devel-opment Bank (ADB) had already waded in to fund the Melamchi project—unaware of the quicksand, or believing they could find a way through it.

The ADB tried to get around the problem of intractable local management of the water system in the valley by requiring that an established foreign water man-agement company be hired to manage the system before construction of the Melamchi tunnel began. Although some people refer to this as water privatiza-tion, strictly speaking the foreign company doesn't own anything; but it may have the opportunity to make a lot of money while doing the work.

The process of selecting the foreign company took several years, by which time King Gyanendra (the brother who had succeeded his murdered sibling) had been deposed and the former Maoist insurgents had become part of the government. To the relief of many in all parties, a Maoist minister finally put the kibosh on the highly controversial choice of the British water management company Severn Trent.

The Maoist minister, Hisila Yami, said "we want ADB, we just don't want Severn Trent." The choice had been widely criticized, both because the company had some blots on its record and because no other company had bid for the contract. Pos-sibly no one else wanted to get involved in Kathmandu's water problems; none-theless the lack of transparency in the bidding process had been criticized. The agreement was also distasteful to members of the government because the contract had no guarantees and no penalties for failure.

Thanks largely to the Maoists who joined the government after a peace treaty in 2006, Nepal managed to maneuver through this episode in the Melamchi quagmire, getting rid of Severn Trent while calming the nerves of ADB project managers to keep them and their money involved. So, by the summer of 2007, the Melamchi water diversion looked back on track.

I was finally getting a handle on some of the confusing information about the Melamchi project.

Separating the Melamchi tunnel from Kathmandu's water infrastructure, as would happen after the banishment of Severn Trent, made some sense. They are entirely different problems and unrelated engineering challenges. But making Melamchi contingent on repaired pipes makes a lot of sense too. Melamchi is a waste of money and a travesty unless leaks are minimal.

On the other hand, who is going to want to focus on nitty-gritty repairs when the big money is in Melamchi? In a poor country where no one ever knows what is going to happen next month with the government, where there is a culture of patronage and no developed sense of working for the common good through government, big wads of money injected into the system can be a big problem. With so little certainty and continuity, the guiding principle becomes getting while the getting is good.

Thus, giving the apparent drudgery of managing the water utility to a foreign and ideally disinterested entity with a competent track record looks reasonable. But even if Severn Trent had worked out to be that competent foreign company, such an approach was going to cost a lot more than a good home-grown version.

After the ADB was convinced to continue its involvement without Severn Trent, a new public company took over management of Kathmandu's water infrastructure. In the original plan, the British company Severn Trent would have supervised this organization, called Kathmandu Upatyaka Khanepani Limited (KUKL) or Kathamandu Valley Drinking Water, Limited. Now the added layer of foreign management was gone.

Nonetheless—as KUKL's new managing director, Rudra Gautam, told me early in 2010—his job was supposed to be held by a foreigner for the first six years. He said ADB and Nepali officials eventually realized that plan was not working; they saw that it made no sense to have a foreigner overseeing the day-to-day operations of an all-Nepali staff, especially one who knew little of the country and did not speak the language. Instead, Gautam would have international experts to advise him.

KUKL is not wedded to the big-money projects. It's a public company, overseen by a board, whose mission is to manage the infrastructure that is in place, now and in the future. The company's offices are in the old Nepal Water Supply Corporation building that I had visited once early in 2007 when its employees were on strike, protesting the potential loss of jobs under the new management.

The building had a new coat of paint along with its new signs. And Rudra Gautam constructed a waiting room for customers who came to complain about service so they would not have to just stand around in the hall. In his assistant's office, where I waited for a few minutes before seeing him, there were two bottles of Aqua Smile, a popular brand of bottled water: a reminder that even here, in the offices of the water management utility, there is no good drinking water on tap.

Rudra Gautam seemed to be striving to do some of the things that were needed twenty years ago, and that has meant opposing entrenched patterns. He said he started by trying to cut the bloated staff. He inherited 1,400 employees from the Nepal Water Supply Corporation; Rudra said he needed only nine hundred. The trade union wanted to keep all employees, many of whom had been hired as part of the old patronage system. When I spoke with him, three hundred positions had been eliminated through voluntary retirements.

In the past, Rudra told me, the general manager of the water company would typically make a handsome "contribution" to the reigning minister's party in order to be appointed to his job. Then he would seek to extract the money he paid for his position from some other source —often from his own underlings. When a new minister came into office, he would "request" the head of the water company to hire a certain number of his followers. Rudra said KUKL's board appointed him, so he's not beholden to a minister. His new position was a significant promotion; before his appointment as KUKL's manager, Rudra had been an engineer in a branch of the valley's water utility.

Many management decisions used to be made because of pressure from unions affiliated with political parties. Rudra told me he was trying to change that. He ignored strike threats. He said he tried to treat employees with respect, but that includes reminding them that half of the responsibility for making the organization work for customers now falls on them. He denied requests from the ministries, whether to hire people or to give away his allocation of petrol: "I say please help me by not asking for these things. If I take one request there will be a flood, so I take none."

Rudra said he was exhorting staff to work harder. He was aiming to get at least four hours a day of real work from the staff. "I love the staff. I punish the staff if they violate the rules. Just this morning I fired two members of the staff. They had been asked many times to improve."

"I have to be very tactful," he said. "I have to maintain my morality. If I make biased decisions, if I'm involved in some corruption, they will not obey me. It's challenging to be good among the bad. Easy to be bad among the bad."

Rudra's waiting room often filled with people who came to complain that they did not have water. "I have to talk to them. They are aggressive. I try to cool them down. At least they realize I am doing something for them. What they care about is equitable distribution. And my policy is equitable distribution. One person may be getting four hours a day and his neighbor is not."

The government does not have a land use policy, Rudra lamented. The city has some zoning laws, but they are rarely enforced. He said there's been too much growth, too much building: "Someone can build a house on some land and come to the office and bang the table and demand water. There is no policy to control this. And there is no policy because there is no stable government."

On a whiteboard in his office he had drawn a triangle, annotated with arrows and goals. This was a diagram of his first two-year plan for improving service to customers, stabilizing finances, and training the staff. One of the projects the staff had begun was a house-by-house survey. Rudra said there had not been such a survey for more than a decade, and thus there were no systematic records about water users. There were illegal connections that were not recorded. The survey is just a questionnaire, not a physical inspection, but Rudra hoped the survey would discover both illegal connections and "bypass connections," those that are set up to go around meters.

This seemed like progress. In the past, employees rarely left the office; now at least they were out visiting customers. I saw evidence of this every day as I walked around my neighborhood: the letters KUKL were chalked on iron gates along with a date.

"Whether Melamchi comes or not, at least we have to manage, we have to increase sources of water, have a better infrastructure." Rudra was planning to increase the water in the system with deeper tube wells and efficient use of surface water until Melamchi water might arrive. And he planned to get all customers connected to the system with meters, avoiding illegal connections.

Until the pipes are better, and until the magical infusions of water from Melamchi arrive, I asked him, what about encouraging rainwater harvesting? Rudra agreed this is another solution, but with a tinge of boredom. He was much more enthusiastic about other improvements. He said surface leaks, which in the past had run for weeks, would be taken care of in twenty-four hours. He established a phone number so that customers could report them.

Rudra was not overseeing any physical improvements to the system. That is being handled by a different office, through which money from ADB is channeled. The company that would design a new system of pipes had yet to be chosen when I talked to him in early 2010. That office, unlike Rudra's, is not answerable to the public.

Rudra said he got the complaints from the public, and he was trying to make the organization work better while struggling to get more support from the government. "In the last 10 to 12 years this organization was run by political interests. The plan for Melamchi is very delayed. But there were no improvements made in this organization because they were waiting for Melamchi.

"The government is not even serious now," he complained. They say Melamchi is where we are going. But they have no idea what we are doing here at KUKL, dealing with the people. Either support us or you take responsibility, I told them. I am happy I told them the truth. If I speak in a low voice no one can hear."

In the spring of 2011, employee unions padlocked Rudra's offices, demanding that he resign. The unions accused him of ineffective and inequitable management; staff said that KUKL was on the verge of becoming a failed institution. A chorus of editorial writers chimed in, saying there had been minimal progress on infrastructure improvements during his tenure. A month later, Rudra resigned. A further editorial urged the KUKL board to appoint "someone with a vision" as his replacement. The appointment "should not be based on political influence or pressure from the Asian Development Bank; it should be based on open competition. Only the best and most capable candidate should be chosen to manage KUKL and the drinking water system of the Kathmandu Valley."

In late November 2001, on my first trip to Nepal, I was a dazed and rather sick trekker as I waited for a bus back to Kathmandu. I had no idea I was only yards away from the confluence of the Melamchi Khola and the Indrawati River —and also very near the future site of the Melamchi project field office. And I had no inkling that any of this would come to mean something to me. After learning about the fits and starts of the Melamchi scheme over the years, I ventured again into the valley east of Kathmandu to try to see for myself what progress was being made toward the much-heralded tunnel.

The Melamchi Khola is a pristine little river, though not fed by a glacier. It springs from the high snowy mountains of the Jugal Himal, south of Langtang National Park, the trekking region I visited back in 2001. Its source lies at 19,000 feet, or 5,875 meters, and it flows for about 25 miles before joining the Indrawati River near the small town of Melamchi Pul Bazaar where I waited for the bus. The proposed tunnel would carry this river's sparkling clean water into Kathmandu's leaking metropolitan system, and let it flush out the polluted Bagmati.

In the steep valley watered by the Melamchi Khola, about 70,000 Nepalis grow rice, wheat, lentils, potatoes, millet, and vegetables, almost all the staple crops of Nepal. The weather is similar to the Kathmandu Valley's: hot summers and chilly winters, but it rarely drops to freezing, and there's always plenty of daytime sun. Hillside forests are being thinned—just as they are throughout this nation of almost thirty million—because people in the hills still depend on wood for fuel as well as for building their houses.

The area is relatively accessible compared to others where development projects are underway in this mountainous land; still, it has taken twenty years to carve out the dirt roads needed to bring heavy equipment into the mountains so that workers can blast through solid rock at several points simultaneously to begin the tunnel. It's an arduous job, certainly, but there have been repeated delays.

In January 2009 I decided to walk into the hills for a few days with my reliable young trekking companion Cholendra Karki and his brother. I wanted to explore the parts of the region I had missed on my earlier trek. These hills on the edge of Langtang National Park were no longer just the endpoint of a popular trek; they had become a political hotspot because of the on-again off-again Melamchi project.

Our destination was Tarke Gyang, a village in Helambu, the northern and higher end of the valley through which the Melamchi Khola flows. Travel books sometimes refer to the farmers who live in this region as Sherpa, like the people who live in the Khumbu and Rolwaling valleys; but they adamantly call themselves Hyolmo. To them the Melamchi is the Hyolmo river. Like the Sherpa, they are Tibetan Buddhists and their ancestors lived on the Tibetan side of the border.

The first day we walked up into the hills, passing Sundarijal, or "beautiful water," a pilgrimage spot in the hills northeast of Kathmandu and also the proposed endpoint of the Melamchi tunnel. The next day we took a long downhill walk from the top of the ridge to the bottom of the valley where the Melamchi flows. We cut cross-country through farmland. As typically happens when one leaves the main trekking routes, there is a sudden sense of remoteness—not just from Kathmandu and the places Westerners frequent, but from the present time. The trails were often just gullies carved by the monsoon. Men and boys guided ox-driven plows over the dry ground.

There was no place to stop for lunch. There were a few farmhouses that might have given us tea, but everyone was out working in the fields. The day was almost hot as we walked down the rugged gulleys under the intense sun. The previous night up on the ridge had been cold and clear. We stopped for a rest and nibbled the nuts and biscuits and dry noodles we carried. We walked for several more hours, eventually following the Talamarang Khola down a long rock-strewn slope, the site of a recent landslide, to a village of the same name.

At Talamarang we joined the main road, built to open up the region to development and to Melamchi. The town has some simple hotels and restaurants and plenty of noise and garbage, a shabby sense of progress. The hotel where I stayed was dark, which was perhaps a blessing because it wasn't very clean. The darkness was compounded by the electricity outages, which were reaching sixteen hours a day nationwide in early 2009.

The next morning we walked up the road for a few hours. Below us flowed the Melamchi Khola. This was the dry season and the river was depleted, both from lack of rain and from many small irrigation diversions upriver. I wondered what would happen to the farmers here when and if the water started going to Kathmandu. The hillside terraces were dense with cultivated fields. Would the diversions deprive them of water they needed for these crops? And would the concrete tanks of the small fish farm we passed have enough fresh water to continue raising rainbow trout? And what about the wild fish in this part of the river after more than three quarters of its water is diverted to Kathmandu? Small tributaries will join the river and increase the flow as the Melamchi approaches its confluence with the Indrawati, but for this stretch it will be just a trickle.

Two bridges on the newly graded dirt road had been washed out already. We detoured around them via the hillsides and the rocky river channel. At the end of our morning's walk we discovered a recently built guesthouse, a complete contrast to the one the night before in Talamarang. The simple inn was immaculate and bright. The name of the village was Thimbu, but the original Thimbu was higher up on the hill. This part of the village had spilled downward to meet the new road and embrace the business it might bring. The innkeeper's family used to have an old, run-down little inn up on the hill, in the old part of the village, which trekkers occasionally visited. New business had already begun to arrive in the form of Melamchi project workers and an occasional trekker like me.

Beyond Thimbu, the road had been carved out next to the river in order to carry workers and construction equipment up to the "intake point," the point on the river that had been chosen as the starting point of the tunnel. I wanted to see this intake point, even though I was unaware then how much in dispute it was and would become. After lunch Cholendra and I walked up the deserted road; it was rough but still passable for sturdy vehicles. We saw no people. There were some simple structures that could accommodate workmen, but no one had taken up residence yet.

There was more water in the upper river channel than there was below. The floodplain was wider and flatter; I could see what a lovely little Himalayan river the Melamchi is even in the dry winter, hugged by steep canyons on both sides, its rippling waters shallow in these reaches but clear as they hurried over the rocks and gravel.

Less than a half hour's stroll up the road we saw that monsoon floods had undone the workmen's labor. There were wire cages full of rocks, secured in stair steps up the hillside to reinforce the loose soil, but still the hillside had collapsed in places and parts of the road had dissolved into the river. We clambered past the first big gap in the road and proceeded on our walk. I wondered how they were going to reinforce this road adequately to prevent damage each monsoon.

Eventually the road disappeared as we neared a sharp curve in the river. The monsoon had completely washed it out at this point. There was just the river licking the hillside with very little dry ground to walk on. We could not be sure, since there was no one to tell us, but we didn't think we had reached the intake spot. Later I found out it was another fifteen-minute walk beyond the bend, near the confluence of the Melamchi Khola and a side stream called the Ribarma Khola.

Even without seeing the exact spot, I had gotten what I wanted. I had seen the terrain from which the water for Kathmandu was supposed to come: the steep wooded hillsides, still remote and quiet, a waterfall even this far into the dry season plunging down to a river full of rushing water. But perhaps there was not enough of that rushing water for both the farmers who depended on the Melamchi and the clamoring city on the other side of the mountain.

Back at the inn in Thimbu, the owner, Dawa Lama, said he had never in his life seen such a monsoon as the one that came after the road construction was finished. That was soon after he had bought this riverside plot of land for his small inn. The last part of the road up toward the disputed intake point had been finished during the previous year. Then the summer's monsoon floods broke the road in three places and filled its upper portions with rocks and sand, reclaiming it as floodplain.

As he kneaded chapattis for us, Dawa said the water rose almost to the top of its deep channel just feet from the inn. It didn't flood his house, but he was afraid of what might happen if such storms came again. Outside the lodge one could sit in the sun above the steep-sided channel and watch the winter river hurrying on its way. All night I was comforted by the roar and whoosh of the water.

Dawa, a Tamang with a round friendly face, did not seem to have strong feelings about the Melamchi project. A good portion of his income depends on the Melamchi staff using his little hotel, because Thimbu isn't likely to become a trekking destination. He told us his father used to own a water mill a short way up the valley. It was torn down to make way for the road. The father filed a claim to be reimbursed for his loss. Ten years later, according to his son, he still had not received any money.

Dawa Lama said the Melamchi project had authorized a payment of seventy thousand rupees, about a thousand dollars, for his father, who was still hoping to receive the money soon. Salil Devkota, an environmental engineer who handled public relations for the Melamchi project for seven years, said that he did not have authority to pay out money promptly for such claims; they were subject to a long process through the government.

Sometimes villagers blocked the roads to protest such delayed payments or to register other complaints; these roadblocks created expensive delays on road construction. Salil said villagers "know how to milk the cow," relating a story about a roadblock set up by a boy who wanted money for a children's group. The project gave him five thousand rupees for nets and balls; that got the road opened again. Apparently small disbursements could be handled without approval from Kathmandu.

This dance between the project and the villagers went on for five years; villagers tried to gain some benefit from the big-money invasion of the valley. They did so in ways that were certainly irritating from the point of view of the project managers, whose work might be halted for days or weeks. According to Salil, sometimes villagers wanted handsome sums because the road was passing by their homes. One stretch I had seen in an otherwise blacktopped length of the road was bare dirt. The village homeowners along that stretch did not want to give permission for the road to be blacktopped without receiving compensation. Salil said the project had allocated about five million dollars for social and environmental improvements for the local population, about 2 percent of the overall budget.

From the bottom of the canyon where the Melamchi River flows to the top of the hill on the eastern bank is upward of three thousand feet of friable soil compacted into steep hillsides. Early the next morning we started up the hill, passing many farms and some small villages. We reached our destination, Tarke Gyang, a quiet village on the next ridge, by mid-afternoon. A Hyolmo village in Helambu, Tarke Gyang is a popular stop back on the main trekking route.

Our reward after the long climb was a large, comfortable old lodge in an open meadow at the top of a hill, surrounded by trees. Cholendra knew the lodge. He liked its dining room, where we sat on cushions along the edge of a richly polished wood floor. The food was delicious, the *sauni* gracious. The dining room was filled with trekkers even in January. The sauni, like Dawa Lama, did not express any strong feelings about the Melamchi project.

The next day, after a visit to the gompa, we had to leave in order to reach Kathmandu by the following day. We plunged down the trail from this haven and its fresh breezes, back to the dusty roads and garbage-strewn villages below. As we got closer to Melamchi Pul Bazaar, we walked down another dirt road, but this one had nothing to do with the Melamchi water supply project. It was built by villagers who had collected money from each household to carve out a road from Melamchi Pul Bazaar up into the hills.

Construction had taken less than a year. Each house contributed ten to fifteen thousand rupees, about a hundred and fifty to two hundred dollars. A few more substantial donations came from relatives in Kathmandu. It was unpleasantly dusty to walk down in the dry season, but I was impressed by how fast the work had been done. It may be decades before this one gets a blacktop, but perhaps because it's locally managed and locally owned, it will be repaired faster than the ones that are part of Melamchi, should the next monsoon damage it.

I spent the night in Thimbu again a few months later, and trod the steep path up the hill to the east a second time with some of the staff members from the Melamchi project. Our destination this time was a remote village called Kharchung. The Melamchi staff members—specialists in communications and conflict resolution—had been visiting small villages like Kharchung. On this visit they were supposed to meet with villagers to discuss their current demands, which included moving the site of the intake and changing the project's name to Hyolmo. Both these demands, Melamchi officials told me, were impossible to satisfy.

The atmosphere at the top of the hill in Kharchung was tense. Only a handful of men had shown up from Kharchung and nearby villages, even though it was a Saturday. They kept to themselves, not even greeting the Melamchi staff. Salil Devkota told me that the men had been instructed by their leaders, who now lived in Kathmandu, to boycott the meeting.

Other villagers seemed nonplussed and set about cooking lunch for everyone who had made the long morning walk up the hill. We waited for our food in separate, largely silent huddles. I was puzzled as usual about what was going on, even though a young local woman who worked as a "facilitator" for the project tried to answer some of my questions. I wondered why these men had made the long walk when they suspected this would happen. They had to, as I understood it. It was part of their job to make the gesture even if the other parties were not interested in coming to the table that day.

I later found out that the villagers had become suspicious of the Melamchi staff's motives. The staff had interviewed villagers to find out the cost of their land and homes and livestock, getting information that would allow the project to come up with estimates of how much compensation they might have to award the villagers. Even though the farmland in the hills would not likely be touched directly by the project, all people within the "project-affected" districts might be entitled to compensation. The villagers and their stakeholder committee leaders were afraid the project could take advantage of them because they did not really know how much their land was worth.

Relations between locals and the Melamchi project deteriorated. In August 2009 villagers placed multiple padlocks on the gates at the field office near Melamchi Pul Bazaar. Each padlock represented a group of protesters who were trying to halt any further work until local demands were met.

In September 2009 I went to a meeting in Kathmandu organized by the Melamchi project and the public works ministry. The meeting was intended to allow representatives of local groups from the Melamchi region to air their grievances. The padlocks were still in place. The current prime minister had recently held a groundbreaking ceremony on the Kathmandu side of the hills, in Sundarijal, instead of on the Melamchi side. This had offended some people. Salil Devkota told me he had advised government officials against holding the ceremony in Sundarijal for precisely this reason.

I reached the hotel where the meeting was held in the afternoon. Dozens of men and a few women were on a shady terrace having lunch, scooping rice and miscellaneous oily curries from chafing dishes. The secretary from the Ministry of Public Works would arrive soon to hear the participants air their grievances during the afternoon session.

Inside the meeting room upstairs, tables were ordered around the perimeter of the room in a big rectangle so that participants were facing each other. The tables were draped with orange pleated shams and in the center there was an arrangement of coral gladiolas. There were no fans; I was wilting from the heat. I fanned myself with my notebook and sweated, while some Nepalis wore vests and topis, even scarves and jackets, despite the heat.

The representatives rose in turn to speak. Many identified themselves according to political affiliation. Speakers complained that the government had not included them in decisions about the project. Some were worried about the environment, about whether there would be enough water left in the river, about villagers being adequately compensated, about whether the project would bring them jobs.

A trim man with a chiseled face, mustache, and graying hair named Shrestha said he wished there had been a meeting like this earlier. He added that he was ready to give the water to Kathmandu but wanted some benefit from the project because so much money was being spent on it and the demands of the locals were small in comparison. Then he threatened that if the government did not fulfill the demands of local people they would not let them have the water.

A clean-cut young Maoist named Thapa in a plaid short-sleeved shirt spoke slowly and a little hesitantly at first, but his voice got louder and stronger as he said he wanted representation for indigenous people. He complained that the ministry had told villagers the disputed intake point could be changed while the Melamchi office said that was impossible. Why the contradiction? He wanted the intake changed.

Another speaker in a black vest from a right-wing party asked why the meeting was being held in Kathmandu instead of Melamchi. Why didn't the project leaders come to Melamchi to talk to them? And why had it taken seventeen years? "It's too late," he said. Someone should have responded earlier to their grievances.

Others were more conciliatory, saying agreement would not be difficult to achieve if the government would be more forthcoming with information and compensation. The main thing was that the local people benefit from the project, even if the name and the intake were not changed.

The secretary from the ministry spoke, agreeing the meeting was long overdue and should have been in Melamchi. He said the project must succeed at any cost and the government would solve people's problems. He urged the local committee members to help solve the problems, to provide mutual understanding. He said that money was available for people affected by the project. There seemed to be no representatives from the Asian Development Bank at the meeting.

As the secretary finished speaking a man yelled out, "You say it all so easily, but where is the commitment?" The secretary tried to say some soothing words, then gave up and smiled. He listened for a few more minutes, then started to ease out of the room. He had been at the meeting for two hours.

Men began to yell angrily: "If he is leaving, we are leaving. He has authority to make decisions and he's gone." The secretary's premature departure seemed to have canceled any goodwill his words might have garnered.

I talked to a German consultant who had recently been contracted to help manage the project. He told me that as bad as this project was, considering its delays, confusion, and contradictory demands, things had been much worse in Nigeria, where he worked before coming to Nepal. There, people were hostile and foreign workers were not safe. The army settled most issues with guns. He appreciated Nepal's democracy, no matter how chaotic. Even with the many changes of government, there is some rudimentary order and continuity. Perhaps it can't be called progress, but it can't be called a failed or police state either. The delays from the padlocking of the project field office were costing the country about eighteen thousand dollars a day. Even the Chinese contractor waiting to start work on the tunnel was dismayed at the idea of charging this poor country that kind of money, the German management consultant told me.

As the meeting broke up, I met a man named Ngawang Lama, one of the leaders of the Hyolmo group demanding that the intake point be moved. He was from Tarke Gyang, the pleasant village I had stayed in earlier that year. He said 1,500 Hyolmo families lived within five hundred meters of the river; for their sake the intake had to be moved.

Some months later I arranged for a translator so I could talk with Ngawang and find out if his group had made any progress in negotiations with the Melamchi project. My friend Anil and I met Ngawang at the entrance to the famous Boud-hanath stupa. During the short walk back to his home, Ngawang spoke animat-edly to Anil. On the noisy street I was unable to catch the drift and Anil, who knew little about the Melamchi project, seemed confused about what Ngawang was telling him.

Ngawang's flat was in a fairly new apartment building; the furnishings were simple and sparse, but the rooms were very clean. We sat on the floor in our sweaters and jackets in the unheated blue-walled room. I started asking questions about what had happened since I met Ngawang in September. Ngawang talked and my friend Anil translated.

He told us about something that happened the previous week. I tried to get Ngawang to back up to the meeting where I had met him, but it was difficult to get this challenging circle of question and answer on a more chronological course. Then Anil said something about Ngawang having been released from custody the day before. That explained why neither Anil nor I had been able to reach Ngawang until that morning. His mobile phone had been switched off; I stopped asking questions and just listened.

Ngawang and some other members of what they had named the "Hyolmo Drinking Water Concerned Committee" discovered that work had resumed on the roads. Apparently the contractor who had built the roads was tackling the various washed-out areas I had seen when I visited. Without these repairs, the road was useless not only for building the tunnel but to local buses and other travelers.

The men went to talk to the contractor. Ngawang said his group was not armed and had no intention of fighting, but other men waiting in the bushes started throwing rocks. Ngawang showed us a scar on his scalp. Frightened, Ngawang and some of his colleagues ran. They left the area and came to Kathmandu on motor-bikes via back roads.

Ngawang filed a police report near his home in Bouddha, saying he had been attacked without provocation. The police told him they would send the report to the Melamchi police office. Ngawang and some friends then went to the Melamchi project office in Kathmandu to talk to the project manager about why work had started up again; the agreement they had made in August with Melamchi man-agers had still not been honored.

The meeting was inconclusive; afterward a plainclothes police officer arrived and asked Ngawang to come with him. Ngawang was held overnight in Kathmandu. Meanwhile his family, a lawyer, and even a member of parliament talked to the police, saying Ngawang was the injured party, having been attacked by what they called "goondas," paid hoodlums. The police said they were going to take Ngawang to Melamchi the next night.

At this point in the story, as I kept trying to get the sequence of events straightened out, Ngawang started to cry. Being taken by police or army at night, in a country that had until recently been embroiled in a civil war, was terrifying. Ngawang's wife came into the room and also started to cry as they talked to Anil, who understood instantly the nature of their fears. Many people taken at night, whether by the government or the Maoists, had never returned home. Even those who had not been affiliated with any party had been tortured and killed during the insurgency. Maybe their bodies were found later, maybe they simply "disappeared."

Shankar Limbu, an attorney who works for indigenous people's rights in Nepal, had known the Hyolmo group for five years. He corroborated Ngawang's story. To protect Ngawang as best they could, his supporters obtained a document signed by the police saying they had him in custody and were taking him to Melamchi. Along with half a dozen officers, the police inspector left with Ngawang around two the next morning. His wife came to the station to find him gone and followed in a taxi.

Ngawang was held for two more days in Melamchi and finally released after Shankar Limbu went to the supreme court of Nepal with a writ of habeas corpus. But before his release Ngawang was threatened that he might be held financially responsible for any damage to the contractor's equipment that occurred during the scuffle the previous week, when apparently some equipment had been set on fire. Ngawang said he hadn't set any fires and didn't know who did.

Ngawang was still frightened, and he believed that that was the point: to intimidate him, as the leader of the Hyolmo group, and to silence him. He said he was not a violent man. He had a plastics recycling business in Kathmandu, and his two teenage children were in school. I met them before I left his apartment: a girl and a boy, bright and polite, they greeted me in English and smiled warmly. Ngawang said he would leave the work to others for now. He said he did not want to stop the Melamchi project, but he wanted indigenous rights recognized.

Limited or clumsy communication seemed to be at least one of the problems between the Hyolmo group and the Melamchi project. In a part of the world where many big projects, especially dams in India, have gone forward while the people displaced by such dams have not been compensated, there is great fear and suspicion. People in Kathmandu learned about the Melamchi project over many years, when it was in the talking stage. But in the villages of Helambu and Sindupalchowk, the district that is home to Melamchi Pul Bazaar, there were few newspapers, and not that many people could read Nepali. So the locals learned about things largely through rumor.

Villages in these districts are widely scattered; at the beginning of the project probably no one on staff knew how and where to interact with local people. But many years into the project, the lack of successful communication seemed troubling. I was puzzled to learn that Ngawang Lama, an intelligent and courageous man, was still confused about important details of the project. Even Shankar Limbu, the indigenous people's rights lawyer, believed there could be a large dam as part of the project, with inundation and displacement of people, when this is not the case.

Ngawang said that Hyolmo villagers began to voice their concerns as soon as they heard about the project, but were continually reassured that the tunnel was years away and their issues would be dealt with later. In 2004 Shankar Limbu gave a presentation at the head offices of the Asian Development Bank in Manila, outlining various concerns of the local people. Chief among those concerns was the placement of the "intake," the beginning point of the tunnel that would carry water to Kathmandu.

Hyolmo villagers live up in the hills, not along the riverside; yet they were worried about thousands of their people being displaced. They seemed not to believe reassurances that there were no plans for a dam on the river that would inundate large areas. Their fears seemed a little inconsistent and confused. But there was nothing confused about their assertion of their right to understand the details of the project and to have some input in the planning process, essentially to have their rights as the region's indigenous people recognized.

Shankar Limbu intimated the fight was not over. Nepal was at a crucial turning point in its governance, and there was much debate over federalism and representation of the country's many ethnic groups. The Melamchi project was caught up in this debate. Indigenous people throughout the country were organizing. They had long felt oppressed by the Hindu hierarchy that has ruled the country since Prithvi Narayan Shah invaded the valley in 1768 and started forming a nation. Perhaps this wider movement will weigh in on the side of the Hyolmo people and stall this high-profile project once again.

Soon after I met with Ngawang, I talked with an ADB project officer. I told her I had trouble understanding how the Hyolmo people could still be in doubt about crucial details of the Melamchi project. Lakshmi Sharma, the project officer for Melamchi, suggested that enemies of Melamchi "might have poisoned their minds to kill the project." Or perhaps it was simply a communication error; she said she too was puzzled.

Many observers criticize the two big development banks for being in the business of pushing burdensome loans on poor countries for ill-advised and unnecessarily costly projects. But many of the people working for these banks see themselves as public servants.

I asked Sharma why ADB was sticking with the project after all the many frustrating delays. "We have started something," she replied. "We have to be accountable and make it happen. It's a matter of doing the work together. We have to be responsible to the country. It's actually the government's money. We are just giving our helping hands."

"Water is really critical for Kathmandu to survive," she added. "Without water Kathmandu's development is hampered. It's the capital. How can it lead without water? If no water, then no Kathmandu. It's the life of the people."

Sharma said the leaking distribution system in the valley would be improved to acceptable standards before Melamchi water comes down the tunnel. I asked her whether she felt sure the ADB could accomplish a task the World Bank had given up on in 2002. She laughed and agreed: "It's really a mess. It's difficult. But it can be fixed. Why not?"

I admit that I deplore the idea of harnessing the lovely Melamchi River and squandering it on Kathmandu's uncertain system. But I see the romanticism of this attitude, and its hypocrisy. For decades I lived across the bay from San Francisco, in Oakland and Berkeley; I was a customer of what we fondly referred to as East Bay MUD (Municipal Utilities District). I happily drank water from the dammed Mokelumne River. There was no mud in the water that came from California's Sierra Nevada. The Mokelumne was snow-fed like the Hyolmo-Melamchi, and its name indigenous, from the language of one of the tribes of forgotten Indians, the people who once lived throughout what is now California. The water still tastes good, much better than the treated water from the big blue bottles the guard brought up to my apartment in Kathmandu.

Yet none of the objections to the Melamchi diversion seem like romantic desires to keep this river as it is, just for the sake of keeping it. They have to do with cost, efficiency, logic, and a keen desire to make the best use of resources. And now, after all the many delays in tackling the problems of water distribution and sewage management in the Kathmandu Valley, some of the water experts who opposed Melamchi have ceased to do so actively.

UN Habitat's Roshan Shrestha says the country is in too deep now to stop Melamchi; Nepal has taken too many loans already, so there's no turning back. But he would like Melamchi to be part of a rational plan. "My concern is that still we are only talking about Melamchi. We need it, but even after that we will have water problems. We have to work on proper management, and how to stop overextraction of groundwater. All over the world people are talking about wastewater recycling. Where is this? Or eco-san toilets? We have no plans for this sort of thing."

Another expert says water harvesting and groundwater recharge are finally becoming a priority for the government. Without groundwater recharge,

Kathmandu—built on a lakebed, under whose surface depleting aquifers are like empty beehives—will sink. The government could start by requiring big hotels to institute rainwater harvesting. They are drilling the deepest wells to extract groundwater for the showers and flush toilets tens of thousands of tourists want.

Rainwater harvesting is already spreading in the valley as small-scale efforts proliferate. Perhaps what Dipak Gyawali suggested—ameliorate the water shortage with all the less expensive interventions and then build Melamchi if necessary—will happen, but not in the order he suggested. Perhaps everything will lurch along together. More and more neighborhoods will try what the residents of Chyasal have done with rainwater harvesting, or what the community around Alko Hiti has done to take advantage of its still-prolific *dhunge dhara*. Maybe the city's leaking pipes will be replaced before Melamchi arrives, which is likely to be delayed beyond the 2013 date ADB projected in 2010, when tunnel construction finally got underway.

The China Railway 15 Bureau Group, a Chinese government enterprise, won the bid to build the Melamchi tunnel. The Chinese are very good at building tunnels. I passed through an impressive one going from the Lhasa airport into Tibet's capital city. The trip used to take at least three times as long because vehicles had to go around a mountain range. The wide, well-lighted tunnel simply goes right through it now. But the mountain ranges of the Tibetan plateau are older and more stable. The hills of the young and ever-shifting Himalaya may prove more challenging. And the project will likely continue to be buffeted by the politics of a country that lacks the strong central authority to which the Chinese construction company is accustomed.

# 8

# More River Blues

In the past, there were no latrines, no toilets. But people
did not pollute the river. People went to the fields and hills.
Now there is filth all over the country—around springs, rivers,
everywhere. Sin has advanced and Dharma has been left
behind. I feel that sin has won. Will it always be like this,
or will this be reversed one day?
—Hira Devi Dahal, elderly resident of
Sundarijal, Kathmandu

It was not entirely clear to me why Ngawang Lama and his group wanted the intake point—the head of the Melamchi tunnel—moved to the spot that was proposed back in the early 1990s, when the World Bank was funding the revamping of Kathmandu's water supply. But I learned that Norwegian engineers, who were then consultants on the project, had originally placed that intake upstream to provide for a hydropower plant. They saw the Melamchi project as a good opportunity to get more for the same investment of money.

The Norwegians had proposed to place the intake several miles above the spot that Cholendra and I almost reached as we walked up the damaged access road. Using that intake point, called Nukute, would have allowed for a twenty-five-megawatt hydropower plant in Sundarijal, where the tunnel ended. The higher intake could give an additional three hundred meters of "head"—water pressure to generate electricity.

When the Asian Development Bank took over the project, they scuttled the hydropower component. After this, the Norwegians pulled out. The proposed twenty-five megawatts of electricity would have been welcome in a country that is likely to see power outages for at least another decade. Now, with the planned intake point lower on the river, hydropower is not possible because there would be insufficient water pressure.

The ADB's reasons for dropping the hydro component are a little vague. Ratna Sansar Shrestha dismisses the economic and environmental costs the organization cites as its rationale for dropping hydropower. Ratna is a water resources specialist who is well known for wanting the Melamchi project to include hydropower. He is one of three members of the Regulatory Commission for Water Supply that over-sees tariffs and quality of service throughout Nepal.

To be charitable to the ADB, he says, "working with Nepal's bureaucracy is not easy." Hydropower projects require negotiating with an entirely different ministry from the one that oversees water supply. Cutting out the hydropower component also cut out half the administrative red tape on a project that has been drowning in it for years.

I went to visit Ratna at his office one day. A wiry, energetic man, slight with broad shoulders, he looked more elegant than the average well-groomed Kath-mandu water expert, wearing a brown suit with a cream-colored shirt and a silk paisley tie. Closely trimmed salt-and-pepper hair framed his dark, square face. He has a thin mustache and prominent eyes that gaze intently. A ready laugh punctu-ates his quick and sometimes colorful speech. He also moves quickly and purpose-fully, like a man who has a lot of things he wants to get done.

On the one hand, Ratna says, Melamchi is the wrong water supply project for the Kathmandu Valley. He agrees with critics who say there are many things that can be done in the valley to fix the system so it won't lose water. He acknowledges these would cost a tenth as much as the water-diversion project. "Melamchi is too big a project for Kathmandu. But the 'hydrocrats' don't like the alternatives; they like big projects for obvious reasons. They can get bigger cuts. This is the wrong project, but because of their cut they are pushing it forward. It's that simple."

On the other hand, if a big project is going to happen, let it get the most bang for the buck. "I want a bigger project," Ratna says, "not just a water supply for the pampered people of Kathmandu valley." And he wants not just twenty-five mega-watts of hydropower, but 265 megawatts for this power-starved region. He says he can do that and keep the proposed tunnel where the ADB wants it to be. "Now we say let it continue as is, but let us take water from the original place and get fifty megawatts of hydropower from Thimbu."

Ratna's guru is one of the Norwegian engineers who worked on the project in the 1990s, and now with the benefit of Google Earth has seen ways to use Nepal's rivers and gradients to an even better advantage than he had proposed earlier.

From the beginning, plans for the Melamchi water diversion included bringing water from two rivers to the east of Helambu: the Yangri and the Larke. But con-struction of tunnels to carry water from those two rivers was not scheduled to happen until years after completion of the Melamchi tunnel, when the valley's water needs would increase. Ratna thinks the work should be done simultaneously and that there should be an additional tunnel to bring water from a fourth river, the Balephi. The water would be used for Kathmandu, of course; it would also generate fifty megawatts of hydropower in Thimbu, and another 215 megawatts at a pair of generating facilities on the Bagmati a few miles south of Kathmandu. Then the water would be available for irrigation for many miles downriver.

The much-disputed intake point for the Melamchi tunnel would remain the same. The water from all four rivers would be sent through a penstock from

Nukute to the first hydropower plant at Thimbu. Ratna's plan would work only if all four rivers contribute their water, so the tunnels would have to be constructed simultaneously.

"You have to be kidding," I interrupted as he outlined this plan. "They can't even get one tunnel built!" He smiled as if he were expecting this objection. "You are right, we are talking about a series of tunnels." But the work does not need to take decades, he maintains. For a hydropower project he worked on called Kimti, Ratna says construction of a twenty-seven-kilometer tunnel was complete in a little more than three years. I recalled the dusty road I walked down returning to Melamchi Pul Bazaar, built in a fraction of the time the Melamchi project took to construct the road to Thimbu.

In Ratna's plan, water would leave the plant at Thimbu, go down the tunnel to Sundarijal, and on to the Bagmati, which according to Ratna would "get a new lease on life." The river could be used for fast, cheap transport between Kathmandu and the *tarai*, the agricultural heart of Nepal. There would be enough irrigation water downstream for three or more crops a year, including the dry season. "If they would do this Kathmandu could have gravity flow for 24 hours a day—no water tankers, no pumps to carry water to the roof, no illegal connections and pumping from aquifers," Ratna said enthusiastically. He says he is not looking for a job on such a project, but would be happy to give advice pro bono.

ADB maintains that any such changes in plans will cause delays, though delays have been the hallmark of the project already. Ratna says he outlined his proposal in recent years to various prime ministers as the government shifted from one party to another, from kingdom to republic. All were interested in the ideas, but they told him he would have to persuade ADB. ADB has said that the government needs to make up its mind. Ratna thinks Nepal's politicians and bureaucrats don't want to say things to upset white foreigners.

So perhaps however good this plan, it will languish; meanwhile the Chinese contractor will struggle to stay on schedule with the single Melamchi tunnel.

And there would still be sewage in the Bagmati, even with a brilliant hydropower scheme. "But all the sewage will still go down the Bagmati and on to the Ganga," I ventured. "No, it cannot," Ratna replied. "If all the sewage from Bagmati goes down to my plant, it will be destroyed."

At this point Ratna had to take a phone call, which left me to think about what he had said. Ratna's office was on the ground floor, next to a daycare center. I became aware of children's excited voices. How many more generations might have to wait for a city with good water and sewage services? I wondered.

What Ngawang Lama and his group said they wanted—to move the intake point back to the one originally proposed at Nukute—was impossible. Once the hydropower component was removed, ADB had to move the intake. There would be too much water pressure and so there would have to be energy-dissipating equipment placed at the end of the tunnel if Nukute were the starting point. That made no sense.

Yet including hydropower in the Melamchi project would benefit the Hyolmo people and all others in the watersheds of the rivers affected without their having to fight so hard for recognition. Water supply projects are not yet required by any law to compensate local people, but hydropower developers have to pay royalties to the people affected by their projects. Nepal's laws require this. Such royalties could allow the Melamchi region to develop, providing schools and hospitals, probably far more than what the Melamchi project's social uplift program might provide them.

After meeting with Ajaya Dixit and Dipak Gyawali at Nepal Water Conservation Foundation one warm winter day, I took a taxi back to the other side of the city. Ten years ago this trip would have taken twenty minutes; now it takes an hour. Gridlock brings squadrons of motorbikes and flocks of little white taxis to a standstill at various intersections and bottlenecks. I sit back and muse.

The dry winter air is dirty. Grit hits my eyes. But in spite of the haze, to the north I can see the white peaks of Langtang and Shisha Pangma, the Himalayan "water tower."

Helmeted motorbike drivers ease around the cars and taxis; my driver switches off his engine to save petrol, which is often in short supply in the valley. Taxis and motorbikes sometimes line up for miles, waiting for hours for a little gas. I wonder how much wear and tear all the extra starts put on his starter and whether he's really saving money.

We start crossing the bridge over the Bagmati; midway over, gridlock again. Women crossing the bridge on foot pull scarves over their mouths and noses to block the stench from the ruined river. I do the same inside the taxi.

The whole conundrum seems to be laid out right here before my eyes. Unseasonable warmth: a sign of climate change? The so-called water tower gleaming above, South Asia's snowy, frozen sponge, now threatened by climate change. Unmanaged sewage. No urban planning. Intelligent people in the offices of nonprofits with good ideas. The paralysis of policy that is like the traffic gridlock. The perhaps needless and politically driven shortages of petrol, electricity, water.

The Kathmandu Valley I had lived in for the past few years was reaching a crisis point that mirrored the nation's precarious government. Streets clogged with undisciplined drivers, air quality like Los Angeles before regulations, roads full of potholes, car horns damaging one's eardrums, hideous new construction, no city planning. For now, life lurched on. For now, the tourists were still coming.

Yet I can visualize the Kathmandu I never saw, the remains of which still inspire thousands of visitors. I can just wipe out the noisy, importunate crush of cars and motorbikes and all the boxy concrete buildings that may someday crumble in the big earthquake that is overdue, and instead hear the quiet and see only trees and green fields of mustard and wheat or rice paddies surrounding the red brick city centers of Kathmandu and Patan, stretching to the hills that ring the valley, the shoulders of the Himalaya draped in white. And the rivers that wind through Kathmandu, the Bagmati and Bishnumati, I can almost imagine them clean.

There are many people in Kathmandu who can easily remember clean rivers. Lajana Manandhar is one of them. She played in the Bishnumati as a child. A slender, attractive woman now in her forties, she directs an organization that works to gain safe water and better living conditions for some of the poorest people in the valley. I was just as interested in her memories of a cleaner Kathmandu as I was in hearing about the work she does now. So one winter day we took a walk down the bank of the Bishnumati, near an old part of the city where she lived when she was growing up. "Look at this, it's all sewage now. It used to be clean water, a clean river. Can you see that small temple—I used to swim there," she said, pointing across the river. "Now it's a road."

Lajana said she and her friends made regular visits to this area where there is a triad of temples. She is Newar and a Buddhist, but also follows some Hindu traditions. "Young little girls like nine, ten, eleven years. We used to hold hands. We were scared of the flood and we all held hands so we wouldn't be swept away." The water came to mid-thigh on the girls, and in monsoon it was deeper. "It was fun. We used to enjoy being in the river, crossing the river, worshipping at the temple. I don't think now the children are following that tradition. There is no water."

By twenty years ago, the water had gotten far too dirty for swimming. I ask her if she thinks she will ever see in her lifetime the river she knew as a child. "Well, it will never be like before. Roads and houses have encroached. But at least they could make it as clean as it was before. Maybe twenty, thirty years," she trails off. "So smelly, so dirty." She looks around again. "I remember in that area we used to dig the sand and build small temples. But now there's no sand."

I returned to the same spot near the three temples another winter day. Smells assaulted me as I walked. Acrid smoke blew from the Newar cremation ground just above on a rise at the confluence of the Bishnumati and a filthy little stream that was once the Bhacha Khusi, flowing out of the forest of a hill called Nagarjun.

Out in front of the Shobha Bhagwati temple I caught the earthy smell of a big black ram, unaware of its impending death at some upcoming sacrifice. The almost sweet smell of old garbage mixed with that of the animal.

The far more repulsive smell of sewage hit me as I walked closer to the side stream. A cow with a wounded and horribly swollen jaw approached, the jaw so distended I wondered if it was partly a birth defect. How can so huge an infection not make the cow stumble and die instead of what she does, grazing on the stubble of dry grass, inspecting some garbage, then wandering down to the river for a drink of water before crossing it in a way that appears routine, purposeful.

I stand and gaze down the nasty rivulet that passes under a small footbridge over the erstwhile Bhacha Khusi. On the bank I see a lump and soon realize it's a blond sow. Her bloated body is so still I think the wretched animal must be dead. Then I see an ear twitch. And then I make out two piglets by her side, almost indistinguishable from the garbage.

Boys keep yelling hello at me as if the repetition and volume will finally get my attention. I ignore them. A small girl comes up behind me and quietly says hello. I smile at her and answer.

In the river there are blobs of plastic bags like those I saw in Delhi, resembling balloons with the air mostly let out of them; but the water here is slower and shallower than the Yamuna, so they clump in a mass. Dust from the unpaved roads on both sides of the river gathers thickly on them, creating a series of monochromatic, asymmetrical rafts on top of the water. A single white egret walks on one of these rafts and pecks around for something worth eating.

The Nepalis are busy with their own affairs and move on by me with only an occasional glance. They are as oblivious to the wounded cow as they seem to be to the filth in the river and the stench that is making my mouth feel dirty. I want to spit. The smoke from the cremation ground blackens as the heart of the body burns. I rinse my mouth with my own saliva and spit to get rid of the taste from the sewer.

I can imagine this was once a peaceful place to say goodbye to someone beloved, to commit a family member to the flames and then to the fresh water coming down the hill. At the bottom of the steps that lead from the Newar temple down to the Bhacha Khusi is a remnant of that stream, which seems to have been about twelve feet wide. Now it's a dry and rocky waste of garbage. Only a little water

The Bhacha Khusi flows with filth

comes from upstream; the flow that reaches the Bishnumati a hundred feet below is mostly coming from an open pipe in the middle of the streambed. It may be the end of a storm drain, or a pipe that was intended as a storm drain. Now it's sewage.

If Lajana Manandhar could swim here thirty years ago, why does it seem so difficult to retrieve that? There have been various clean-up-the-Bagmati schemes, just as there have been ones to clean up the Ganga and the Yamuna. They are ongoing now. All the smaller rivers in the valley drain into the Bagmati, which of course carries all the valley's muck south into the tarai and on through Bihar to the Ganga, which receives the effluent from Delhi and Varanasi and Patna long before the Bagmati reaches her with Kathmandu's crud.

Whether or not Melamchi includes hydropower, how is Kathmandu going to deal with its sewage?

In a social setting one evening in Kathmandu, I chatted with a water engineer who worked for one of the big development banks. I started talking about the plight of the Bagmati, as I often do.

"Pave it over," the young engineer suggested, perhaps unwinding after a long day's work, sipping a glass of wine. My face no doubt registered pure shock. This engineer, raised Hindu, had proposed a heresy.

"But it's too important culturally," I sputtered. "Yes, but at what cost?" the engineer shot back. In fact, the Bagmati is a sewer these days. So why not treat it like one and pave it over? I started to see the logic at least. Unsurprisingly this engineer did not want to be quoted making such a blasphemous proposal, but the idea continued to reverberate in the back of my mind.

What is the cost of cleaning up the Bagmati? It's a multifaceted question. I got some ideas about the size of the problem from the current director of the sewage department of KUKL, the water management company. Kiran Amatya started managing the sewage department in the new KUKL in 2008. First he outlined some history.

Throughout the city there are storm sewers, many from the Rana era, intended to carry monsoon runoff to the various rivers and streams in the city. As Kathmandu grew and developed, toilets became common. But unless the owner of the toilet installed a septic tank, the toilets most likely drained into the storm sewers.

During the period when the World Bank was trying to help sort out the confusion of Kathmandu's underground pipes, several sewage treatment projects were begun under the auspices of international aid organizations. One project involved a sewer line running parallel to the Bishnumati River. The line was meant to intercept the sewage coming from homes and hotels chiefly on the east side of the river, carry it downriver by gravity, and then pump it up to a triad of ponds where the sewage would be treated naturally by bacteria. Eventually the clean water would be released back into the river.

This method of treatment is called slow filtration and it requires a lot of land for large ponds; it was appropriate for Nepal because it costs a good deal less than the sewage treatment plants common in the developed world. But the project never worked. In the 1970s, when the treatment system was constructed, said Kiran Amatya, there was not yet enough water in the system to keep the sewage flowing. Homes had little water supply yet, and people were unaccustomed to using much water in any case. The system had been built on foreign assumptions about water consumption.

Soon the interceptor pipe was clogged with sewage because there wasn't enough water flowing through the system to flush out the pipes. The pump broke for similar reasons. The ponds were essentially never used, and the land sits there, now surrounded by homes whose owners at this point would find a functioning sewage treatment facility near their property distasteful.

Two similar treatment facilities were built around the same time but have ceased to function as intended. Even if all three of the facilities built more than a generation ago—when the population of the valley was about a third what it is today—were working well, they would treat less than a quarter of the two hundred million liters a day of sewage the valley produces.

Thus, almost all of the valley's sewage goes to the rivers via storm sewers. When new homes are built, their sewage pipes carry waste away from their premises. Kiran says that's about all his department can do now—keep the city clean, if not the rivers, by maintaining the pipes that carry sewage away from dwellings.

A slender man, wearing a caramel-colored leather jacket, Kiran spent two hours explaining the ins and outs of the failed system. His office was well heated with a kerosene stove and his assistant brought me tea as Kiran showed me, with the help of Google Earth, where the sewage treatment plants had been placed. Kiran, who has a long, narrow face and a strong thatch of coal black hair, did not seem a gloomy man; but he was very frustrated by his inability to accomplish more than piecemeal solutions. He saw no way out. He said the city and its services were essentially trapped by the utter lack of city planning in the valley. Basically without a complete rehaul of the city, there isn't much that can be done but to keep sewage out of neighborhoods.

The land that might have been used for sewage treatment ponds like those near the Bishnumati is gone. The valley is clogged with construction, haphazardly laid out because so many different people own small plots. Land for sewage treatment is needed in logical places, Kiran said, allowing large interceptor pipes to follow the course of rivers like the Bishnumati downhill, using gravity to reach treatment ponds. Most of this land has been covered by houses.

I had stood on the roof of Dipak Gyawali's house in Patan one day as he showed me his rainwater harvesting system. We tried to find a gap in the houses, a mile or so to the north, where the Bagmati flowed. Once he could see the river from his house, but now it was impossible to distinguish where the houses stopped on one side of the river and began on the other, so thoroughly has the floodplain been built up. Kiran said there are old sewage pipes in that area along the Bagmati, deeply buried. Now houses have been built on top of those pipes, so it's no longer possible to rehabilitate them.

At some point, a functioning city government might have to buy out some landowners and just go ahead and build treatment systems. Ironically in such a dysfunctional city, the price of land in the Kathmandu Valley has risen dramatically in recent years. It's no longer within reach of the enterprising individual who hopes to build his family a house. It's in the hands of speculators to a great extent. Perhaps, if the dysfunction continues, the land value will drop and the government can buy land and start to institute a system. But instead of an overall plan for the city, there is a plan to build another ring road outside the existing one, encroaching on more hillside land, more agricultural land. Kiran is afraid this will simply ruin the 30 percent of the valley that has not yet been engulfed.

"Sewage is more an opportunity than a problem," says Ratna Sansar Shrestha, who believes there is enough land left in the city for innovative approaches. Starting over and doing things right and planning the city from scratch might be nice, but things are the way they are. Dealing with the current situation means thinking outside of the usual boxes.

And as he sees it, good use of sewage could go hand in hand with the revamping of the Melamchi project that he advocates. From a selfish point of view, he points out, his proposed hydropower plant doesn't have to clean up the filth from the Bagmati until just before it enters the turbines. The water would have to be de-silted anyway in order to preserve the power plant he would like to see built, so the silt and other wastes could be removed at the same time at a plant designed for both purposes. And down in the area south of the valley there is plenty of land for such a treatment plant.

"So that's great for Bihar and the Ganga and beyond," I pointed out as we continued our discussion in his office that day. "Meantime we still have this sad, stinking city here." He agreed that from the point of view of Kathmandu, and people who used to be able to enjoy and revere the Bagmati, this would indeed be selfish. So the treatment could be done in the city, Ratna said. He admitted that he doesn't have a plan yet, but "we can come up with one," based on the notion that sewage is an opportunity. With the right treatment, it can become biogas and fertilizer.

He says the money for establishing biogas plants could come from the hydropower developer; it would not have to come in the form of loans from an entity like the Asian Development Bank. The hydropower developer will have to invest in treatment anyway so that clean water enters the plant. So why not put the treatment plant upriver, in the city? I point out the lack of land that daunts Kiran Amatya. Ratna says he's not that pessimistic. For starters, landless people who have built slums along the river can be resettled. Such dwellings are illegal. And biogas plants do not need as much room as do sewage treatment ponds.

And it would not all fall on the government. He envisions pilot programs, partnerships with big hotels. Many multistory buildings with hundreds of toilets now dump their refuse into the rivers if their septic tanks can't handle the waste. Help them install biogas they can use to offset energy costs, Ratna says. The infrastructure for biogas is not very expensive. Several different biogas companies could undertake this. It would not take twenty years to get the sewage out of Kathmandu's rivers.

Initiatives like this would require a mental leap, a new approach to problem solving to get beyond conventional ideas about how sewage should be handled. The technology is not complicated. The financing may even be feasible. What is missing is the will to act on the part of the government. But by 2010 there was beginning to be talk about private approaches to sewage treatment in Kathmandu. And on the fringes of the valley, enterprising people were pursuing solutions involving biogas and other low-cost methods. Even the Asian Development Bank—perhaps beginning to move out of the bigger-is-always-better paradigm—was helping them with small grants.

All these were hopeful signs. But still the main directions and the central assumptions about how to solve the problem of sewage in Kathmandu, as in Delhi, involved treatment plants that needed a lot of water and required moving sewage long distances in pipes. All this would call for help from foreign donors, again leaving thousands of villages in hock for Kathmandu's convenience.

There's a canal not far from where I lived in a neighborhood known as Lazimpat. It was once a small river called the Tukucha Khola. I crossed it several times a week on my way to yoga class. I held my breath because it's a sewer. Nonetheless I often saw children and animals down by the water.

Emanating from the canal is a stench much like what I smelled near the Bishnumati River where the polluted side stream with its sewage joined the river. Following the Tukucha Khola north one day, I discovered that it has been roughly paved over in places with big slabs of concrete. This made me think of the engineer who shocked me by suggesting the most sensible solution for the Bagmati would be to pave it over. The pavement over the Tukucha doesn't keep the smell contained. Big pastel houses are right on the edge of it; perhaps their owners built them when there was less sewage in the channel.

On another walk, I saw a garbage collector throwing plastic bags from his handcart into the Tukucha, just before the dark water goes underground near the perimeter of the modern Narayanhiti palace. Until recently it was the king's residence and is now a museum surrounded by acres of what could be a wonderful city park someday.

On those walks east of the Tukucha, on my usual route to yoga, I skirted the edge of the very ancient capital and palaces of the Lichchhavi, perhaps of the Kirats as well. The Tukucha Khola was more or less the western boundary of that capital.

Artifacts of ancient life in the valley lie buried or are churned up by heedless new construction in these neighborhoods. It's all too easy in a country as poor and ungoverned as Nepal to get around whatever regulations there might be to protect its artistic and archaeological heritage. Secrets that architect and historian Sudarshan Tiwari would like to know about are lost each week.

The Tukucha travels south, sometimes above, sometimes below ground, to meet the Bagmati. Near that confluence—if one can call this sad prayag, this unholy veni, a confluence—lives a man named Huta Ram Baidya.

The house he was born in and worked in for much of his eighty-nine years sits very much within smelling distance of the Tukucha. Now the house is crowded by concrete buildings and many lanes of traffic rushing to or from the Patan side of the river.

Huta Ram was the first agricultural engineer in Nepal and he founded the Nepalese Society of Agricultural Engineers. The grandfather of Nepal's environmentalists, he has spent the second half of his life advocating for the river and what he calls the "Bagmati culture"—the rituals and rhythms of life in the Kathmandu Valley, centered on the Bagmati.

I went to visit him one day. His "office," adjacent to his home, was outside, covered with a corrugated roof. There were several dark and crowded little rooms nearby, but his desk was in the breezeway outside, piled high with notebooks and publications. When I arrived, unannounced, he was trying to screw a cardboard sign into a vertical beam. He wore big, dark-rimmed plastic glasses with lenses so thick they distorted his face, which was fringed with white, a beard and a fluff of hair behind his bald head. I tried to help him, but the wood was too hard. Finally he gave up and, smiling, asked me why I had come. I explained that I wanted to understand his ideas about saving the Bagmati.

"Bagmati is in the coffin," he said. "But Christ was in the coffin. Somebody, somehow opened the tomb. That *may* happen to Bagmati.

"So many things come to my mind when people come and ask me about Bagmati. I first saw Bagmati when I was four years old. Bagmati has been in my mind all the time. People come and ask me with the preconception that if we have enough money we can save Bagmati. So have money first and then we can get enough men to work on Bagmati and save Bagmati. But that is wrong. We start with resources. Land. Bagmati is land in economic terms.

"First do the simplest thing with the least money and quickly save the land. Save the land means make the best use of the land you have with the least money. Don't do large projects. There are many small things you can do on the river. And at least go ahead one step towards saving Bagmati and Bagmati culture. With money they are not going to save the culture of this valley."

Based on what I had seen, I could not fault anything he said. Yet my mind was on the filth in the river. It seemed to me that first you had to get it out, stop putting it in, then set about restoring the river. But this seemed not to be the topic Huta Ram wanted to dwell on this particular sunny day. He was charming but a little gnomic as much in his speech as in his weathered face and slightly hunched, thin body. He continued in this vein, making analogies I could not quite follow between the river and a sick person. What does a sick person need? Clean water.

The cars on the road outside honked incessantly. Finally Huta Ram gave up trying to explain his vision to me and went to get his photo albums. He had hundreds of photos. He showed me ones of people swimming in the Bagmati near the bridge over to Patan. The riverbed was much wider than it is now.

He showed me a photo of a diagram, demonstrating the kind of simple restoration projects he wanted to do. I gathered he had tried to implement such projects but no one had followed through on them. The picture showed a low, inexpensive dam, built before the monsoon rains, which would slow the water so that the river would deposit sand and silt. This would restore the riverbed, and vegetation would return. He envisioned fish and ducks in these ponds, sustenance for people.

It looked lovely, but in no way could I imagine how this would work until the river no longer carried sewage and garbage. When Kiran Amatya toured me through the valley's failed sewage system on Google Earth, he had pointed to a place where there was a sewage interceptor line near the Bagmati. It had been built to intercept the flow of the Tukucha, not far from where I was sitting with Huta Ram.

Polluted riverbed in Kathmandu (photo courtesy of Ginna Allison)

But too much besides water and sewage got into that line. Dust and solid waste—like the plastic and other trash I saw the garbage collector throw into the Tukucha near the palace—had blocked the line, making the interceptor useless. Kiran said the line was buried deep under the houses that have encroached on the Bagmati floodplain. His workers could not clean it. So whatever flows down the Tukucha goes directly to the Bagmati, like everything from the Bishnumati. I did not see how Huta Ram's vision could take form until this stopped.

Some years ago Ajaya Dixit and a group of environmentalists penned a request that after their deaths their ashes *not* be consigned to the filthy Bagmati, the equivalent for Kathmandu Valley Hindus of the Ganga, and one of that holy river's tributaries. Consignment of a loved one's ashes to these rivers is an essential ritual and comfort for Hindu families, whether or not they believe it confers a good rebirth.

The Kathmandu Valley, once a bowl of water, is now filling up with concrete and asphalt, covering up some of the best agricultural land in the nation, the farmland that generations of Newars protected. I think of another valley that has not been sufficiently protected, parts of it now hidden by pavement and houses: the vast Central Valley of California, America's salad bowl, a paradise of orchards. The Central Valley still produces staggering amounts of produce, but coastal cities have sprawled into it as they have grown and land prices have spiraled. Farmers could make more money for their families by selling out to subdivisions than by continuing to cultivate.

As Kathmandu succumbs to concrete, its people can no longer see it. The visual experience of the city has all but been lost. The signposts that throughout the day remind the people of the valley who they are—the temples and shrines and *stupas* and *hiti* with flowing water—are crowded out by big buildings. Money is coming into Kathmandu from somewhere for high-rises and multistory condominiums.

At the end of a lecture in Kathmandu about Nepal's changing human geography, a seventy-two-year-old man said: "I can do nothing but cry." Sixty years ago as a boy he had made daily trips to the Bagmati, to wash his face, to place tika on his forehead, and start the day right. "The water was clear. I could see grains of sand, and mica shining in the water."

That's the kind of water I too had seen, walking along the banks of the Melamchi River one sunny winter afternoon.

# 9

# Belji of Dhulikhel

> It's possible that as a nation we've exhausted our quota of heroes
> for this century, but while we wait for shiny new ones to come
> along, we have to limit the damage. We have to support
> our small heroes.
> —Arundhati Roy, *The Greater Common Good*

I first heard of Bel Prasad Shrestha five years before I met him. An article in the *Nepali Times* lauded his efforts to establish a water system in the town of Dhulikhel while he was its mayor. I clipped it and set it aside. Fifteen miles from Kathmandu was a municipal utility that put Kathmandu's to shame. I wanted to know more.

Perhaps I saved Bel Prasad for last, expecting the visit to Dhulikhel to be a pleasant excursion—a hopeful encounter that would show me that the breakdown of urban management I saw every day in Kathmandu was not an inevitable part of development in Nepal. After all those discouraging discussions about Melamchi and about Kathmandu sewage and water supply problems, perhaps I was going to meet a Newar who had a gift for water like his ancient forebears.

I went to Dhulikhel the day before May Day, 2010, when Nepal's Maoists were planning to outdo their usual May Day celebrations with protests all over the city. They were massing their cadres in Kathmandu, ostensibly to pressure the prime minister of another party to resign.

On a Friday morning I set out with my friend Ram, a Kathmandu taxi driver who was always available when I needed to venture out on a longer excursion. The shocks on his little white Maruti Suzuki were shot, as they were on most taxis in Kathmandu, but Ram was a good driver who knew all the roads and backroads.

Aside from worries about being able to return to the city in the face of demonstrations and roadblocks—or perhaps the complete countrywide shutdown that the Maoists were threatening—Dhulikhel was a green and quiet escape, a fine place to wait out urban riots if any were to materialize. And I found a charming host in Bel Prasad, a unique and now elderly gentleman who had straddled the wide gulf between the rural Nepal of his childhood and the world he had seen in visits to Europe, America, and Japan. He had knit ideas from these two worlds together in ways both clever and inspiring. Here was a social entrepreneur whose approach to problems like water scarcity and lack of sanitation would seem to point the way for others.

When I reached the Himalayan Horizon Hotel, the popular lodge he has owned for several decades, Bel Prasad was not there. I worried he had forgotten our appointment. A man who worked at the hotel accompanied us up the hill to find Bel Prasad, who was overseeing work on a new family home that morning. He was going to move out of the rooms in the hotel where his family had lived for years and into a spacious, hilltop house.

The weather lower in the valley was getting hot, but Dhulikhel was a thousand feet higher and ten degrees cooler than Kathmandu. Bel Prasad wore a natty outfit that looked like a leisure suit a yacht captain might wear—cream-colored slacks and shirt, a pale yellow windbreaker, and a cap with a bill. Actually these were mourning clothes. Bel Prasad's mother had died a few months earlier, and though stark white is the typical color of mourning in South Asia, light colors like this were also appropriate. He would wear such colors for a full year.

Bel Prasad showed me around his new home, after which we made a plan for the day. He couldn't walk very far because of trouble with his ankles, so we rode the taxi back to the hotel. We settled down with some tea in the upstairs dining room overlooking a lawn and the hills and mountains beyond. I asked Bel Prasad to tell me about his life.

The Dhulikhel in which Bel Prasad was born in 1940 was a village whose way of life he calls "primitive": it had local healers but no doctors; it had only a few old standpipes from the Rana era along with some *dhunge dhara*, ponds, and wells; and it lacked toilets. "But we had some good things," he said. "We can see the mountains, the climate is good." Dhulikhel is located east of Kathmandu, where the valley starts lifting up toward the middle hills. Now it's less than an hour's ride if the traffic is not too bad. When Bel Prasad was young there was a partially paved road that came only halfway from Kathmandu. After that travelers usually had to walk because there was no regular transportation and the road was bad.

And there were no schools. The Ranas were still ruling Nepal when he was a boy and they had forbidden schools. When he was nine years old, Bel Prasad's father, a cloth merchant, told him "school is coming." He had no idea what that meant. What was school, and how was it coming? It was moving to Dhulikhel? The teacher wore clothes from India and eyeglasses, which Bel Prasad had never seen before.

He was the first in his family to go to school, and his father exhorted him to do well. His parents had helped start the school and bought books printed in India for the students, who sat outside under the trees for their classes. School was a puzzle for Bel Prasad, who had not learned to read or write. He had been out in the hills tending cows and goats. Slowly he got the hang of it.

Bel Prasad stayed in school and eventually went to the engineering campus in Kathmandu, where he trained to be an overseer. He got a job with the Nepal government and also opened a small bookstore in Dhulikhel, the first of its kind in the area. After six years he decided that he wanted to go back to school. He went to Kathmandu, studied history, and got a BA.

The Peace Corps had begun working in Nepal. A volunteer in Dhulikhel recommended that Bel Prasad teach Nepali to new Peace Corps volunteers. Bel Prasad traveled to the United States to teach an incoming crop of young Peace Corps volunteers at the University of California at Davis. He says his life was never the same. "I learned so much about myself, and I wanted to change my small Dhulikhel."

He could see the connection between the education system in Nepal, one which taught children merely to memorize, to be "parrots," and the continuing stagnation of his country. He saw that Americans were "so free. They can do so many things—drive, sing, dance, swim—free like a bird. Here is like handicapped person. Here they said 'don't do, don't do.'" Parents and teachers wanted to keep children traditional. Bel Prasad, or BP as the Americans called him, saw that if people cannot be free to do things on their own initiative, to think on their own, they cannot grow. And his town's problems would never be solved. "If everything stays the same who will bring water to Dhulikhel, who will make toilets?"

By then "hippies" from the United States and Europe were settling in Kathmandu and a few had discovered the rural quiet of Dhulikhel. Bel Prasad's father wanted him to come home. He was worried about foreigners having a bad influence on Dhulikhel. Bel Prasad was the oldest son; his father felt his place was near his family.

Adjusting to life back home was hard, but Bel Prasad coped by starting new enterprises. The first was a small lodge he opened in 1969. He had one room with four beds he could rent to travelers. "I had six spoons, six glasses, a burner and a bucket." And he built Dhulikhel's first toilet. "My grandmother was not happy that I put a toilet in the house. God doesn't like a toilet, she said. It's smelly. Not clean. That's why nobody puts toilets next to temples."

The lodge expanded. Bel Prasad housed groups of Peace Corps trainees and his lodge became a training center as well as a favorite stopover for foreign guests. After he mastered the twenty-four-hour-a-day responsibility of a guesthouse and savored the satisfaction of having his own restaurant, he started looking for new challenges.

"I listened to what they talked about in my restaurant. They talked about past events. Not about ideas. I was tired of listening to them. They asked about life in America. My shirt, my trousers. I was tired of telling them. I tried to stay a little away."

He had a daughter, but there was no school for her. People didn't like girls and boys to mingle in school, because there was so much fear at the time of inter-caste marriages. So Bel Prasad decided to start a school for girls. And in the back of his mind, he was still worrying about the lack of toilets in his town.

He talked to a hermit, a holy man. He wanted to understand how the hermit lived on the little rice and money that people gave him and why people trusted him. The holy man had been a friend of Bel Prasad's first teacher, whom the town had honored by erecting his statue. Bel Prasad told the holy man that he wanted to put his statue at the first girls' school in the town. "He was excited. He asked me what he should do. I told him to collect some funds."

Volunteer teachers started holding classes outside, near the main street, much as Bel Prasad's first school had begun. The girls came in the afternoon after they finished working in their houses and fields. "The girls were excited, they learned faster than the boys. And many boys came to see the girls." But there was no school building and there was a shortage of land for building one.

People had been using a piece of land at the edge of the village as an outdoor toilet area. Bel Prasad announced that he wanted to use part of the land for a school and then abolish the outdoor toilet. He thought this strategy just might make Dhulikhelis change their habits. Some women asked him to make the toilet area cleaner, which gave him the opportunity he needed. "So I said: see my toilet. See, it's not smelly. I am a university graduate, the son of a businessman. I am doing it. You can do it too. You can make a toilet. I can subsidize if you make a toilet in your house. They started thinking it was not a bad idea."

His family and others donated some additional land. He put up a photo of the hermit, calling him the founder of the school for girls, and asked people to help. They did. "Tourists and Indian visitors pulled out handfuls of money for the school." Soon the hermit was unhappy because it was taking so long to build the school; but this did not daunt Bel Prasad. "I want everyone unhappy with me—it helps to get things done."

Bel Prasad maneuvered so that people in the community started competing to donate money for the school, and eventually there was enough for construction. On the toilet front, some German donors offered cement and pans to build the squat toilets. A few Dhulikhel residents then went to India to learn about the "sulabh" toilets Bindeshwar Pathak was promoting in Bihar.

Soon the town had a few toilets, but there wasn't much water for flushing—not even for the water-thrifty sulabh toilets. A drought came; the only water was what was in the ponds. "My wife worked very hard carrying water back and forth." Aside from his wife's discomfort, Bel Prasad was happy. "I want problems to make people think. Once people start realizing problems, then they think of ways to solve. This is the best way. I get very excited."

Bel Prasad laughed at himself as he remembered those days and his genial face opened up as his eyes crinkled closed. His prominent nose was framed, when he smiled, by the crescent moon of his mouth below and a pair of arched white eyebrows above. His white hair was a halo above a broad forehead. He says he wondered during this time if maybe Americans faced the same sorts of problems he had encountered before they had water and sewage systems that worked. Maybe there were leaders in America who had to employ the sorts of strategies he had been using. That idea seems to have comforted him.

At that time Dhulikhel was a district headquarters, but there was talk of moving the headquarters to another town nearby because of Dhulikhel's extreme water shortage. Now here was a really big problem—and Bel Prasad loved it. "Big pressure on the people to make them think about water."

Bel Prasad invited a foreign friend to come to Dhulikhel and advise them about how to solve their water shortage. The friend offered fifty thousand rupees. A man from Dhulikhel said: "We don't want money from you, we want water." Bel Prasad was surprised but happy. "I thought it was very good—a Nepali thinking of water, not money." The foreigner was agreeable, indicating that once the people of Dhulikhel could find a water source, he would try to help them set up a water system. "We made a team to look for the water source. I didn't want to be the head. Everyone looking at me." Bel Prasad wanted someone else as a figurehead for the water supply project, the same way he had wanted the hermit out in front during the girls' school campaign.

"I used up four, five pair of shoes looking for water. Dhulikhel is located at 5,000 feet. Someone said I know a place higher where there is water. He showed us a small stream, beautiful clean water. We prayed, please God let us bring this water to Dhulikhel."

Bel Prasad requested help from a German government representative he knew, which angered someone in Nepal's government. "'You are not supposed to talk to the Germans. Who do you think you are?' So I asked him to please write a letter to the Germans asking for help for Dhulikhel. He said no. I said okay, we will all resign from our posts on the town council and say you are not helping, that you are not supporting a water supply for Dhulikhel." Eventually the government official made the request.

"We took the Germans to the water source. They liked it. They made a design. Then we made an agreement with the government to put a water system in Dhulikhel." Part of that agreement included the germ of the unique management system that has served the town well for twenty years. The people who drink the water manage the system.

But plenty of challenges remained before the local users group could take over management of the system. All along the eight miles from the water source to Dhulikhel where the pipe would travel, local people complained that they didn't want the pipe coming through their land. Each obstacle required a different solution: the group building the water system for Dhulikhel had to build a separate water system for a neighboring village, then a school for another one. One woman had already planted her rice that monsoon season and couldn't bear to see it dug up for the pipe. "We told her we would pay for the rice. She was not happy. 'I don't want to remove it,' she said. I said okay we will do it and plant it again and you won't have to see it. She said okay."

The problems arose, and were solved, one by one: funds, right of way, technicians. And the pipes had to be purchased in Germany, because those manufactured in Nepal and India were not sturdy enough for the intense water pressure in the new system they were planning.

"Once the water was here most people forgot they had promised to pay for it. They said in Kathmandu they're paying only 15 rupees a month. You want to charge us 35. The German and Nepali governments gave money to build the water system. Why do you need money to operate it?

"We told them that in Kathmandu they get dirty water for two hours in the morning, maybe two in the evening, maybe some days no water. But here you will get clean water nonstop. Women can sleep, not get up at three to go get water." Sounding a little like a public broadcasting tout, Bel Prasad told people in Dhulikhel that the daily cost of the service was only half a cup of tea. For the price of a coke they would get ten days. For the price of a beer, they would have a month of service. So give up one beer, and the water is free for a month, he told them.

"So are you ready now to pay for water? Please raise a hand. They raised their hands and then clapped." They paid their deposits and their connection fees. They got meters. I saw one of those meters, still in service. With a glass face a few inches in diameter, it looked very much like the one I had seen at my own home in California.

Bel Prasad had taken his time telling this story, and I had asked him questions, so by now I was hungry. We ordered lunch. As Bel Prasad continued talking he broke off little pieces of a grilled cheese sandwich, smacking a little as he took a bite. He ate slowly, while I wolfed down paratha and curd. In the occasional silences while we ate, I could hear the rhythmic horns of trucks on the road. They were passing through town just down the hill from the hotel on the main and only highway from Kathmandu into the eastern hills. The two-lane road was slowly going a little deeper into the wooded hillsides each decade, getting closer to the Khumbu.

I had once taken that road to its end and started walking, trying out the longer, slower route into the trekking region instead of flying. I hoped the road would never penetrate the Khumbu the way roads have almost encircled the Annapurna region, the other most renowned trekking region in the nation. But Dhulikhel is no longer a trekking village like those in the Annapurna region that are now losing their charm. It is a regional hub of activities, and the road is necessary if noisy.

In 1986 Bel Prasad was elected Dhulikhel's mayor. Once construction of the water system was underway he started trying to find a way to bring a modern hospital to Dhulikhel. That dream materialized as a handsome campus with a medical school spread itself on the hills south of the road. Then a popular university also started up, called Kathmandu University, drawing students and professors from the metropolis but making its home in Dhulikhel. Neither of these developments could have happened without the water supply. And Dhulikhel remained district headquarters. "Everyone was so happy. Can use water for toilets. Women can sleep. Stomachs are healthy because water is clean."

Bel Prasad believes it's good to borrow ideas from anywhere; but the initiator of a project has to be local. After he finished his story, Bel Prasad took me up the hill just outside town to see the water treatment plant and to meet Rameshwar Parajuli, the manager of Dhulikhel's water system. Parajuli commutes from Kathmandu to his job; he's been doing this since 1986, when the water system was under construction.

The day I visited, Parajuli was busy trying to solve a problem. The gravel beds that filter sediment from the stream water were going to need extra cleaning because of changes in the system. In order to get a stronger discharge of water down to Dhulikhel, a sedimentation tank closer to the source was now being bypassed. All the filtering would be done here at the plant, so the gravel beds would get dirtier and need to be cleaned more often. Digging up the beds requires strength. But most of the strong, young men had gone "to the Gulf," to countries like Dubai and Iraq to do construction work. There was little work in Nepal and what there was typically paid low wages.

Parajuli discussed the issue with Bel Prasad. Bel Prasad is no longer mayor, but he is the town's senior statesman and people still turn to him for advice. While foreigners call Bel Prasad "BP," Nepalis call him Belji—the suffix being an honorific in Nepali and Hindi, conveying affection and respect—or Bel dai, older brother.

We walked around the treatment plant. Parajuli showed me the stair-stepped gravel beds of the filtration system. It's a redundant system, so that one series of rock and gravel beds can be in use while the other is closed off for cleaning. After one month the gravel beds develop a buildup of algae, so the water is switched to the parallel set of filters. The gravel beds trap particles and clean the water, which during the monsoon is especially turbid. These are the chief impurities, since the stream that is the source of this water, unlike urban ones, has relatively little human or animal contamination.

The water passes horizontally through three gravel beds—the first with stones the size of tennis balls, the last with much finer gravel—then through a slow-sand filter. The water percolates through more than a meter of sand, which cleanses it of suspended solids and bacteria. After this cleansing the water goes to a rectangular, covered reservoir and a small amount of chlorine is added for a final purification.

The system is run by gravity. The water descends several hundred feet from its source eight miles away to reach the plant, then continues down to the town. At the outset when there were fewer users, there was such good pressure in the system the pipes could send water up to rooftop tanks at each home without electric pumps.

The grounds surrounding the treatment plant are like a garden. Rosebushes scent the air, honeysuckle clings to the fence, and pine trees circle the perimeter of the property. We walked slowly back to where Ram was waiting with the car, just outside the gate. On the way I saw a faded sign in English and Nepali: Dhulikhel Water Supply Project—German Nepal Cooperation.

Over the years Dhulikhel has outgrown what was once a twenty-four-hour, on-demand water supply. The town of six thousand now has double that population depending on the plant. Once there was a continuous flow of water in the pipes; now it's released from the tank for several hours twice a day.

The diminished eight-hour availability, half in the morning, half in the evening, makes Dhulikhel's system more typical of South Asian water systems. Still, the water is of good quality and so far there are no waterless days. A well-trained staff has kept the twenty-year-old system functioning as intended. Parajuli told me he has visited other water projects in Nepal that cannot say the same. The Dhulikhel community takes pride in the system. "They let us know even if they see a little dampness on the ground. So we catch leaks early."

The cost has gone up from 75 to 105 rupees a month for basic household use. That's still a pretty good deal, about a dollar and a half a month. Dhulikhel's hotels—of which there are now many since the town has become a tourist haven, conference center, and weekend getaway destination for Kathmandu residents with money —pay more. They not only use more, they pay at a higher rate, which helps to sustain the system for everyone.

But there are still conflicts with neighbors. One pragmatic decision made at the outset eventually backfired. Dhulikhel needed to be a municipality in order to get foreign funds, but its population back in the 1980s was not sufficient for that designation. The town had only six thousand people and needed ten thousand to be a municipality, so Dhulikhel incorporated some adjoining rural areas that were not to be recipients in the water allocation scheme. As time passed, some of these areas asked to be connected to the system, but the water supply for Dhulikhel alone was already shrinking. The users' committee stalled on a decision.

People in villages that had been incorporated into Dhulikhel's municipality held a demonstration in 2007, demanding water. Later, in retaliation for what they saw as inaction on the part of Dhulikhel's water officials, they cut the water line.

Dhulikhel was without water for two weeks. The act of vandalism plunged the grown-up, developed Dhulikhel with its modern hospital and university back into the dark age of water insufficiency. Residents had to go to streams and standpipes again, filling buckets for kitchen use, forgoing showers, carrying laundry out into the woods to wash.

Eventually the town and the village reached a compromise; the water lines were repaired and water was shared.

Bel Prasad said he wasn't surprised that the conflict with the neighboring village led to a crisis. Starting a system is one thing, while managing it for several decades is another. Bringing a hospital and a university to the town without increasing the supply compounded the problem. The system was superb in 1997, unparalleled in all but a few cities in South Asia, but by 2007 water availability had diminished both because of the inclusion of the outlying area and the increased demand from Dhulikhel itself. Still, the water was reliable and clean, something people in Kathmandu can only dream about. Bel Prasad believes people even now aren't so worried about vicissitudes in the amount of water they get. What's important is that it's there every day.

Because of the increasing population in the area, Dhulikhel plans to participate in a bigger project in concert with neighboring municipalities. Together they will install larger pipes to siphon water from the source, and the expanded system will require a new management structure. A board will oversee the whole system, in turn empowering Dhulikhel's users' committee and whatever management structures the other municipalities choose.

Upstream villages will receive royalties from the towns in the system, along with schools and other infrastructure. There will likely be an array of problems to solve with this new system. The population is increasing in the villages higher up in the hills. Local people are worried they won't have enough water for their own villages. And maybe the sources are shrinking, maybe global warming is affecting the streams.

Construction on the new project was set to begin in late 2012 with a loan from the Asian Development Bank, which has recently begun to foster smaller-scale endeavors. The project is to be completed by 2017. Perhaps then, Dhulikhel will again have water flowing in its pipes twenty-four hours a day. Before then, the current eight hours a day may shrink. Maybe Dhulikhel residents will receive water only every other day. But Parajuli says the people understand why this is happening. Like Belji, he says people who can remember the shortages of the past know that even the reduced service is better than relying on hand-carried water from streams, springs, and wells.

Before I left Dhulikhel I met some of the members of the users' committee. Parajuli had called an emergency meeting to get approval from the committee for spending extra money for workers to clean the gravel beds.

The nine members of the committee are elected by fellow citizens who own private water taps. There are 1,500 taps in Dhulikhel now, and that determines the number of votes: one tap, one vote. By law, there must be at least two women on the committee.

Yog Lakshmi Shrestha has been on the committee for six years. She was halfway through her second term and said she would run again; maybe eventually she would run for chair. Yog Lakshmi lives at one of the higher elevations in Dhulikhel, so she gets less water than people in other parts of the town. Instead of several hours, sometimes her neighborhood has a flow for only half an hour each time the water is released from the tank. She runs to let people know when water is coming so her neighborhood can get its share.

I asked Yog Lakshmi why she wanted to be on the board. "It's social work. In Nepal most women need water. Men don't work in the kitchen, so they are not so interested in these things." Yog Lakshmi also teaches fifth-grade English and social studies. She can remember the old days, before the water system. I asked her what she does with the extra time, not having to fetch water. She did not always understand my questions without help from Parajuli, but she could answer in English. "Now it's easy. I can study. I can sleep more. We used to spend all our time washing clothes and getting water. I used to get up at three to get water. Then after work I went again."

"Why do you think Kathmandu can't solve its water problems?" I asked Bel Prasad before I left Dhulikhel. "When you go to meetings where people are making decisions, what do you see on the tables?" he asked in response. "You see Coke and bottled water. The people who really need water are not at the meetings." Bel Prasad advises other towns in Nepal not to go to the government or the World Bank if they need water. Waiting for "God or the government" to help will not get them water. "If they are thirsty and they need water, they can pay for it and then they will have it. Unless they do this they won't get water."

Later I asked Dipak Gyawali why other towns in Nepal have not developed a system like Dhulikhel's. They lack the kind of charismatic leadership Bel Prasad provided, he said, and the commercial culture is different in Dhulikhel. "Dhulikheli shopkeepers don't trust the government to do anything." Dipak was part of the team that recommended handing over the German-funded project to the local users' committee.

Shedding some more historical light on the Dhulikhel phenomenon, Dipak said the townspeople decided that if their water project were put in the hands of a users' committee, they would all be willing to pay the amount the committee set. But not if it was to be managed by Nepal Water Supply Corporation, as all the other water projects in Nepal had been.

Bel Prasad sets a high standard for anyone who would follow in his footsteps. He has not slipped into the kind of self-serving performance that is so common among officials, here and elsewhere in the world. And he does not congratulate himself. He implied that other towns in Nepal could accomplish what Dhulikhel has if people take the initiative and set a fair price for the service. I asked him what role his own leadership played. Yes, he agreed, leaders are necessary, "to make people feel the importance of the goal. I was mayor for sixteen years. I put water at the root, not at the leaves. I was not a big boss. I went to the people to see where I should work."

That night I stayed at the Himalayan Horizon, curled with a blanket under my chin in the chill mountain air. I woke to a view of the hills under a light fog the next morning. Belji introduced me to some other people in town that morning, and I went to see the hospital.

We heard the roads were still clear, but I was feeling anxious. Ram and I ate a quick lunch, then headed back to Kathmandu. It was a Saturday and there was little traffic on the roads. Ram had no difficulty reaching Kathmandu and getting across town on the roads he chose, but we knew the May Day demonstrations were underway toward the center of the city.

I reached my apartment with relief. The countrywide shutdown the Maoists had been threatening started that night. No vehicles moved on the roads of Nepal for six days, except a few ambulances and the motorbikes of Maoist organizers.

Dhulikhel's water system, even with its diminished supply after ten years of population growth, is admirable. But sewage treatment here lags behind. It got off to a good start when Belji started constructing toilets in the town. Later when the new hospital was built, the first reed bed wastewater treatment system in Nepal was created at the bottom of the hill behind the hospital. This is an inexpensive system, a step above simple settling ponds and more aesthetically desirable.

That system is still working well to handle sewage from the hospital, but the rest of the town is only beginning to progress beyond the toilets Bel Prasad helped to establish. Some people have septic tanks, and there are still some functioning sulabh toilets, but a lot of sewage is going to nearby rivers. Bel Prasad said the municipality was "working on it."

But down the hill on the road back toward Kathmandu, several Newar villages are installing their own reed bed treatment systems, using the Dhulikhel Hospital's system as a model. I arranged to see some of these very small but successful sewage treatment systems. Ram came to pick me up in his taxi one morning and we went across town to fetch Mingma Gyaltsen Sherpa, who was to be my guide.

Mingma was working under the auspices of UN Habitat, helping villages near Dhulikhel set up sewage management systems while he finished a PhD from a university in Bangkok. (This was the second Mingma Gyaltsen Sherpa who helped me understand water in Nepal. He shares his name with my friend Mingma from the Khumbu; the latter translated the stories of Sherpa farmers who witnessed the Dig Tsho flood in 1995.) This Mingma was born in Solu, the region just south of the Khumbu, which together form the district Solukhumbu. We discovered as we chatted in the taxi that I had walked through his village the previous year after visiting my friend Cholendra's village, a day's walk from the main trekking route.

First on our itinerary that day was Shrikandapur, a village whose reed bed sewage system had been in operation for a year, following the model that had worked so well at the hospital campus in Dhulikhel. The village is situated on a small hill so the town's sewer line can take advantage of gravity. Sewage used to go directly to the nearby Punyamata River.

Purna Karmacharya came down the path from the village to meet Mingma and me; he is both the chair of the town committee that set up the system and the plant's only employee. After he retired from teaching social studies, Purna turned to community activism. People in Shrikandapur, a village of farmers and a few merchants, had begun to think it was bad to keep sending the town's sewage to the river.

The Punyamata is holy just like the Bagmati, said Purna; putting dirty water into it was a sin. The townspeople wanted to find a way to send treated effluent to the river, and they liked the idea that they could use biogas for cooking. Biogas systems have been used in villages in Nepal, where cow dung is available and supplies of wood may be dwindling, for some years now. Propane canisters and kerosene have replaced wood in some places, but biogas is a cheaper alternative to those fossil fuel sources, and recent technology allows human feces to be put to good use.

Wastewater that comes down the line under Shrikandapur's hill is first channeled through a trap that catches plastic and other garbage that could clog the system. It then passes into a biogas chamber where it is digested by anaerobic bacteria to produce methane gas. From there the water can be channeled into one of six different filter beds. Right now the town needs only two of them, but can use the others as more people join the system.

Purna said only about half the households have connected to the sewage system so far. Some are not able to connect because their houses are too low on the hill, below the sewer line, so they would have to continue using pit latrines. Others were in the process of connecting to the system, but at least the sludge from their pit latrines could now go into the biogas digester instead of into the river as before.

The bulk of the discharge into the system occurs in the mornings and evenings when people are busy at home bathing, cooking, using the toilets—most of which are the simple pour-flush variety, not commodes. Gray and black water are mixed in this system. Sludge from biogas is automatically pressed down into a concrete tank while the water passes on down to the reed beds. The sludge is available to be used as fertilizer after it is dried.

The "reed beds" are square, slightly sunken pits about fifteen feet on a side, covered with gravel, reminiscent of the gravel pits I saw in Dhulikhel that filtered out impurities at the beginning of the water cycle, before the water went down to the town. The reed bed pits that aren't in use yet are just smooth gravel. Those that are in use show a crop of tall grasses, somewhat like pampas grass, growing up between the rocks. The wastewater moves through the gravel, guided by gravity, while the roots of the reeds oxygenate and cleanse it. Aerobic bacteria eat the waste.

The cleansed water is caught in a small canal and channeled down the slope toward the Punyamata. A thick plastic lining keeps the wastewater in the gravel pits to avoid contaminating groundwater. There is no odor. I would not have known this was a sewage treatment plant without someone telling me. Only near the domes of the biogas chamber is there a vague odor of sewage, but nothing like what one smells crossing the bridge above the Bagmati.

Purna says maintenance of the reed beds is simple. All he has to do is cut the reeds and wash the gravel from time to time. He said people from other towns, both near and far away, had been coming to look at Shrikandapur's system and he hoped they would learn from what his village is doing. The bulk of the money to set up the system, which cost about $100,000, came from UN Habitat. The village itself paid for less than 10 percent of the construction but donated land for the site; the municipality also contributed. The villagers who use the sewage system pay nothing.

The biogas system was a pilot program; when I visited only a few houses were using the gas in their homes. As the system expands, more homes can run gas lines into their kitchens. Shrikandapur started out seeking only a sanitary sewage system, but the potential for biogas may lead the town to become a sewage entrepreneur. Purna Karmacharya said the town was planning to use some land it owns on the other side of the river to build a septic waste treatment plant. This would allow it to collect sewage from other towns on the floodplain. Shrikandapur would produce biogas from the sludge, charging fees both for taking in the sludge and for providing biogas to people in and near the village.

Reed bed sewage treatment at Shrikandapur

The villagers who currently use the biogas in Shrikandapur pay 350 rupees a month for the fuel.

At the end of the tour Purna took us down to the Punyamata so I could see the treated water going into the river. A pipe discharged just a trickle at midday. As I watched the clean water joining the small river, I saw that the water coming from upstream looked murky and pieces of garbage were floating in it. Purna acknowledged that this is sadly true. Most of the pollution comes from Banepa, he says, a large town down the hill from Dhulikhel, which was once its rival for the privilege of being district headquarters. Banepa has not managed its sewage very well. It plans to do so with help from the Asian Development Bank, but that could take a long time.

A smaller town upstream called Nala was much closer to setting up a system like Shrikandapur's. Mingma was working with that town as well, so we bid goodbye to Purna, got back in Ram's taxi, and headed up the road. In the taxi Mingma told me that efforts like those in Shrikandapur and Nala—very much like Dhulikhel's celebrated water system—require a community champion. There has to be someone like Bel Prasad Shrestha or Purna Karmacharya with an "ability to visualize the future," as Mingma put it; someone who can focus people's attention, persuade them to join an effort, show them it's in their best interest. An effort like this requires someone who won't get distracted and drop the ball.

Nala is larger than Shrikandapur. It is famous for its potatoes, Mingma told me, though we both agreed the potatoes up in Solukhumbu are tastier than those grown in the lower Himalaya. Nala has several historic temples, and legend has it that when the Shahs launched their conquest of the valley back in 1768, people in Nala dug a tunnel under the road to hide in and that's how the town got its name, since "nala" means drain in Newar. The tunnel didn't help. Nala fell to the Shah invaders just as the kingdoms of Kathmandu, Patan, and Bhaktapur did.

One problem slowing Nala's progress toward sewage management was lack of public land. Unlike Shrikandapur, which had enough open space down the hill for its system, either Nala had to buy land or landowners in town had to contribute it. Mingma told townspeople they could not get everything from donors. The town had to contribute money and land.

Twenty percent of the houses had no toilets whatsoever; those residents had to resort to the river or to land outside the village. Those who had toilets mostly had single-pit latrines (not the two-pit sulabh toilets) that had to be de-sludged. Often the sludge also ended up in the river.

As we walked down the sloping stone streets of Nala, we stopped from time to time to talk to a resident. I had to bend over to get through the small doorways. We stopped to look at a toilet with a septic tank behind one house. It was a pour-flush toilet used by two households. The lady of the house told us the last time they emptied it was one year ago.

Another woman, Krishna Lakshmi Shrestha, showed us into the ground floor of her small house, situated on a dark, narrow street near one of the main squares. The toilet cubicle was set in the back of the small ground-floor space, which was dark because there were no windows on this level. The house was a typical brick Newar home with a narrow stairway leading up to the higher stories; the delicately carved wooden grilles so typical of Newar houses covered the windows on the upper floors.

Krishna Lakshmi, a tiny, elderly woman, was pleased about the project and eager to talk. She said she had attended planning meetings to discuss the new sewage system. She wanted a new system because the wastewater from her kitchen one floor up was running into the street. It flowed directly down into the narrow sloping passageway outside her house before joining the storm water system. She told Mingma she had wanted to cover up that drain but the community would not allow it. Mingma explained that enclosing such drains was forbidden because people might start discharging their sewage into them. He said that some people used to wait for a big rain and then open up the pipe from their toilets and let the sewage pass out into the storm drains. Here in Nala there is a plan for separate lines, one for gray water and storm water, one for sewage. The smaller, more compact village of Shrikandapur combined the two.

Lakshmi Sharma of the Asian Development Bank told me the bank was no longer funding storm water projects in the region because of this misuse. She said people who were supposed to have septic tanks were often just connecting their sewage to storm water systems and the municipalities were not policing this violation. "So we have stopped supporting storm water drainage because we know it will become illegal sewer lines later." Instead the bank was beginning to help set up reed bed systems in small towns. Unlike Kathmandu, she said, these towns still had enough land to accommodate the size of settling ponds they needed. The bank was funding these efforts as an experiment, to find out for itself how well they worked.

Mingma said discussions were underway with the town's users' committee to determine an appropriate payment for residents who were going to use the sewer system. The town is divided into wards, each with a users' subcommittee. People will have to pay for their own connections to both the storm water and sewage systems.

Mingma was working with the user groups on a final report and the plans were almost complete. The wastewater system could be built in a year, but the storm water system would take longer because the town's brick and stone streets would have to be dug up. Money was allocated to build toilets for the 20 percent of the houses that lacked them, and for hygiene campaigns to educate people about waterborne disease.

People were very friendly in these villages, frequently greeting us with smiles and "namaste," much as they do in more remote villages. Kathmandu isn't unfriendly. It's just that everyone is in more and more of a rush, and foreigners are not a big deal because they're all over town. Here a visitor is unique. Mingma asked me if I wanted to try the yogurt. "It's really good," he said brightly. As we savored our little pots of buffalo milk yogurt outside a small shop, Shyam Sundar Shrestha, the chair of the users' committee, joined us. I asked him why his village needed the new system. He said the main reason is that they would stop polluting the river. And it would be good to get biogas and manure for the fields as well.

Nala chose the reed bed system like Shrikandapur. Maybe later, if the village was offered some financial help, it would start a pilot project so that some houses could participate in a urine-diversion program as well. Many houses in town were still using an old style of urine management that preceded toilets in the valley. This was essentially a spot on the ground floor of the old Newar homes that people could use as a urinal. In some cases there might be a pile of straw. Ash would be sprinkled on the area. Once the smell became too strong, the pile would be taken out and spread on the fields, where the phosphate-rich fertilizer would not be wasted.

Progress in delivering public services sometimes seems paralyzed in the capital; but at the fringes of the valley, there is movement. Roshan Raj Shrestha, who pioneered the reed bed system for treating wastewater and is working to expand it throughout Nepal, says there is plenty of talent in Nepal for innovation and good management of resources. "We just need some political stability and then we can jump very fast."

My visit to Dhulikhel brought home what Ajaya Dixit said a few months earlier, explaining why "a century since modern technology was introduced in Nepal we still haven't figured out a way of using those systems to deliver basic services."

Ajaya agreed that Dhulikhel is an exception. He sees it as evidence that once the right structure is in place, with the right kind of decision making and transparency, some of the contradictions between Western technology and Nepali social and political structures are eased. I thought of the nine-year-old Bel Prasad. What is school? he asked. How is it coming? He was at first baffled by the whole process. A gifted individual, he caught on, and took his whole town with him.

How many Beljis will it take to get Nepal on the right road? The kind of leadership he offered has not emerged at the national level and may never prevail. For now Nepal lingers in a kind of twilight under the heavy hand of the Maoists who once promised much but have delivered little social progress. They have shown instead plenty of ability to manipulate a weak democracy. There is political resistance to pouring any more resources into the overpopulated and attention-stealing capital. Combined with general instability, this leaves potential foreign donors unable to spend money to rescue Kathmandu from decades of bad management.

Journalists, who might explore public policy issues to the benefit of Nepal's citizens, are not safe in the country right now. Many are afraid to expose corrupt practices or are not allowed to do so even though there has been constitutional freedom of speech in Nepal since the early 1990s. Meanwhile, a sense of public responsibility thrives mainly in a few pockets, chiefly the progressive, nongovernmental groups working on water and sanitation, and various other social and environmental issues.

When we were talking about the dhunge dhara, the Ranas, and Melamchi, I asked Ajaya if Kathmandu really needed a centralized system. He talked for a while about the Mississippi River and the US Army Corps of Engineers, the Bureau of Reclamation, the Ganga canal and the British colonial system. What these had in common was that the agencies had redefined the hydrological system for their own benefit. Anything that did not belong to the state and generate revenue for it was wasted, subject to "reclamation."

"You can't isolate what is happening here from that historical process," Ajaya said. "It's the same paradigm." That paradigm is expertise-led, finance-dominated, technological, centralized, and bureaucratic; in it, local ideas and needs are disregarded. The water development system that came to India was the U.S. model of big dams—big river valley transformations for flood control and hydropower. India had been following this model for fifty years. Nepal is smaller, its progress looks nothing like India's, but the guiding philosophy is the same: trying to fit the Western model into the South Asian context.

"Melamchi is an extension of the U.S. model," Ajaya concluded. "You have a desert, you have scarcity, the solution is to bring more water. The lesson of the *dhunge dhara*," he said, coming back to Kathmandu, "was that it was decentralized. Supply and demand were decentralized."

"And communally maintained," I interjected. "Yes, but not bureaucratically maintained. The users themselves assigned roles, clothed in religious ritual to some extent. The water system was not transferred to an external, alien organization. There is some element of hierarchy and centralization even in a decentralized system. Perhaps the lesson is to see how the interest of the users is reflected and articulated."

Here in Kathmandu, and in most of Nepal, "that space has to be created," said Ajaya. But in Dhulikhel the "space" for the local people's interests and abilities was created, thanks to Bel Prasad's efforts. And it helped that it was a small town, growing slowly. A place like Dhulikhel could simply slip out of the gilded net of foreign money to some extent. Too small to figure in big loan schemes, the town could go its own way.

Twenty years ago Ajaya and his colleagues argued for decentralizing water in Kathmandu. Divide the city up into maybe ten distribution sectors, they said, with different supply systems. Try to match demand and supply in the sectors, using whatever water was available. Some would have enough, some would not. In that case "you could deal with the shortage at the sector level," not try to take on the entire valley all at once.

Ajaya says the World Bank and the ADB argued against the idea because it lacked economy of scale. For the thirst of the whole valley, "the only solution" seems to be torrents of Melamchi water shunted through a tunnel.

# 10

# The Sorrows of Bihar

*The more the levees confined the river, the more destructive it became when they failed.*
—John McPhee, *Atchafalaya*

Before I went to Bihar I knew little about embankments. I had seen levees in California, in the Sacramento–San Joaquin Delta and in the Central Valley. There, if you stand on an embankment on one side of the river, you can look across and see a matching embankment. Some have been set back from the river a half of a mile or so; but even then it is easy to grasp in a glance the relatively linear triad of a river and its pair of embankments.

The first embankment I saw in Bihar after miles of bumping along in the back seat of a gray Tata Sumo SUV in the April heat was a steep-sided loaf of sand, maybe twelve feet above the adjacent land. I scrambled to the top and looked around. The Kamla River glared below, reflecting a hazy but intense sun. It flowed lazily between the embankment and a wide stretch of sand a few inches above the water. Together the water and the sandbank narrowed as they receded into the distance.

I didn't see another embankment. I was disoriented by the incessantly jarring ride and the heat, but I recall asking where the other embankment was. A gesture directed my eyes toward the horizon of low trees and brush and sandy soil. Nothing was very distinguishable in the monochromatic haze of dust and heat.

Over the next two days my eyes and brain continued to struggle in vain to make sense of what I was seeing by comparing the north Indian state of Bihar to California. California rivers are powerful and can flood portions of the flat Central Valley, but they are in no way comparable to the rivers that rush out of the towering Himalaya. The Sierra Nevada ranges from five to twelve thousand feet. At twelve thousand feet in the Himalaya, one is still in the "middle hills," where in spring there are forests of rhododendron trees blooming. There's another fifteen thousand feet to go for those who plan to scale one of the higher peaks, and like many trekkers I have climbed to eighteen thousand feet just to get a good view of those peaks.

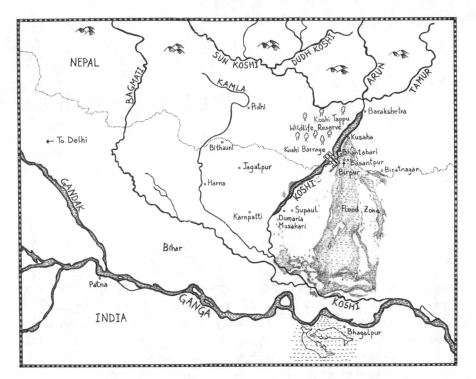

*Map 5 Bihar and the Koshi River Basin. Created by Kanchan Burathoki*

Down on the Gangetic plain in Bihar the Himalaya are too far away to see, but their presence is palpable in the legendary wildness of the rivers, as powerful as the mountains are high. During the monsoon, they toss and turn in their channels, fighting to release the energy stored in them by their precipitous descent from dozens of glaciers two vertical miles higher, carrying torrents of rain and tons of silt.

Bihar or vihar, in Sanskrit, means monastery, a place of refuge. The Buddha once wandered in this Indian state. He was born just over the border in what is now Nepal and later attained enlightenment in Bodh Gaya, south of present-day Patna, Bihar's capital. He walked long distances through the well-populated areas along the Ganga to teach. Patna—now the overwhelmingly noisy and dirty capital of Bihar—was then known as Pataliputra and later became the capital of the powerful and enlightened King Ashoka. He followed some of the Buddha's precepts to establish a peaceful, prosperous realm for a time in this region. Two thousand years ago there was no border, and a culture called Mithila flourished on the plains below the Himalayan foothills. A version of that culture's language, Maithili, is still widely spoken here.

The Buddha and his followers rested during the monsoon, staying in one place because of the incessant rains and flooded roads. They adapted, just as did the farmers, who had learned to live with monsoon-swollen rivers in these plains. After the rains ended, the Buddha and his followers returned to their wandering and teaching. And the farmers would go plant their *rabi*—post-monsoon—crops in the moist fresh soil. There's evidence that rice, wheat, barley, and lentils have been cultivated on these plains since at least 2500 BCE.

After India's independence from Britain, the effort of adaptation began to look less attractive than politicians' claims that they could end the inconvenience of floods. Indian politicians made promises to the people to gain votes, while engineers convinced themselves they could control nature and end the yearly annoyance of a few weeks of flooding.

There were only a hundred miles of embankments in Bihar when the British left. Now there are upward of two thousand miles of them. Critics of these embankments say they have done little to protect land and people from floods, and in most cases have made life far more difficult than it was before. Where there were six million acres of flood-prone area in 1952, now there are seventeen million. And where there was inconvenience there is now a yearly disaster.

Even though the rivers are far stronger than the embankments humans have built here, these low-tech innocuous-looking lumps in the landscape have nonetheless caused no end of grief to the people of North Bihar, blocking the rivers' drainage and robbing its soil of nutrients. They have contained floodwaters some of the time but have also blocked the yearly replenishment of the sweet Himalayan silt that once flowed at the top of the floods and spread over the land, a yearly vitamin and mineral bath to nourish the next season's crops.

In the southern part of Nepal, just north of Bihar—in fact the same landscape and culture, bifurcated by a national boundary—the rivers flatten out and get wider and shallower. Their rapid descent stops here at the bottom of the Himalaya, and they begin to dump their load of silt on the plains, making this some of the most fertile land in South Asia. The strip of low-lying land in southern Nepal where the rivers change their character is called the *tarai*. It produces more than 50 percent of that nation's crops. Bihar too was once a rich agricultural land, not the province of legendary *dacoits* (bandits) that it has become, of political corruption, disease, and low literacy rates. The embankments that have blocked the mountain silt and kept the water from finding its own path to the Ganga have steadily made this once rich land the poorest state in India.

In August 2008 the Koshi River overwhelmed one of these embankments, as this and other rivers in Bihar had done countless times in the past half-century. The flurry of news reports, both Indian and international, presented the catastrophe as a natural disaster. There was no discussion of the inherent fallibility of embankments. There was no mention of the fact that when the flood occurred, the river was nowhere near its highest point. Rarely was a water expert asked to explain why over the years this river alone had already broken through its embankments many times before. The answer, some would say, is simple: these unruly Himala-yan rivers go where they need to go. Trying to put the Koshi in a straitjacket had made the August 2008 jump out of that constraint virtually inevitable.

Like the hurricane-spurred levee and flood wall failures in New Orleans in 2005, the massive 2008 inundation of Bihar and parts of Nepal had been predicted for years. So had the less severe inundations that have occurred almost every year in Bihar, lying as it does where the powerful rivers out of the Nepal Himalaya con-verge and do a kind of dance on their way to the Ganga, creating a landlocked inland delta, vaster than many in the world where great rivers meet the ocean.

The people who made these predictions, and who warn that the same threat hangs over Bihar every monsoon season, are not in power. They are not even members of mainstream media as the Cassandras of New Orleans were. They are members of civil society in India and Nepal: advocates, intellectuals, former engineers, and people who work for the social good.

They say the embankments have harmed far more than they have helped people in Bihar, turning a reasonably prosperous and certainly self-sufficient agri-cultural area into the most backward and impoverished state in India. Before the embankments were built the rivers rose during the monsoon and spilled a few inches of water over the land every year, distributed silt, then subsided. People were a little inconvenienced but had adapted to the yearly floods for centuries, probably millennia.

After the embankments came, the entire drainage system was disrupted, leaving some areas "protected," others deep in water for months. People lost the productivity of their land and were not compensated. Laborers lost livelihoods. People started migrating to cities for work, others scratched by. And everyone who lived outside the embankments came to live in fear of a breach, which would bring not the few inches of floodwater that used to spill out of the unjacketed river channels, but a torrent of water and sand that could sweep them away, destroy their houses, and make their farmland useless for years.

The Cassandra of Bihar is Dinesh Kumar Mishra. And the main reason I went to Bihar in April 2007—more than a year before the devastating August 2008 Koshi flood—was to meet him and to see the embankments that water specialists in Nepal and India had told me are the central water management problem in India's poorest state.

After that visit I felt bewildered, uncertain how to write about north Bihar and the embankments. I did not trust my understanding of what I had seen and I wondered if this grand scheme of mine—to try to understand the Ganga water-shed from Gangotri to the Sundarban—was perhaps a mistake. For one thing, my knowledge of agriculture was limited to driving down country roads or interstates where I would see linear row after row of mechanically plowed, sowed, tilled, and harvested crops: low broccoli, tall corn, mounded strawberries, bunched lettuce, endless rippling wheat. That was agriculture to me and that was what farmland looked like; in California and wherever else in the United States I had lived and traveled there was the same symmetrical, almost mathematical, and easily perceived order. I had not been in South Asia quite long enough, even though I had lived for many months in the meandering lanes of Kathmandu, to lose such expectations.

My bewilderment in Bihar was not for lack of well-meaning assistance throughout the trip. I did not venture across the Nepal border alone. My young Nepali helper, Pushpa Hamal, smoothed over some of the confusion of traveling in a part of India where tourists and Westerners rarely go. Pushpa had recently finished college and came highly recommended by the director of the Fulbright program in Nepal. This would be his first trip into India; he had taught himself Hindi and wanted to practice. He bought us both broad-brimmed hats made of nettle fiber to protect us from the intense sun before we set out by bus on the first leg of our overland journey from Kathmandu.

From the moment we arrived at the Patna train station, Dinesh Mishra's trusted friends took care of us and tried to help me see as much of north Bihar as might be possible given limited time and bad roads. Mishra himself was traveling when I arrived. I would meet him finally at the end of my visit to Bihar.

Mishra's friends were kind and they thought perhaps I might write something that would boost their own efforts to help the farmers of north Bihar. I didn't know how to tell them I was trying to understand the dynamics of the whole watershed, the entire drainage basin of the Ganga of which they were only a small, however important, part. And that I was then to write a long meandering account that would lack the timeliness and immediacy of what they were hoping for.

One of the many donor-funded nongovernmental groups active in Bihar provided the Tata Sumo, a sturdy vehicle with air-conditioning. We left early in the morning with Ajaya Jha, who worked for the organization called Ghoghardiha Prakhand Swarajya Vikas Sangh or GPSVS. Our driver was a genial Maithili-speaking man named Chandra. We headed out to see the embankments on the Kamla first.

Ajaya Jha was doing social work, but his passion was religious philosophy. His square, friendly face was so much like that of a good friend in Kathmandu that I felt reassured from the outset. He wore a short-sleeved shirt of brown and white checks with pens in the pocket. His mustache and dark-rimmed glasses made him look very serious when he was not smiling. As we bumped along all morning on the dusty two-lane road that was the major thoroughfare from Patna to the north, Ajaya earnestly asked me to tell him if I had any suggestions for how they might do things differently in Bihar in terms of water management, based on my knowledge of water management in the West. I guessed perhaps he misunderstood that I was a journalist, that in fact I was the student and he the teacher. Perhaps he thought I was some sort of water specialist, or that being a journalist looking into these issues presupposed specialized training.

I mentioned my inadequacies. But later, perhaps fearing he might think I was being churlish by withholding my observations, I ventured to tell him that one thing that had been recommended in California after flooding in the Central

Valley was pushing embankments back farther from the main channel to give the river more room to drain. Ajaya listened politely, still treating me like some kind of expert, as many foreign visitors to Bihar perhaps are. Later, of course, I finally grasped that these embankments were already pushed back from the rivers' immense channels and that's why I could not see both of them at once. Pushing them any farther back would in some cases land them in another watershed, before which they might collide with some other river's embankment, so criss-crossed with obstructions is this land now.

After I climbed up that first embankment on the Kamla River, we crossed the river and traveled north along the western Kamla embankment to reach a tiny village called Harna.

We talked to a group of weathered farmers. Bare-chested, white *dhotis* folded around their loins and legs, the elderly men had gathered in the shade of a thatched open-sided shelter on the edge of the embankment. Below us were ponds still full of murky green water six months after the monsoon. Beyond that we could see cultivated fields.

I asked Ajaya Jha to translate. One farmer, very thin like all the others, was seventy-year-old Basu Thakur, who had lived here all his life. He told us that at first he felt the embankments were good and had been built for their safety. "But now they are a danger to us."

Before the embankments the people of Harna had lived on land that is now inside the Kamla embankments. There were mango orchards once. After the embankments were built and the monsoon floods could no longer spread out over the entire floodplain, their houses and trees were washed away. The government gave some of them land outside the embankment. But starting in 1962 the embankment breached periodically and flooded that land too.

Sometimes they had to move up onto the embankment. For a while the breaches were small and not a great threat. But since 1987, as the Kamla's bed has risen from silt deposits, there have been bigger breaches every year and living on the embankments has become the norm for them. It is their security. They built houses on it, but then the government took them down to build the road on the embankment. We could hear tractors rumbling in the distance. Workmen were reinforcing the embankment we had driven down.

I asked what solution Basu Thakur would like from the government. At first he mused that all the rivers wander their own way here and he did not want the government to do anything. "God will take care of it." Later, perhaps warming to having an audience, he said he wanted the government to move the river, change its course. The Kamla is moving closer to their embankment each year, threatening them even more. Other farmers chimed in, "Yes, change the course of the river." If this is not possible, then make the space between the embankments wider; or dredge the channel, make the river deeper. The men seemed to argue among themselves about the best solution.

"Either break it or make it stronger," Ajaya summed up the comments. Take it down or build it higher. We rode farther down the embankment. Tractors fitted with shovels buzzed diagonally up the side of the embankment at one spot, carrying loads of dirt from the floodplain to make the embankment higher.

We stopped at another village called Bithauni. Villagers gathered around while one told a story about a boy who went out to cut grass and was stranded in a tree on an island in the river for three days after the floods swept in. Three thousand people, most of them landless, lived there. About half of them live right on the embankment. They said the embankment we were standing on does not help them because five miles away the embankment ends, so the water comes on both sides of the embankment anyway. But there has been a large investment, and the embankment is good for transportation. "So improve it. At least then we will have some benefit from it," they said.

"It does not matter what we want," one said, "it is what the government wants." The *mukiah*, or headman of the village, said he too wanted the embankment to be strengthened, even though he admitted "sometimes it saves us, sometimes it harms us." A child's bronchial cough punctuated the silence. Just down the road, the tractors continued to rumble.

Whatever added protection from floods the raised embankments may offer, they are clearly helpful for traffic, in particular for the government's jeeps and other vehicles. Riding the embankments was a challenging, uncomfortable way to travel for someone used to real roads; but on the other hand, had they not been there I could not have seen so much of north Bihar in the short time I was there. I later learned that villagers used to be hired to cut the earth for the embankments, but that has been contracted out to tractor owners from other states. So now engineers and contractors are the only ones who benefit from raising the embankments: they and the politicians who may receive kickbacks from them.

We rode down to the end of the embankment. The southern border of Nepal was only ten miles to the north; our driver was listening to Nepali music on the FM radio. In 1956 residents of the nearby village of Pidhi protested the construction of the embankment. The contractors gave up in the face of their opposition. We could still see a low hump in the ground, the remnants of the construction. It would block a very low flow, but not a monsoon flood.

The monsoon water spreads evenly here, villagers said, and does not stay for long. "Why are they raising the embankment to the south if the water will still come in here?" I asked, still puzzled. "Because it is useful as a road. The government wants a road here. So do some of the villagers," Ajaya reminded me. Come in the rainy season, they all urged as we walked back to the jeep. Then perhaps I would understand better where the water goes.

The Kamla was adored for its rich silt. But since the embankments came, the yearly nourishment has ended, except here near Pidhi.

We spent a pleasant night at the GPSVS headquarters in Jagatpur in the district of Darbhanga. The organization offers health and maternity care to villagers in the area. Then the next day, after a generous breakfast of rice, vegetables, curry, and tea, my companions and I drove toward the east for a couple of hours to a spot where several worm-like embankments converged. These mounds were smaller than the one we rode on the day before, about half the height. I saw no water. I wondered why they came together here, and why I didn't see the river that could flood this summer. I was still looking for two parallel lines with water in between.

"Where is the other embankment?" I asked again. Ajaya gestured toward the horizon as he had on the first embankment I saw the previous day. "Some kilometers over there." I looked and saw only the flat earth spread out around us. In fact, this was the spot where the embankments of two rivers meet—the eastern Bhutahi Balan embankment and the western Koshi embankment. The Bhutahi, which had been a relatively benign river before the Koshi was embanked, had to be fitted with a jacket once it was no longer able to drain freely into the Koshi, for it became a threat to the Koshi's western embankment.

At the junction, three barefoot little girls with no mirth on their faces were fetching water from a hand pump, elevated on a tall pipe. They filled different sizes of old plastic bottles. One wore a scarf around her head. Another looked as though her hair had not been combed in many weeks. Nearby were a few small homes, with thatched roofs and walls of cane and twigs.

We drove out along one spoke of the radiating embankments. Embankments serve as roads, and roads as de facto embankments all over north Bihar. Soon I noticed a train moving along atop another mound. Trains created a whole new set of mounds. I gave up trying to figure out which were the paired river embankments. Even if somehow the many flood control embankments in Bihar were torn down, which some people advocate, there would still be roads and railway lines directing water away from where it wants to go. Most of the roads and railways cross the state from east to west, while the rivers are trying to flow southward toward the Ganga.

We cut across the sandy open land, off the embankment, and got stuck. Some local men showed up to help our intrepid driver extricate the vehicle. Then we headed off along another spoke of the converged embankments, toward the Koshi. Barely clothed children looked at us dully when we stopped. I asked fewer questions the second day. I just looked out the window of the jeep as we bounced down atop the embankments.

In an hour or so I saw piles of wheat seeming to move across the surface of the river toward us. The river, an arm of the Koshi, flowed alongside the embankment below us. People were slowly wading across it, the water reaching the armpits of some of the villagers, the chins of others. They all kept the bundles of wheat they carried on their heads dry. We could see bright green fields interspersed with the rusty red of ripening grain on the other side. These people were among the unlucky ones whose ancestral fields had been trapped inside the embankments of the Koshi, spaced from seven to ten miles apart. After the monsoon floods subside, farming families go back to their old fields, those that are not still inundated, and plant their winter wheat. The men, women, and children I saw carrying the wheat are among a million people who still live inside the embankments of the Koshi.

While I was in north Bihar I attended a women's meeting in a village called Lachi Katuka. One woman, Parvati Devi, lost her five-year-old daughter to diarrhea and consequent dehydration after a flood. She said the local doctor gave her some glucose solution that had already expired. Whether the expired remedy hurt the child or she was already too sick to be helped was not clear, but the child was a flood victim as much as anyone who had drowned.

Another woman, Punam Devi, told the group that her husband goes to Punjab to work now. She said it was no longer possible to grow crops on their land. There's too much water on the land in the monsoon, the time for growing rice. And then the ground stays too wet afterward for wheat. "The land is either drowned or dry," said another, older woman, Sukia Devi. Before there were lots of crops, she said. The villagers had constructed their own drainage systems, so the land did not become waterlogged before the embankments were built. While these women spoke to me through an interpreter, children coughed, their lungs badly congested. Outside in the sun it was 105 degrees.

In Patna I spoke with *Times of India* reporter Pranav Chaudhary, who chronicles floods in Bihar each monsoon. He told me that over the past half century a whole

Farm families carry wheat across the Koshi

Villagers in Bihar save their cattle (photo courtesy of Eklavya Prasad)

new economy has evolved in North Bihar; and it depends on the floods. A nexus of contractors who build boats, provide timber, or make stone chips, sand, and cement turn a profit every year. "Everybody loves a good flood," he said.

Over the years, the people have become more and more dependent on relief, Chaudhary told me. They have come to expect the government or the nongovernmental groups to give them food or build new houses, so relief dependency has developed out of the yearly cycle of floods and flood relief. He said people are in such dire need of help, politicians are afraid of them.

Chaudhary says some people die in the floods, but thousands more Biharis die slow deaths all year round, living a life of bare survival. A chronic disease called *kala-azar*, which can be fatal, is common. Also known as leishmaniasis, the disease stems from filthy living conditions. Transmitted by sand flies, it causes fever, weight loss, and anemia. It's more prevalent than AIDS, he said, which is also common because the men who leave home for work pick it up from casual sex in other parts of India. All the able-bodied men leave to work during the four months when their land is waterlogged, disrupting family life. Many boys go to work in the carpet factories in neighboring Uttar Pradesh. Girls who are left behind are often sexually exploited. Sometimes sex becomes the currency used for repayment of loans.

Dinesh Mishra had been traveling when I arrived in Bihar. I met him at the end of my visit, after his colleagues had taken me out to see the embankments and the villages. About sixty years old, his hair a thick white halo, he inspires deep affection among other men in Bihar who are trying to help farm families in the flood region. They call him *guruji*. Though he had just returned from a trip the day

Villagers wait for rescue from floodwaters (photo courtesy of Eklavya Prasad)

before, he answered all my questions patiently. From time to time his eyes twin-
kled, even though much of what he told me was the stuff of tragedy. We talked for
hours in my hotel room while Pushpa listened and poured us tea.

An earlier flood on the Koshi, a quarter of a century before the dramatic 2008
shift in the river's course, had led Dinesh to change his own course. He was a
structural engineer then, running his own business. He had some experience in
disaster housing and was asked to come see if he could help the people in an area
engulfed by water in North Bihar. The river had bifurcated during that flood, one
arm inside the embankment, one outside.

He declined what looked like an impossible assignment. But he also found he
was unable to continue his own work after seeing the terrible conditions in which
people were living following that flood. The Biharis told him little when he asked
them questions. If they don't want to talk, he concluded, something must be seriously amiss.

So in 1984 he decided to spend a few months visiting libraries and reading old
newspapers. He scanned forty years' worth of news reports. They told him a story
of blunders, uncertainty, and suffering. He dedicated a year to documenting his
research, chronicling a debate on floods that had begun during the era of Muslim
ascendancy in the region, through the British Raj, to independence.

Since then he has been writing about the eight major river basins of North
Bihar and raising questions about embankments and how they are managed and
mismanaged at every forum he can attend. He's able to dedicate himself to this
work so completely because he doesn't have a family to support. A bachelor, he

makes his home with his brother in the nearby Indian state of Jharkhand. He stays with friends in Patna when he needs to remain in Bihar.

Dinesh said for thousands of years the people on these plains knew the inconvenience of floods for a few months, maybe real trouble for a fortnight, but the rest of the year life was good. "When rivers are free to flow, the water spreads everywhere and subsides. When river water encroaches the banks, it spills. The river is like a trough," he explained. "Only the top layer of water goes outside. And that top layer contains the fertilizing silt. That is the silt that travels all over the land. It's very light stuff." The heavier sediments flow along the bottom of the river channel. But when a flood breaches an embankment, all the water rushes through, carrying sand and gravel that can devastate agricultural land.

"Floods in our area are not a disaster," Dinesh continued. "They made it a disaster by building embankments. It's like going to a doctor with cough and cold. He says I don't have medicine for cold but I have medicine for pneumonia so when you contract that come back. We wanted protection only from some inconveniences and they built embankments. They created disaster, so now they give us disaster management."

Dinesh Mishra is also a Sanskrit scholar. Mythology captures the essence of these rivers, but engineers overlooked this folk wisdom to the great detriment of people living in the plains. Sanskrit literature characterizes rivers as female. Some are girls. Some, that have been stabilized, are married rivers, like the great mother Ganga. All the rivers of North Bihar are girls, young and a little wild. Dinesh says people welcomed the floodwaters of the wild Koshi and sang to her; but if she continued to rise, the people would ask her to go back. If she kept rising, women put red *tika* powder into the river, threatening her with the vermillion symbol of the married woman to convince her to behave herself and subside.

In Sanskrit literature the river goddesses even have hairstyles that capture the way they lie on the landscape. Some have a single stream like hair cascading down a woman's back; others have many channels crisscrossing the landscape like braids. "To Hindus these rivers are not just drainage channels," says Dinesh. "They are much more, and we have a relationship, like family. They are the mothers of the world. The rivers flow for the benefit of others. If they stop flowing, become stagnant, their benefit to mankind ceases.

"The British tended to look at rivers and see money," Dinesh says. "How big a boat could navigate a certain river and for how many months a year to ship goods? The British thought if you can bring water and give it to people for irrigation you can make money. So they started making money from Ganga." But even though they tried to make money by diverting water from these rivers for irrigation, the British were daunted by the idea of embanking Himalayan rivers for flood control. They decided the cost to rehabilitate an embankment that would likely breach in a few years made the business unprofitable.

The Koshi River is often called the "sorrow of Bihar." But Dinesh says people here never called the Koshi the sorrow of Bihar before the British came. "The British gave it that name because they couldn't make money from it," he contends. The British caution about embankments on Himalayan rivers persisted for a few years after they left the subcontinent. But a big flood on the Koshi in 1953 changed that. Nehru came and said something must be done to help the people. Nothing can be done but building embankments, politicians told him. Dinesh says engineers, both British and Indian—who had opposed embankments for the entire previous century—fell in line.

I asked him why the people themselves accepted the embankments in the first place. "The way I have understood it," says Dinesh, "politicians told people the British were exploiters and didn't do anything for their betterment. It is our turn to do it. We will do it because we are a welfare state. And they did it and people thought they might get protected. They got trapped into it. They were promised heaven on earth in 1953."

The Koshi may be the least predictable of Bihar's rivers, but there are seven other major rivers that have been embanked, including the Kamla and the Gandak. The debate centered on the Koshi. After the Koshi was embanked there was not much debate about any of the other rivers of Bihar. So the Kamla, the Gandak, and the Bagmati were eventually fitted with their jackets, and then their tributaries had to be jacketed too, because they could no longer drain into the main stems. Water pooled outside the embankments, making the land useless for agriculture. Sluice gates built to allow tributaries to drain through the embankments into the main rivers seldom worked.

The embankments have pitted people against each other. The "insides" and the "outsides" now have different needs, where once they were subject to the same floods and the entire community pulled together. The people inside want the embankments breached so the water can spread. The people outside want the embankments raised.

The world I start to see through Dinesh Mishra's eyes is like some upside-down place Gulliver might have visited, something that the mind of Jonathan Swift might conceive, insane and absurd, in every way the opposite of what makes any sense. Dinesh says Western-trained engineers felt contempt for the people here, and the feeling was mutual. "Someone who has grown up on the river knows much more from the day he is born about how the water behaves. We come and look and we don't see what he sees; we can't understand what the land is saying." Hearing this I feel a little less stupid for not understanding what I was seeing as I gazed out over the floodplains of North Bihar.

Is there a solution? I ask him. "The state will never allow these embankments to die unless there is a revolutionary thinker in politics." His advice for any forward-thinking politician: "Please stop making it worse—don't make any more of them. Don't build them any higher. If an embankment breaches, people living around it should have the authority to say yes or no to plugging it."

Whatever happens, the state should not get involved in removing the embankments because that would open the door to the same corruption and kickbacks

and skimming that the building of the embankments has fed on. "Building embankments is a money-making device; taking them down is also a way to make
money," Dinesh says. "But the river does it for free, so please don't spend money
on doing away with embankments when the river is showing the way. When it
opens a path, it is telling us that it needs to get out from here. In five or ten years
it will take its own course."

# 11

# The Koshi's Revenge

> Who is going to represent the river? The river is a living thing. It
> has a wish. It comes from the Himalaya and wants to reach the
> sea. The river always takes revenge. It's an aggrieved living
> creature. It will always take revenge.
> —Nayeem Wahra, Save the Children, Bangladesh

The Koshi spoke during the monsoon of 2008. She opened a new path, just as
Dinesh Mishra predicted. The river breached an apparently ill-constructed and
certainly ill-maintained embankment. A photo taken as the flood began shows the
ridge of sand dissolving as water poured through a widening gap in the embankment and flowed southeast. In both Nepal and Bihar, villages and farms that had
not seen a flood for the past half century were devastated.

The embankments on the Koshi had already breached seven times at various
spots downriver. This time the entire river below the Siwalik range in Nepal, where
the land flattens, had essentially jumped out of its straitjacket and returned to
one of its old channels—one it had flowed down two centuries ago.

In Nepal the Koshi River is known as the Saptakoshi, or "seven Koshis," because
seven Himalayan rivers merge to create it. The Tamur flows down from Kanchen-
junga in eastern Nepal near its border with Bhutan and India; the Arun comes
down from Tibet. Out of the Khumbu comes the Dudh Koshi, the milky blue
river that entranced me on the way up to Gokyo. The Dudh Koshi joins the Sun
Koshi, which is also fed by the Tama Koshi, which in turn receives water from the
Rolwaling Khola and Tsho Rolpa, the threatening glacial lake I visited during the
monsoon of 2006. From farther west, toward Kathmandu, come the Likhu and
the Indrawati. The latter receives the as yet undiverted waters of the Melamchi
Khola.

These seven tributaries of the Saptakoshi drain more than a third of the Nepal
Himalaya, the wettest and highest of the great range, which includes the Khumbu
and Ngozumpa glaciers. The Koshi drains almost thirty thousand square miles. It
is Nepal's largest river and one of the largest tributaries of the Ganga.

Less than ten miles above the plains, three of these great rivers come together
in a final merging: the Sun Koshi from the west, the Arun from the north, the

Tamur from the east. They meet at a *triveni*, a confluence of three rivers, even more sacred than a *prayag*, the confluence of two rivers.

The Saptakoshi is wild and powerful and carries more sediment than any other river in South Asia. The floodplain of the Koshi is an inland delta where the river has been building land for eons from sediments the Himalaya shed. Water specialist Dipak Gyawali calls the Koshi "a massive conveyor belt" that delivers sediment from the top of the Himalaya. Each year more than a hundred million cubic meters of gravel, sand, and mud flow out of the Barakshetra gorge from the triveni where the rivers meet. During the two hundred years before it was jacketed, the Koshi ranged across a seventy-mile fan, abandoning one channel when it got choked with silt and finding a new one down to the Ganga.

Ajaya Dixit has been trying to understand why the Koshi is so very capricious and dynamic. There are other powerful rivers in Nepal that flow down to meet the Ganga, but they do not rival the Koshi in terms of unpredictability. He has puzzled over this and believes the answer lies in the different personalities of the triad of rivers that merge just above the plains to become the Saptakoshi.

Each has very different characteristics by the measures hydrologists use to define rivers: slope, the amount of sediment the rivers carry, the land area they drain, and the amount of rainfall in those areas and when it falls. Further, the ten miles between their confluence and the plains, where the Saptakoshi's channel flattens out, offer only a very short space for such powerful rivers to assimilate. And in the monsoon the behavior of one river at any given time may be very different from its sisters, because each comes from a distinct area in the Himalaya. For example, there may be a high flow in the sediment-laden Tamur for a day. This might be joined later by a high flow and a load of coarse sediments from the Arun. Then the two of them might subside and twelve hours later the Sun Koshi could send a mass of monsoon rain down the gorge. The dynamics of the mixing and the variability of sediment loads and volumes of water create hydrological behavior that is almost impossible to predict.

Embankments on a river like this, says Ajaya Dixit, could never guarantee protection to surrounding lands, even if they were well constructed and well maintained, which those on the Koshi were not. But they can give people a false sense of security, leaving no one prepared when the breach happens. When the river does break through, its natural caprice is compounded by all the failed attempts to control it.

On August 18, 2008, the river's unpredictability combined with human error to create one of the most devastating floods in South Asian history. For several years before the flood, the braided channels of the Koshi had been moving east, flowing closer and closer to the embankment. Because of the massive amount of silt the river carries, its bed had risen over the years until it was about twelve feet higher than the land outside the embankment. The high, rushing water gobbled its way through the embankment at a point about fifteen miles north of the Indian border, quickly creating a mile-long breach through which the entire river escaped onto the low-lying farmland.

For the previous sixty years the area on the eastern side of the embankment had steadily been built up. It remained agrarian, but towns and roads had encroached on the Koshi's old channels. The water that broke through on August 18 had to find a new way. Taking the path of least resistance, it spread across the land, filled depressions, found some of its old channels, and traveled more than a hundred miles south to the Ganga. On its way it took people and animals, but even months after the flood there was no official death toll. Some said five hundred people drowned; others said thousands more were missing and presumed drowned.

After the 2008 flood, the governments of India and Nepal, as well as the state government of Bihar, could envision nothing more innovative than giving meager aid to the flood refugees, then proceeding to put things back the way they were before. Ajaya Dixit says this was a lost opportunity to begin to correct a terribly misguided endeavor, though he admits that would have been a formidable undertaking. More than three million people lived in the part of Bihar that was flooded in 2008. Moving all of them was not an option. It would have been about as unthinkable as abandoning New Orleans to the Mississippi's muck, though for different reasons.

Months after the monsoon subsided, Indian engineers were able to build a series of temporary dams and then move the minimal, dry-season flow of the Koshi back inside the embankments. By late January the river was back in its former course, and the repair of the breached embankment began.

Five months after the 2008 flood, I visited the Sunsari district of Nepal and the Supaul district of Bihar, the two districts closest to the breach in the Koshi's eastern embankment.

"This area was a garden," a retired Indian engineer in Supaul told me as we sat on a dusty verandah, looking out over a bare, dun-colored schoolyard that had been under water a few months before. "Now it has become a desert."

He said that the deep sand could become useful farmland again, perhaps in twenty-five or thirty years. Plowing under organic material from whatever people could manage to grow in the sand, combined with animal manure, would over time create a soil where crops could grow. He said he had seen the sandy floodplain become productive in his lifetime. But now it would take even longer to restore this land, barring some extraordinary outside intervention like dredging the sand and hauling it away. It seemed an unlikely solution for all those sand-covered acres.

Dinesh Mishra, exasperated, wrote in a letter after the breach: "Thus there have been altogether eight breaches in the Koshi embankments so far, five in India and three in Nepal. Three of the eight breaches have occurred in the western embankment and five on the eastern." Noting the lack of pattern in the river's behavior, he asked: "Why on earth were the embankments built at all? There is no answer."

Before venturing into Bihar, I was a guest for a few days at a bird-watching camp adjacent to Nepal's Koshi Tappu wildlife reserve, a wetland alive with some of Nepal's 860 bird species. This haven north of the breach was a garden of tall, dense trees, largely unaffected by the flood. It was wonderfully peaceful except for the intermittent rumbles and squeaks of trucks and tractors. From early morning until well after dark they carried loads of big river rocks down the nearby embankment to fill up the breach.

A Nepali friend, Shankar Tiwari, invited me to stay at the camp where he worked and offered to help me see what the flood had done on the Nepal side of the border. Shankar is a guide and one of Nepal's most knowledgeable birders. I had met him high in the Himalaya while trekking with friends. Only later over tea in Kathmandu did I learn about his knowledge of birds and about the wildlife reserve near the Koshi embankment's breach. Fit, forty, and handsome with a gentle sense of humor, he was on a mission of his own when we arrived in Koshi Tappu. His boss, the owner of the camp, had sent him to try to iron out some labor problems with the local staff.

It was not yet the season for bird-watchers, so Shankar had time to be my guide while I tried to see the devastation from the Koshi flood. First we walked around the camp and down to the river; out of the corner of his eye Shankar could spot a small bird a hundred feet away. I tried to keep up with him as he pointed out dozens of species. I saw egrets, cormorants, and herons, species familiar to me. But the majority were too small or quick or unfamiliar for me to get a really good look. Shankar patiently repeated the names and pointed the binoculars for me to get a glimpse of the endangered swamp francolin, the white-throated kingfisher, and the ruddy shelduck, also known as the love duck because they're believed to stop eating and die if they lose their mates. Then there were the noisy brown birds that always congregated in groups and in low branches where I could easily see them. They are called jungle babblers, but they're also known as "seven sisters" because they are never alone. I thought it appropriate to find them here near the Saptakoshi.

The villages I saw near the camp were poor. The goods for sale in their evening markets were simple. But the farms strung out along the dirt roads nearby looked productive and orderly. Some houses were tiny; others had two stories, thatched roofs, whitewashed walls, and blue paint on their carved wooden windows. Each farm had outbuildings and small earthen storage silos for rice, along with goats and water buffalo, cows and chickens. Piles of fuel sticks made from dung and straw dried in the sun at each house. Banana plants and bright yellow flowering mustard interspersed with wheat made the land look lush even in the dry season.

Shankar took me to explore the area on a borrowed motorbike. We rode out to the main, paved road and headed south. We soon saw refugee camps on the flat, bare, sandy land near the road. Some of the camps had sturdy gray camp tents donated by one of the international aid organizations that came in to help after the flood, but for the most part there were merely blue plastic tarps spread over bent bamboo, like flimsy blue igloos. These too were donated, whether by the Nepal government or an international aid group, just not so luxurious as the gray camp tents with window flaps and doorways. Off to one side there were woven bamboo structures, small rough booths over pit latrines.

On each side of the road, sand stretched to the horizon. Soon the road ended: a long section of the single east-west highway in Nepal was either under the sand or had washed away in the flood. Vehicles could go no farther; buses had turned around and people who had walked from the other side of the dunes were boarding them.

Shankar steered the bike over the mounds while I walked across the inland beach. People milled around, loading and unloading heavy sandbags and sacks of foodstuffs to cross a short span of water in canoes. A man sat on the ground with snacks and cigarettes spread on a cloth for sale; rumbling bulldozers shoveled sand. Just a few months earlier this sea of sand must have looked like the farmland I had seen from the back of the motorbike as we passed through villages on our way out to the paved road.

We approached a pair of bamboo bridges elevated about six feet above the remaining rivulets of water. A couple of dozen people crowded at the end of one of them. A plump man standing on the bridge had a fistful of rupees in different denominations poking from between his fingers. Some of the people in the knot at the bottom of the bridge—women carrying baskets on their heads, men with bicycles—started yelling at the man, complaining about having to pay just to walk

Tents pitched on embankment shelter Bihari flood victims (photo courtesy of Eklavya Prasad)

across. They might have been going home, or might have been refugees from the camps going into town. This was apparently not a government bridge, but the work of some enterprising individual who was trying to cover the cost of construction—and make as much profit as he could—before the road was repaired and traffic would abandon his bamboo toll bridge. The altercation seemed about to come to blows, but the people lost the argument and paid the toll.

The bridge was about six feet wide: woven bamboo held up by bamboo supports. It was springy to the step, and stretched the length of a city block. Men walked their bicycles and motorbikes over it; women with bundles of driftwood on their heads for cooking fires swayed across in their colorful *kurtas*.

We got back on the motorbike on the other side of the bamboo bridge and Shankar drove on through the small town of Bhantabari before the road swung to the west. We passed a turnoff that went south to India as we approached the Koshi barrage. A barrage is a kind of dam, but unlike a hydropower dam with a deep reservoir behind it, it is constructed in a flat river channel and is not meant to hold back a reserve supply of water. Instead the barrage slows the water down so it can be diverted into canals.

The Koshi barrage forces water to pool and flow into an irrigation canal just above the barrage. The barrage is about a mile across and thirty feet high; all along its length are wide gates that can release the full flow of water when they are all open, as they would be during the monsoon. A paved road, elevated just above the water, crossed the river on the downstream side of the barrage. India completed the structure in 1963, a decade after the embankments, leasing the land from Nepal. India has always been responsible for maintaining and managing both the barrage and the entire length of the embankments in both countries. The breach in 2008 was the first of the eight Koshi embankment failures that had occurred upstream of the barrage.

I watched a group of boys happily take running leaps into the river while Shankar went to ask some women at a nearby temple, from which music emanated loudly, whether people had seen any Ganges river dolphins. They used to congregate in the deep water just below the barrage. Unsurprisingly there were none here now, because the river had only recently been moved back into this channel.

We crossed the river and went north on the west side to a lookout. As we traveled away from the barrage the embankments steadily diverged, fanning out until there were more than five miles of sand and water between them. At the lookout, I tried to imagine this expanse of sand surging with turbid monsoon rains. The other side of the flood channel was barely visible, just a hazy beige blur with a green smudge of trees in the far distance. Shankar was looking for birds, because he was planning to bring another group of keen British birders to the camp in a few weeks and this had been one of the best viewpoints for water birds. He was sad to see that the channels in the wetlands were all changed now; we saw only a few birds.

We returned to the barrage and sped past its mostly open gates to the other side of the river, then went north atop the eastern embankment, the side with the breach. North of the barrage were more than a dozen refugee camps, spread out for mile after mile on the dry land just inside the eastern embankment. Tarps were fraying, and some of the latrine shacks looked like they were about to fall down. But there were children playing and domestic animals roaming around. A giant buzzard rested on the side of the embankment while a group of sari-clad women stared at it, reminded perhaps of all the farm animals that had drowned in the flood and become carrion. The bird spread his six-foot span of wings and flew away.

A *pahadi* woman, a member of one of the ethnic groups from the hills that had migrated to the richer soil of the *tarai*, was walking down the embankment, returning to one of the refugee camps. She told us the warnings had started the day before the flood arrived; people up the road in Kusaha phoned to tell people in the villages that the embankment was going to give way. It rained heavily all night and by morning they were afraid to stay in their homes; she gestured toward some houses we could see a mile or so to the east.

They came up onto the embankment soon after dawn. Some hours later, the water arrived. People were afraid it would not stop rising and they would drown. There was a small army garrison just down the embankment. They were packing up to leave and asked her if she wanted to stay in one of their buildings and look after things. She agreed. Other people came and found another army building locked up. They said "the army left us here alone; we're going to break in and take shelter." She asked for their names but didn't try to stop them, Shankar told me, translating the bulk of what the woman described in rapid, colloquial Nepali.

The woman said her family was repairing their home and would move back soon. She said other people were maintaining a presence in the camps so they could continue to receive aid, even though they had already moved back into their houses.

Just past the last refugee camp we reached the army post; the soldiers let us through so we could travel north on the embankment. Entering the trees of the Koshi Tappu wildlife refuge, we left behind the bare floodplain where the refugee tents huddled. We bounced down the dusty embankment in the fading light for several miles. Trucks filled with rocks came up behind us and we waited for them to pass to avoid the clouds of dust they sent up.

Beyond the damaged village of Kusaha the avenue of trees ended abruptly and we reached another desert. Here was the recent flood's ground zero: the breach. We slid down from the embankment onto the floodplain and into a hive of activity, a chaotic construction zone about two miles from end to end.

All those thousands of Himalayan rocks carried on trucks or dragged behind chugging tractors were here, and men were busy assembling them. They shaped them into what looked like giant buff-colored bricks. Rectangular cages of wire tightly held hundreds of loose rocks together in blocks, each "cage" a couple of feet high, four feet wide, six feet deep. The cages of rocks, called gabions, were being assembled end-to-end, four deep, all along the length of the breach. I assumed all this reinforcement would eventually be covered with the heaps of sand we were struggling across, thereby creating a mound of sand with a rock-hard fist inside. There should be no more breaches right at this spot. The river would have to go find another way out of its straitjacket, once this rip was sewn up.

A couple of hundred people were still working in the late-afternoon light. Dry sand stretched to the western horizon, but far in the distance there was a little glimmer of the now shrunken, dry-season Koshi flowing down to India.

I wanted to see the place where the Koshi emerges from the Barakshetra gorge and flattens onto the plain. The next morning Shankar took me exploring on the bike again. An expert driver, Shankar was a good host and charming companion. He had all the caretaking instincts for which Westerners so love Nepali guides. And his abundant intellectual curiosity kept us chatting over his shoulder for mile after dusty mile, during which we dissected numerous of Nepal's environmental and social problems. He told me I rode the bike so well and lightly that sometimes he had to check to be sure I was still there.

We rode north on the embankment, traveling from the camp toward the mountains. As the land sloped upward, we left the embankment and the flat tarai, traveling up into the bottom rung of the Himalaya, the hills known as the *siwaliks*, to a revered Hindu pilgrimage site.

The road wrapped around the sides of the hills. The Koshi was below us, its loose tresses of meandering channels now caught up between the narrowing sides of the gorge. Soon we saw trucks pulled up one behind the other, apparently driverless. The line of trucks snaked on for mile after mile as their drivers waited to cross the narrow neck of the Koshi ahead of us. This was their only alternative, because of the damaged road at the spot where we had crossed the bamboo bridge. Two small ferries plied the river below, slowly moving one truck at a time across the water.

A driver stood on the road beside his truck, which like most in South Asia was colorfully decorated with red fringe and paintings of gods and goddesses. He told us he had gotten in line six days earlier to reach this spot, just within sight of the ferry far below. There were still a couple of miles of trucks—maybe two days' worth—ahead of him, and a week's worth behind him. These trucks carried non-perishable items of commerce from India and southern Nepal into Kathmandu. There was no other way to transport them. In the distance on the other side of the river we could see another line of trucks waiting to come to our side and head to the east tarai and north India.

Ajaya Dixit says this kind of delay, to which Nepalis are so accustomed because of endless natural and political mishaps, need not happen here again. Even if the embankment were to breach again, he says, the road could be built to survive the flood. But the road repairs that were underway were not employing flood-friendly construction. When the water can pass beneath the road through thirty or forty pipes, simple culverts, the road can withstand the flood.

Any extra expense to rebuild the east-west highway with such openings beneath it could easily be justified by the cost of the delays caused by six months of having to transport goods across the gorge on ferries. In addition, the road would be more ready to withstand floods that climate change may bring. "Unfortunately," says Ajaya, "our civil engineering establishment is rigid, entrenched. Not willing to innovate—all throughout South Asia."

We bumped along the increasingly rough road and left the bike just outside Barakshetra. After we visited the shrines, we walked to the edge of the village so I could gaze up the gorge. Behind the hills was the *triveni* where the Tamur, Arun, and Sun Koshi merge. And just below it, many engineers and politicians have advocated placing an eight-hundred-foot dam to hold back the unruly waters of the Saptakoshi.

As Shankar steered the motorbike back down the road after we left Barakshetra, we could see the Koshi's dark green waters filling the gorge below. The river was calm, showing nothing of her tempestuousness. Flowing easily out of the hills on this hazy February day, the Koshi was just another mountain river.

The next morning Shankar took me to the Koshi barrage, near where I was to meet an Indian social worker, Eklavya Prasad. He would take me down to India and around the flood-ravaged district of Supaul. I bid Shankar goodbye and piled my backpacks into the dusty jeep Eklavya had hired and the driver sped away toward the Indian border.

Busy chatting with Eklavya, whom I had met the year before in Delhi, and amazed we had managed this rendezvous in such a remote spot with relative ease, I paid little attention to the road. Soon the car slowed, and Eklavya spoke with an Indian soldier. "Where's the border, where's the Nepal exit post?" I asked, suddenly paying attention. The soldier shook his head. I said we had to go back and get the Nepalis to make official note of my exit. We drove back ten minutes to the customs office near the barrage, where half a dozen men sat on chairs in the sun. The man who appeared to manage the operation said he had no way to register my exit; this was not a crossing point.

I was stunned. It had never occurred to me to check on this in Kathmandu. There was a border, there was a road, Nepalis and Indians go back and forth. They don't need passports or identification of any sort; for them it's a completely open border. But surely there would be an official who could document that a foreigner was leaving. I was ready to pay a bribe, but that would not have helped. There were no stamps. It was only a customs office, not a passport control post.

The customs manager said we could drive to the Darjeeling border, several hours away, and I could exit there. Eklavya seemed ready to do that if necessary. But I said we would have to come back here and exit the same spot anyway because it was the only way into Bihar, or at least the only efficient way to reach the part of it where Eklavya was working. Dismayed, I felt much as I had at the washed-out bridge on the way to Tsho Rolpa. My trip was over. I had come all this way and couldn't go on.

I had been planning to go on down to Bihar's capital, Patna, to meet with Dinesh Mishra again. From there I planned to take a bus back to the nearest crossing point into Nepal. I could not do that unless I made a legal exit here, for it would become clear when I tried to reenter Nepal that I had not officially left it. I called friends in Kathmandu for advice. One counseled me not to take the risk. But I didn't want to give up.

The Indians did not seem to care about letting me in, even though there was not an official immigration post there either. Nor did they seem to care whether the Nepalis stamped me out or not. So I decided to go on over and return the same way, as if I had never left. My Nepal visa would expire, but I would only be a day late renewing it if I returned to Kathmandu over the coming weekend. We piled back into the jeep.

By the time we reached the Indian border, the guard had changed. This soldier was even less sure than the previous one about what information he should take from us. He and his colleague wrote down the license number of the rented vehicle, which would not be the same one in which we would return here. They glanced at my ten-year visa for India. And we left. I worried about what might happen on the way back if a different Indian soldier wanted to know how long I had been in India and why, but I wasn't willing to give up the trip. It would take too much time, effort, and money to get things lined up again to enter this part of Bihar when Eklavya was working here.

We went a few miles beyond the border to the town of Birpur. Eklavya was affiliated with a small, local organization there—one of many in north Bihar, typically supported by foreign donors, through which flood relief is often channeled.

Taller and plumper than most of the Indians in Bihar, Eklavya's uniform in early February was jeans, a loose sweater or a hooded sweatshirt, and sandals. His face is round and open, his alert eyes friendly behind dark-rimmed rectangular glasses. In the field that week he was unshaven, a fat camera bag slung over his shoulder. I felt immediately comfortable with him as we bumped around the rutted roads of north Bihar. Born after I had already graduated from college half a world away, he nonetheless seemed familiar to me and easy to talk to, and not just because of his fluent, colloquial English. He could instantly fit in at Berkeley, with both the aging white progressives and the upwardly mobile, multicultural young people on the campus now.

Two of the staff from the grassroots group in Birpur joined us for a ride east to the flooded area. Small shelters lined the road. People had constructed them from various combinations of thatch, sheet metal, plastic tarps, and wooden doors scavenged perhaps from their own homes. Some of the doors were locked, other doorways had only curtains hanging limply over them. Laundry dangled outside the one-room homes. There were also some low, tent-like shelters of cane and blue plastic that appeared abandoned.

Flood refugees from Bihar went across the porous border to Nepal for months after the flood to take advantage of the aid from international groups being offered there. Though India had become the economic engine of South Asia and

Nepal's tiny economy was steadily declining, the relief after the Koshi flood was better in the nearby Nepal tarai than what they could get in India, where millions were waiting for government aid. The amount of land and number of people affected in Nepal was far less than in India: sixty-five thousand versus three million flood victims. The area of Supaul we were going to visit was just a fraction of the land in Bihar between the Nepal border and the Ganga that the Koshi had inundated.

The road came to an end at a haphazard-looking construction site. Work was still underway in the late afternoon to install a temporary pontoon bridge over what remained of the river now that most of it had been put back into its pre-flood channel. At this point we were right in the path of the flood, some eighteen miles from where the embankment had breached and belched out tons upon tons of sand over the land.

This area had been protected from floods since the embankments were built in the early 1960s. It had become reasonably prosperous, sustaining landowners, farm laborers, and sharecroppers. There was electricity. There were roads. This is what the defenders of the embankments would point to, saying the embankments had helped some parts of Bihar, and thus they had been worthwhile. For two generations this had been productive farmland, where before it had been the floodplain of the wild and unpredictable Koshi.

But what happened in August of 2008 supports the other argument. It might be fifty years before this land returns to such prosperity now that it's a sea of sand. The people here were blindsided by the flood, having lived for two generations thinking they were safe. They were not prepared in the most minimal way for such a disaster, nor were Indian authorities ready to handle the crisis.

Eklavya spoke with some of the men who were gathered near the road construction. Some wore *lungis*, some wore trousers. All wore long sleeves and a scarf or shawl in the mild chill of winter. Some had returned to their houses and were continuing to repair them, but the land on which they used to grow rice and wheat and vegetables was smothered by four feet of sand. Eklavya asked them what they planned to do. He was compiling a report in order to make recommendations for a foundation interested in funding development in the area. "Try growing different crops," said some. "Migrate," said others.

Hitting the roads and rails in search of work is a desperate solution, but many farmers from north Bihar resort to it. Bihar has become a source of cheap labor for the rest of India, though the displaced farmers sometimes become the victims of prejudice when they leave home. In other parts of India the term *bihari* has become a catchall for an unkempt menial worker, living in a shanty by a road that leads to one of India's international airports. A guesthouse owner in Delhi, an otherwise reasonable and pleasant man, maintained vehemently that the Biharis are lazy and they cheat. In Mumbai Biharis have been killed because of this kind of prejudice.

Eklavya found out that the group of men he was chatting with included retired government officials and contractors who owned land. Only a few of them were farm laborers or sharecroppers with no resources to fall back on. "If I'm looking for what needs to be done I think the poor people are the ones who can point the way to it," he said. "The people who are better off don't have to fight for the basics. They probably didn't lose their houses. They're not the people in the tent camps. They'll fight to get help for de-silting the land. But the poor people will say 'before that I need a house, some utensils so we can eat.'"

The sun was setting. Eklavya arranged to meet with a few of the farmers at noon the next day.

Eklavya Prasad's business card says "practitioner," not social worker or consultant, which would also be accurate. He's focused on putting theory into practice.

As a young man he started to follow in his father's footsteps to become a doctor, but soon realized that was the wrong career for him. He studied social work at a time when it was largely the province of wealthy men's wives who did charity work. He spent a few years feeling out of place in well-paid corporate settings and management positions, but missed the challenge he had felt doing field work with deaf-mute children when he was completing his master's degree at the prestigious Tata Institute of Social Sciences. He spent five years at Delhi's Center for Science and the Environment (CSE), one of India's most effective research and advocacy organizations. He says his years there, working on rainwater harvesting and land use, were pivotal. The founder of CSE was his mentor, and taught him an idea that guides his efforts now: "simplicity is the best weapon to overcome any challenge."

He wanted to work in Bihar, having grown up in the southern part of the state, which was later carved into the state of Jharkhand. "I had seen thousands in abject poverty while people in government and other organizations seemed clueless about what to do. Some of the people working here in Bihar are trying, but they're overwhelmed and confused. They do a little here and a little there but don't have a big picture of how their efforts fall into a vision." In 2004 he went independent. He got some funding and single-handedly launched a campaign in north Bihar called Megh Pyne Abhiyan, using the Maithili words for "cloud" and "water," the Hindi word for "campaign."

Eklavya's main tool is his ability to build relationships. He has no office, no secretary, none of the trappings of what we typically consider an organized effort. He has a couple of cell phones, a camera, a laptop. He rents a vehicle and driver to get him around north Bihar when he commutes there from Delhi, where his wife works for the UN.

He seemed infinitely patient as he coaxed information from groups of lungi-clad farmers, even though every few minutes they would all start talking in unison and he would again have to ask them to speak one at a time. But for incompetent drivers and colleagues who tried his patience, he had less tolerance. When staff at the local NGO started to talk about women's help groups over lunch, he demurred. He says there are a lot of generalizations in development now, a lot of clichés. Women's help groups are good, but people need specific approaches that grow out of the particular needs of the place.

He outlined the typical way development work is done in Bihar. The donor agency establishes a partnership with a local group working in the community; the concept and the plan for execution come from the donor; so the local organization is just a way to get the work done. When a similar need arises after a year or two, the local group looks for a new donor agency because they have not learned skills to do the work themselves. He wanted to turn some of that around, giving people ways to cope with the challenging times. While most aid agencies work during the flood season then leave, Eklavya believes in working hard in north Bihar for the eight months before and after the monsoon, helping people learn to deal with the four months of floods on their own.

That's why he launched his "cloud water" campaign, which aims to give the people who face floods every year a way to provide themselves with safe drinking water, especially during the flood. The rainwater harvesting method he has taught people in Supaul and other districts is simple, practical, eco-friendly, easy to construct, and cheap. basically a sheet of plastic with a hole in the middle strung up over a vessel to capture the water.

Local people can manage it themselves, without experts or outside funding, and it transfers skills to the local community. "Unless we think about doing things differently, nothing will ever change here. The problems will remain. And people will continue to depend on external sources," he says. "How much money did we spend changing their drinking water habits? Nothing. Why did I have to come from Delhi to tell people that rainwater harvesting is possible? If more of the grassroots groups were thinking instead of playing the development game, I wouldn't need to be here.

"People may think I'm hard-hearted because I don't just want to dole out aid," he continues. "I do, when it's necessary, during an emergency. But later, you have to try to get at the underlying problem or it's all just a waste of time."

Eklavya said there was really no place for us to stay in Birpur, so we rode about two and a half hours down to Supaul, the district center, where he had been staying. The drive was time-consuming because the roads were rutted and ragged. We rattled and bounced through the dust mixed with ground fog, which further darkened the minimally electrified rural night.

Somewhere between eight and nine we pulled up to a simple, dimly lit brick house in Supaul. It appeared to still be under construction. There was a clay floor and the framed, wooden windows had no glass. It was winter, but not that cold. This was the home and office of Chandrashekhar, who runs another of the community organizations with which Eklavya collaborates. The two men were clearly fond of each other. Chandrashekhar didn't speak much English. His wife Poonam gave us a good supper, cooked outside, in an open area behind the small house.

Their children had gone to bed on a platform under a mosquito net in the one adjacent room. Husband and wife work together to help villagers in this area.

After we ate Eklavya took me to the little hotel where he was staying. He said it wasn't much but it was clean. There was no water in the pipes in my bathroom the first night, but I had my water bottle for teeth brushing and face splashing. The bed in my small room was a plank with a thin cotton mattress. We were about half a block from a little temple that was still blasting music at ten in the evening through speakers that looked like air raid sirens. Eklavya wasn't sure if it was some special day or this was standard everyday celebration. I said I could probably sleep through it. It was pop music with devotional lyrics. I heard "Hanuman," the name of a Hindu god, in one song.

The music stopped. I slept despite the hard plank until the music started blasting again around 5:30 in the morning. Supaul by daylight was dusty and crowded, a small nondescript north Indian city. Poonam gave us delicious homemade yogurt for breakfast before we retraced our steps to Birpur over a different route to avoid some of the bumps.

Eklavya met with the district officer in Birpur. We sat outside the one-story office building under the trees, at a rough wooden table covered with a red and black printed cloth. An assistant shuffled through a pile of papers, each with a handwritten chart of names and figures. The district officer was a slight young man dressed in a fuzzy orange sweater vest and green checked shirt. A group of men stood around observing the exchange. Mobile phones with flashy ring tones punctuated the calls of birds in the trees.

Eklavya asked questions about the flood victims and wrote it all neatly in a little notebook. The district officer didn't waste energy in unnecessary animation. He gave Eklavya the information he requested from his own little notebook: "one of the most devastated wards, in one of them 19 houses were demolished." People arrived to make appeals to him. He acknowledged their requests wearily. The resources at his disposal were limited.

They talked about alternative crops that could grow in the sandy soil: sugarcane, watermelon, cucumber, sweet potato. But they agreed that people here need other ways to make a living now; clearly farming would not sustain the population.

After this meeting we went back down the road, near where we had been the evening before, to a village called Sitapur. About four thousand people lived there. We parked and walked across the sand to a schoolyard. A group of children fluttered around us as we walked to the shade of a verandah. Then they gathered, circling closer and closer to stare at us. Why are we here? What will we do for them? For their parents?

Two boys immediately offered us water to drink, then chairs, so we could sit on the dusty verandah. One of the boys presented a blue plastic cup filled with water from a pink plastic bucket. In this part of the world an offer of water is a basic courtesy. Not offering is an insult. Two other boys were busy smoothing fresh cement on the sides of the damaged concrete platform around the hand pump.

The boy who took it upon himself to pump the water from the rusted hand pump and offer it to us returned my smile uncertainly, then quickly turned away. A beautiful girl with dark chocolate skin and a frilly pink and white dress stared at me. Other girls wore blue and white kurta sets, fading school uniforms. Eventually the children lost interest and went back to playing. The girls adjusted the white *dupattas* hanging over their shoulders as they chased each other across the rough ground of the schoolyard.

Eklavya extracted information from the men who had come to meet him. He did so with humor, gently scolding when all of them spoke at once. Stretches of orderly silence were punctuated by what sometimes sounded like yelling. Arguments seemed to break out over nothing: the population of a village, the number of kilometers to the breach. The men got bogged down arguing among themselves about whether to count a one-year-old child in the population.

Eklavya asked a tall, nice-looking young man with an air of authority how he had managed to save his motorbike, parked nearby, from the flood. He said: "I put it up on a pile of beds and covered it." He continued to answer Eklavya's questions while others chimed in.

"It's a tedious process, but without it nothing will change," Eklavya said to me, referring to the many hours he spent like this extracting information. To me, a former reporter, the long process of asking questions seemed very familiar, however little I understood of their conversation in Hindi. Eklavya apologized about not translating more for me. I told him he shouldn't worry about that, we could catch up when he finished his work.

An older man in a long-sleeved blue plaid shirt came to sit opposite me. He addressed me in clear English: "Madam, this area has tremendous problems." Eklavya and I quickly told him that I was not an aid worker, just a friend. He returned to speaking with Eklavya in Hindi, punctuated by English for my benefit. Eklavya told me when he brought an Italian colleague to Bihar everyone assumed he was her translator.

The man was a retired electrical engineer, age seventy-seven, named B. N. Deo. He said it would have been better not to build the embankments. On the other hand, he said, "until 1944 there was not much cultivation here. The land was ravaged by the Koshi. But the soil changes. It slowly becomes fertile. After 1950," he continued, "it became a fertile area."

Eklavya asked the men what they saw as a solution. Again they mentioned leaving home to look for work. But in any case the government must keep giving aid because they had no way to feed themselves now that their soil was destroyed. Adding to the chaos, the land records here were lost in the flood. This has happened throughout the flooded region and will mean a painstaking process of reconstructing land ownership without records and without many of the landmarks that existed before the flood, a situation not only confusing but likely open to abuse.

As Eklavya continued to talk with the men in Hindi, I gazed around the bleached-out schoolyard. Opposite the building where we were sitting was a row of classrooms. At first I saw no teacher, then a man walked out of one of the rooms to check on some children. They had lingered to play after they took drinks of water from the pink and blue buckets two boys had left outside the classroom. At least one class was in session.

In the late afternoon Eklavya and I left the vehicle behind and walked over the sand toward a remnant of the river, still flowing even now that the water had been redirected into the embanked channel. The people said the flow was created by seepage from the embankment behind us to the west. They tried to find a boat to take us over to another village just on the other side of the water. Villagers were wading across, saris and pant legs hiked up to keep them dry. The water was only up to their knees; we contemplated joining them.

Eklavya talked to some of the young farmers from nearby Basantpur village. Suresh Mandal recalled the day of the flood. He said late in the day he had gone out to protect his vegetable garden, pointing to the sweep of sand behind us. They had heard about the breach in the early afternoon, but there had been similar warnings in previous years and the water had not come, so they didn't leave this time. Around seven in the evening, as he tended to his vegetables in the dim light, he felt water lapping at his ankles.

The water kept coming. It was ten days before the army came to evacuate them. In the meantime they waited on rooftops, then later took refuge on the irrigation embankment south of here. The villagers breached the irrigation canal to drain some of the water. The floodwaters had hit the embankment, then flowed back toward their villages. Eight feet of water, half a mile wide, covered their homes.

Suresh blamed the government in Nepal for the breach, even though India is responsible for the embankments. Perhaps the government in Delhi seems even more remote to him than the government across the border in Nepal. "Why else did it breach when there were only 4,000 *cumecs* of water in the river?" he wondered. It's a good question, because a high flow in the river would be more like 25,000 cumecs, or cubic meters per second. But the Maoists or whoever else might be in power in Kathmandu are not to blame here. Bad construction and bad maintenance are the more likely culprits.

Others have tried to attribute the flood to climate change, an explanation Ajaya Dixit dismisses as "baloney." On both sides of the border misinformation and rumor lead to conspiracy theories. Some people in Nepal even think the neglect that led to the breach was a deliberate effort by Indians to force the issue of building a controversial dam below Barakshetra to control the Koshi.

While Suresh was talking, a young farmer named Lal Dev Mandal Mohunpur yelled across the river to another man. Eklavya said he was telling the man on the other side of the water to stop fishing on his land. Lal Dev's land was now in the riverbed, covered with water, but it was still his land. Eklavya teased him for trying to keep others from fishing in the river, but later he found out the young farmer had come up with several thousand rupees to buy fish fingerlings to stock a little pond in the river and make a living.

This was the kind of entrepreneurial spirit Eklavya wanted to support, to nurture before it got beaten down, or before the man migrated out for work as a laborer, his family growing to depend on aid as a way of life. "I saw the character of our campaign in him," Eklavya said. "Try and fail and try again. Experiment. If you think whatever you do has to be correct you'll never do anything."

Eklavya said that even in the remoter villages away from the road he saw the dependency beginning, people trying to figure out how to manipulate the disaster relief to their advantage. He didn't blame them. But he wanted to understand how and why it happened so he could perhaps steer the community's leaders in a different direction, before aid became a way of life, self-perpetuating and self-defeating. He thinks the energy spent figuring out how to squeeze the maximum amount of relief out of the aid system when the tap is open can be rechanneled into entrepreneurism.

"I admire these people," Eklavya said, as we walked back across the sand to the jeep to make our trip south for the night. "Their resilience, their ability to survive." But still he was worried about what would happen when the government cut off the money and grains it had given out since the flood.

"I fear that might lead to a very distressed situation. These people have lost everything. Basically we're telling them to start from scratch. And the psychological impact of the flood has not been addressed. One of the farmers said to me: 'What can I do? My wife is slightly mad. She just cries.' So the impact is far more than meets the eye." That was why he was so intent on helping at least some of the three million people in this newly devastated area tap their own talents and resilience before they slipped into the cycle of migration, malnutrition, and disease that prevails in so many other areas of north Bihar.

"It will happen again this year, somewhere. It is guaranteed," Eklavya predicted.

The pop devotional music blasted the quiet the next morning. Eklavya left at seven to return to Birpur. He planned to cross the river and walk over the acres of sand to the villages we had seen in the distance the previous evening. He guessed that the more remote villages might have missed out on extra help from the relief agencies and thought they might be fertile ground for creating cottage industries or experimenting with different crops to keep the men here with their families, rebuilding the land and their lives.

While Eklavya returned to Birpur I stayed in Supaul. I had asked to see the area inside of the Koshi embankments, where Eklavya and Chandrashekhar had been working to introduce safe-drinking-water techniques for the past few years. Here, inside the embankments, people have coped with floods every year, not just every forty, or twenty, or ten. Their problems have been chronic for half a century.

Despite the desperation of people in the flood-prone areas of north Bihar, Eklavya believes many problems that beset them can be solved using local resources, and that people can survive floods with the right training and planning. He says some aid workers are presumptuous when they come here thinking to teach people how to survive. "People who have already been surviving an almost impossible living situation for half a century! What you can do is help them organize their approach, help them make their survival mechanisms more sustainable, more effective. They've been ignored by the government for all that time and so they have become self-reliant, which is the positive part."

Life inside the embankments of all the jacketed rivers in Bihar is marginal and anybody who had much choice got out long ago. Supposedly the Indian government compensated people for their houses after the embankments were built, but not for their land, though in some cases they offered other land for farming. Cash helped some people relocate, or build a house outside the embankments.

Some farming families found that the land the government provided lacked any source of water, so they returned to their own flood-prone land. Others said "this is our land, this is our livelihood, we must stay here." So they stay in their villages for as long as they can in the summer monsoon; the higher the ground the village is on, the longer they can stay. The access roads flood so they are often stranded on little islands. Sometimes they travel by boat to get into town to live with relatives or live on the embankments until the water subsides.

Here inside the embankments Eklavya had decided to follow his mentor's advice with a simple but crucial project: to help the people in some of the regularly inundated villages provide themselves with safe water during the flood season. He had enlisted Chandrashekhar and his group to teach the villagers how to harvest rainwater and how to build wells with walls high enough to keep out dirty floodwater. These were old technologies that people here had lost because they had come to rely on hand pumps to extract shallow groundwater. During the floods, these pumps were typically under several feet of water or damaged by the flood, and wells with low walls were contaminated with floodwater. They also taught them how to make simple and inexpensive water filters with clay, sand, and charcoal. With safe drinking water during the flood season, people had a chance to avoid intestinal and other waterborne diseases.

Chandrashekhar and his colleague Suman, my designated interpreter, directed the driver down the bumpy roads outside town to where the first embankment was inaugurated in 1954. We walked over to see the historic spot, marked by a low pillar that looked to me more like a large gravestone.

When we returned to the rented vehicle an altercation developed between the driver and the rest of the group. Apparently the driver was refusing to go any farther, having just learned we wanted to go inside the embankments. Perhaps he wanted more money because of the potential damage to his vehicle on the rough roads. Finally he agreed to go on, though I was not sure why. We veered off the

Raised water well inside the Koshi embankments

main road, up over the embankment, and down onto a thin strip of dirt elevated above the surrounding fields. The driver maneuvered around the deep holes as best he could. As we crossed the embankment Suman told me that people build thatch huts here on the embankment and live there for three or four months a year when the water is deep. When the water subsides they return to plant their *rabi* crops. The water does not necessarily subside quickly.

At the first village we visited, sixty-five-year-old Sukdev Yadav told us he had been a farmer all his life. When he was young he also worked as a laborer, helping to build the very embankment that would make his life even more difficult. He said he didn't get much money for the work. He wore a *dhoti* and nothing else over a thin but strong body. He was balding and his dry, dark skin was slack; his ribs were visible and a few of his teeth were missing.

I wanted to ask this man more questions, but Suman didn't understand what I wanted, and people were crowding close around us, making it difficult to talk. Suman was well-meaning but unable to convey to me with much nuance what we were seeing, and he understood only half of my questions. Communication became impressionistic. My trip into the land between the embankments was sometimes like watching a complicated movie with bad subtitles.

As we left this village two men piled into the dusty luggage area in the back of the vehicle. Their names were Ahmed Hussein and Mohammed Yasin and they were going to take us to see their village, Karnpatti. We were about forty-five miles below the barrage here; seven miles separated the east and west embankments. The jeep stopped at the edge of the water, where one of four narrow arms of the rediverted Koshi was slowly eating the land, scalloping the edge of the ground we stood on.

Ahmed and Mohammed pointed in the distance across the broad floodplain to their village, which they can reach by boat. They said people had enjoyed the last summer because there was no flood. The Koshi was other people's problem for a change. They were able to plant a crop during the monsoon; but they had recently lost their winter wheat crop when the engineers changed the Koshi's course. They said no one warned them the water was coming so soon.

We left the edge of the river and crossed through green fields of mustard and wheat stretched out on both sides of the dirt road. These were signs of a pretty fine year here inside the embankments. Having had their first respite in forty years, residents are not happy about the embankment being repaired. From the car windows we could see many other small villages scattered over the floodplain, surrounded by clumps of trees and banana plants. These villages are among the 380 settlements, with a total population approaching one million people, located between the eastern and western Koshi embankments. The people live on and farm a quarter of a million regularly flooded acres in this limbo between the embankments.

We stopped near a field where half a dozen women in bright orange, yellow, and green saris were weeding. Suman said landowners prefer to hire women for this work because they can pay them less than men would demand: thirty rupees for a full day's work, less than a dollar. Many of their husbands have left for work in the cities in any case.

There are no secondary schools inside the embankments, but there are primary schools. Chandrashekhar went into a classroom in one of them and led the youngsters in a kind of song, a call and response chant. It seems they always sang it before meals. The gist of it was about sharing food together, sharing everything, not leaving anyone behind. Just before Chandrashekhar went in to take charge of the class, a male teacher had stormed out, perhaps driven to his wits' end for the moment by some children's bad behavior.

From the outside, the village of Dumaria Musahari looked green and pleasant. I knew the word *musa* means rat, but did not make a connection between that and the name of the village.

When we arrived at this village a man greeted us and took us around to an open area under a tree. Boys brought chairs for Chandrashekhar and me; women spread a cotton covering with blue stripes and bold yellow and white flowers over a platform for the other members of our group. A new house was under construction nearby, the lattice of the roof awaiting its final layer of thatch.

Forty families lived in this village. Twenty-five young children gathered around to stare. They looked away when I smiled at them, but later I saw them teasing each other and grinning. There is a primary school about a mile away in another village and the older children attend. But it seemed the women did not want to send the younger ones that far away.

The women and girls wore bright-colored saris and dresses; some of them held babies. There were few men, mostly teenage boys. My guess that the men were away, working for wages, was soon confirmed. More than half the men, even in this productive season, were away looking for work. Those who remained were working land owned by people from a different village.

Since the embankments were built, this village has been completely flooded four times. Villagers moved their homes to higher ground each time. The land around the current site has been steadily eroding each flood season, but this past monsoon there was no erosion, and the land was covered with green crops. The people here were happy to see the river go away and leave them alone for a while.

Half a dozen women sang a traditional song about the Koshi, perhaps an ancient one, from long before the embankments were built. Suman translated the lyrics for me. "We worship Koshi as our mother. Your water is very clean, it flows very fast. We are your offspring. Please don't hurt us, because we are your children and we worship you. You made this land. Crops can grow here with your help. We survive with your blessing. You come only in the rainy season. The rest of the year you sleep. We would like to worship you with a boat of gold, but we do not have one. Stay with us always. We bless you in our hearts."

After the song the man who greeted us, Ram Dev, took us around to another part of the village to his house. Scruffy chickens clucked and scattered as we walked. Dirty puppies trotted around hopefully, their ribs sticking out and their bellies concave. Flies descended on us.

Ram Dev had been working with Chandrashekhar on rainwater harvesting and the *mataka* water filter. I asked to see where they kept their filter. They started to bring a disassembled one to show me. I learned they weren't using it in the winter because the clay jar kept the water too cold. This time of year they drank the groundwater without filtering. We sat down under a shelter. Boys brought us all plates of beaten rice and yogurt and sweets. I was full from the sweets and bananas we had on an earlier stop half an hour before. I asked Suman to tell them to please give my food to one of the children.

Later I learned the people in Dumaria Musahari were *musahars*, near the bottom of the group of castes formerly called "untouchables," now commonly called *dalit*. There are more than a million musahars in north Bihar, the name having apparently derived from the group's practice of catching rats to eat. Although officially castes have been abolished and discrimination based on caste is now outlawed, the aversion to the members of this caste persists. The Indian government, as part of its affirmative action efforts, refers to these groups as the "scheduled" castes and both the state and central government have allocated money to improve their lot.

Ironically, around the time of my visit, the government of Bihar had decided to turn a vice into a virtue and help some musahars raise rats commercially. Apparently in parts of north Bihar the meat is considered a delicacy, tastier than chicken. Some activist musahars deplored this, saying they wanted to learn how to use a computer mouse, not raise rats.

That afternoon I did not know about this historical discrimination. These villagers were, as far as I knew, simply among those who had been unlucky enough to have houses or land inside the embankments, and they seemed sufficiently marginalized by that happenstance alone.

Back in the vehicle Chandrashekhar and the driver were having another altercation. Apparently the driver was again trying to extract more money for the afternoon's ride. I wanted to simplify the whole thing by just paying him some extra myself, and quietly offered this to Suman, who gently declined.

When I learned that the driver had grown up inside the embankments, I wished I could talk to him more. He said he now lived with his wife and two children in town. Feisty as he was when talking with the Indian men, he was very pleasant to me and tried to explain to me what we were seeing with his limited English.

We continued our ride over the uneven embankments, passing through a shopping crossroads, a cluster of thatch huts where the embankment intersected with a spur that jutted out into the floodplain. There were booths selling food, candy, cigarettes, and SIM cards for cell phones, as well as a little pharmacy selling drugs. All this construction is illegal because the embankment is government property.

We went farther down the embankment and got out. Chandrashekhar told the driver to swing around and meet us at the crossroads again. We walked on down the dusty road, soft silt filling the crevices between my toes as I waded through it in sandals. I was thoroughly shaken up by the rough ride, and increasingly weary from the blur of confusing information. I wasn't sure where we were going or why we were walking, but it felt good to be out of the vehicle, on my own feet. My attention was lapsing.

Chandrashekhar paused next to two children sitting on the ground at the edge of the embankment. One of them looked a little strange, as though she might be mentally ill or mentally retarded. She could have been six or eight or ten. Her face looked old somehow, a little wizened, a little vacant, her eyes on the verge of fear. She was thin and her head was balding; one of the bare patches of scalp was streaked with mud. Her dark skin was a little chalky, or just covered with dust; she wore a torn and dirty dress with short sleeves. Chandrashekhar said "look at this," shaking his head slightly from side to side. He looked exasperated, sad, and worried. I had no way to ask him the kinds of questions I wanted to ask: what he knew about the child, whether the hair loss I saw was a symptom of malnutrition or some other disease.

We went a few yards farther on the embankment and then down a little path into a small village. Suman translated for Chandrashekhar, pointing to its one-room thatch structures, elevated slightly above the ground on stilts. He said up to ten people live inside each of them. Inside one hut a man was sitting on the floor. The room was just a low shelter for sitting or sleeping, one could not stand up inside. He stared out at us blankly, with an expression similar to that on the face of the girl who was losing her hair.

A group of women and children gathered around us and followed as we walked briefly around the small village. Suman pointed to an even smaller structure, a thatch tent, right on the ground. Whoever built the dwelling had even fewer resources than the people who had built the low one-room cottages on stilts with wooden planks for flooring.

They told me the village had been here for twenty-five years; twenty-five years ago these families lived inside the embankment, where they might have earned money doing farm labor. Now they have no place to go. The government could throw them out at any time.

These people were also *musahars*, less fortunate than those of Dumaria Musahari. Chandrashekhar's group, under Eklavya's guidance, had introduced monsoon season rainwater harvesting here so the people can have safe drinking water when everything else is polluted.

The village stunned me so much I forgot to ask Suman its name. I didn't think of trying to take photos, though usually I did so. I remember one young woman amid the small group of villagers. She was holding a baby. They both looked fairly healthy. The baby was small, its skin dark and shiny. Its head and face were perfectly round and doll-like, its tiny face accentuated by intense eyes and very curly black eyelashes. The baby looked at me. I wanted to smile, but I couldn't. The mother was pretty, her face round like the baby's. Her expression was welcoming but seemed to say "why are you here?" Perhaps she was wondering whether my white face meant they were going to receive some new help.

We left quickly, following Chandrashekhar's lead. We walked down the embankment in silence toward where the driver was waiting for us. "What do you think of all this?" Suman asked me after we had walked for a few minutes through the dust on our way back to the crossroads and the shopping area. "It's very distressing. It shouldn't be this way," I answered. I heard impatience in my tone but didn't know how to take it back, to explain to Suman what I was really feeling. I was rattled by what I had seen, and frustrated that I couldn't really talk to him or Chandrashekhar, as if talking could somehow relieve my distress or theirs.

Back in the jeep Chandrashekhar told Suman to ask me if I wanted to see anything else. I said no. Riding back down the rutted embankment, we jostled against each other sharply. I wanted the day to end. Eklavya told me later that seeing this village had convinced him he had to try to help people here. Yet he said he has mixed feelings about showing up in these villages in a land cruiser with a driver, not to mention with a foreigner in tow. It raises expectations that something is going to happen, that someone has a solution.

All the people who live inside the Koshi embankments struggle to survive the floods, hunger, and disease. They cope each year with a diminished growing season, as well as the lack of schools, medical care, electricity, and transportation. Their chances of a better life are slim. But those who were already marginalized, even before engineers arrived with plans for flood control, those are the people who have been pushed farther to the edge, quite literally, as they huddle on the side of the embankment that was built to promote agricultural prosperity in north Bihar.

# 12

# The Engineers

Humans need to take the river itself into consideration as a
participant while intervening in its process.
—Devashis Chatterjee, Geological Survey of India, retired

I made it out of Bihar with no trouble, much to my surprise and Eklavya's. On the drive back north he was busy talking on his cell phone and I was busy talking to another man in the car about ecological toilets. I had considered trying to camouflage my pale foreignness and look a little more South Asian with some sunglasses and a scarf around my head, but before I could do so we sailed right past the guard post near the border. The guards did not signal the driver to stop.

I was back in Nepal. I tried to insist that Eklavya and his colleagues go on back to Birpur and their work, since I could speak a little Nepali and manage to get the transportation I needed. But good hosts that they were, they stayed and drank tea in Bhantabari until they could see me off in a bus.

The suffering I saw in Bihar would seem to dictate the answer: remove the embankments as soon as possible. Let them melt back into a swirl of earth and water during each year's monsoon floods. There will be some floods, but far fewer disasters, and the subsiding waters will leave rich silt to replenish the land. That is Dinesh Mishra's solution. He knows it's not likely to happen in his lifetime.

As long as the public can be made to believe that what happens in Bihar is a natural disaster, probably little will change. Backing out of this man-made disaster, even with the visionary leadership Dinesh Mishra hopes will come someday, will be monumentally difficult. Even if the government decided to breach the embankments and let the rivers breathe, what about the highways and the rail corridors? Everything would have to be replanned, reengineered, retrofitted. And the work would have to be done well, not shoddily as has so often been the case in this region. People in the flood-prone areas may be trapped for another few generations; they may be out on the roads and railways all over India looking for work, their families sometimes still perched on the very embankments that contributed to the ruination of their lands.

Admitting that the embankments were a mistake is tricky for the government. There are constituents who want to keep them, for one thing. And many parts of

Bihar are not flood prone, so it's not an issue for them. Building a statewide consensus would be difficult. And technocrats are invested in technical solutions. Thus when the embankments breach, they are repaired; when the riverbed rises, the embankments are built higher.

After the 2008 flood there were hints that officials were not entirely convinced of the wisdom of just putting things back the way they were. One official said "we don't have permission to think otherwise." The chief minister of Bihar, Nitish Kumar, indicated that he was sympathetic to some of the arguments put forward by opponents of the embankments, but he was a politician and subject to many pressures from many sources, including, certainly, the state's engineers. I had hoped to meet with Kumar; Dinesh Mishra's friends tried unsuccessfully to set up an appointment for me with this busy chief minister.

Since before India's independence, and thus before any of the embankments, there has been talk of a dam on the Saptakoshi in Nepal as the ultimate answer to floods in Bihar. Some engineers thought the embankments built in the 1960s were only a temporary solution until Nepal and India could agree about a dam. Even if the dam were to help, it could only be a partial solution since the Koshi is just one of the powerful Himalayan rivers that traverse north Bihar on their way to the Ganga. Engineers and politicians on both sides of the border say this dam would control the river's water and sediment and keep the embankments from breaching. The dam would be built in the foothills of the Himalaya just up the gorge from the Hindu holy site at Barakshetra. The dam was first proposed in 1937 and has been talked about ever since, but nothing concrete has yet come of the idea. The 2008 flood on the Koshi may finally push authorities to agree about how to make it a reality. Indian officials and members of the new Maoist government in Nepal met to talk about the Koshi dam after the 2008 flood: thorny questions about how to build it, who will pay for it, who will control it, and who will reap cash profits from it remain to be settled.

A dam was the last step in attempts to harness another tempestuous river in India. Built in Orissa, a large state on India's eastern coast, that dam's history provides a rather gloomy precedent for what could happen in Nepal and Bihar.

The flood-control strategy of the colonial British government in Orissa was dictated by a desire to extract more revenues from the land via taxes, according to Rohan D'Souza, an historian at Delhi's Jawaharlal Nehru University. To collect more in taxes the colonial government needed to increase farmers' yield, and they wanted to standardize that yield and make it predictable. They thought they could accomplish all this by regulating floodwaters, thus insulating the land from the variability brought about by the natural action of land and water, hill slope and rain. In this process the annual flooding of the Orissa delta was redefined as a natural calamity rather than a hydrologic process.

From the early nineteenth century onward the British built embankments on the Mahanadi River and its tributaries, rivers that originated in the hills of India's

Eastern *ghats*, not the high Himalaya. The British had been reluctant to try to tame the Himalayan rivers, but seem to have thought the Mahanadi was not so capricious. D'Souza argues in his book *Drowned and Dammed* that the resulting dislocation of people from their traditional and relatively harmonious relationship with the land was a requirement of political and economic control.

A flood in 1927 "was a catastrophe waiting to happen," according to D'Souza. "The haphazard embankment construction and the canal lines had substantially altered the delta's natural drainage pattern." As in Bihar, "some lands were dramatically sunk below the river's flood line." The flood-control measures had made the land and people in the delta more vulnerable, and thus more than ever they depended on structures to protect them from the increasingly intense floods.

After the 1927 flood, a government-sponsored investigation concluded that the flood was the result of misguided flood protection that thwarted the natural process of drainage and land building in the river delta. But the investigating committee avoided its own conclusions and said that the embankments were not to be removed.

D'Souza says removing the embankments would have been anathema because the economy and the social relationships that had evolved during a century of embankments in the Orissa river delta were completely different from what had prevailed before. Most strikingly there were now two sharply divided zones, those "protected" and those not protected by the embankments, which created two opposed communities, similar to what critics of the embankments in Bihar have deplored. Strong vested interests wanted to keep the embankments, and taking them out would have exacerbated flooding for some people.

Thus there was instead a ramping up of engineering interventions, like treating a drunk by giving him more alcohol. Embankments were made higher and stronger. Later a big dam was built on the Mahanadi as India imported the idea of "multipurpose river valley development" from the United States. The Tennessee Valley Authority (TVA) became the model for flood control in India. The dam went into operation in 1957.

A devastating flood during the August monsoon of 1982 disrupted the lives of ten million people. Eight thousand villages were submerged, embankments were breached in hundreds of places, roads were washed away. The dam did not, after all, banish floods forever. To this day, TVA remains the model for many projects in India.

The senior engineers of Bihar who have watched the state's attempts at flood management over the years may disagree on details, but all those with whom I spoke support engineering interventions.

Dr. T. Prasad says he's not opposed to embankments. A retired professor of engineering, educated in India and the United States, he used to teach how to build them. But he says exclusive reliance on them is "against science." He is adamant about the lack of science in the planning of Bihar's embankments: "it was a quackery of engineering science."

He says the current problems in Bihar are the result of letting politics, rather than science, dominate water resources. A compact, precise man with gray hair, he grew up near the Nepal border and like the other children in his village looked forward to the annual floods that came on gradually and steadily receded. He says before the embankments were built his district was productive, but now it imports rice instead of exporting it, and fields stay inundated for months.

Prasad doesn't trust politicians to have any sympathy for the suffering of the poor farmers in Bihar, but he continues to believe technology has the answer for that suffering. He has long advocated a dam on the Saptakoshi in Nepal as a way to control the river's annual flood. If there were a dam in Nepal the embankments would not breach because the monsoon waters could be held back and released gradually, he believes. Embankments that were no longer needed would eventually wear away; the few that were still needed could be maintained properly. Short of this more scientific management of the Koshi, he sees no solution to floods and poor drainage in Bihar.

He contends the first step, not the last one, should have been a dam in Nepal, and then embankments could be built only in the most vulnerable areas because "if you prevent a river from flooding you are killing the drainage function of the river." It's not too late for dams in Nepal to be the solution, he contends. "This will regulate water and help flood control and irrigation in India," and make hydropower that Nepal can sell to India, which he thinks will catapult that poor country into prosperity.

The ceiling fan in Prasad's office went off while we were talking and I felt a prickling of heat on my back, even though the morning was still relatively cool. Electricity failed frequently in Patna. Bihar has a deficit of electricity, as does Nepal, but the power from a dam on the Koshi would probably go to Delhi first. India would likely build and control the dam as it has the barrage at the border and the embankments in Nepal.

C. P. Sinha is another former engineering professor in Bihar, who like others went by his last name preceded by initials only. Pushpa and I went to visit him at his home. He said embankments were a temporary solution for the Koshi, chosen when negotiations with Nepal over building a dam on the river got stalled half a century ago. According to Sinha, embankments are never a good solution on a river that carries so much silt and shifts its course. But because Bihar's rivers all originate in Nepal, India is helpless to control the water resources.

He responded to each of my questions with a loud, drawn-out *"huh?"* and looked steadily at Pushpa instead of me when he answered. His English was excellent, and his hearing seemed more than adequate. Pushpa, who accompanied me wherever I went, just in case he might need to translate Hindi for me, told me later that *huh?* is even more impolite in this part of the world than in mine. Still, Sinha was considerably more polite than another engineer, who came to the hotel to talk to me at the request of Dinesh Mishra's colleagues. He declined even to shake my hand, instead shaking Pushpa's and treating him like an old friend. His comments echoed those of the other engineers.

I'm not sure which aspect of me was most offensive: female, American, or journalist asking questions about Indian flood-control policy. But having gotten the kernel of information I sought from C. P. Sinha, I just smiled and encouraged Pushpa to ask some questions as I dipped into a lovely tray of sweets and fresh fruit while sipping my tea, all made perhaps by some female hand in the house.

As we were leaving, Sinha, a tall man who was wearing a loose white *salwar kameez*, proudly showed us a wooden bookcase that covered the entire far wall of the sitting room. It was a handsome piece of furniture made of blond hardwood, at least seven feet tall and about eight wide. It held family mementos and photos. He said he had designed the carving across the top, which depicted the water cycle. The semi-abstract bas-relief of sun, clouds, rain, irrigation water, and an ear of wheat reminded me a little of the WPA style of the 1930s. It was a lyrical view of the land before embankments. I saw no dams etched in the wood.

I also spoke with S. C. Sinha (no relation as far as I know), who retired from his post as chief engineer in Bihar's irrigation department in 1994. He says the work of sixty years has led to only 40-percent flood protection for north Bihar, the most flood-prone area in India. He agrees the plan was technically flawed from the start. Embankments were cheap and easy and at the time they made the people feel safe. But embankments were built for small flood discharges; they cannot be built for the so-called hundred-year floods, let alone larger ones. The people had a false sense of security for a while; now they live in fear because the embankments have breached so often.

Still, Sinha said that people in north Bihar lived miserably before the embankments and that's why they were constructed. A year before the Koshi flood, when I spoke with him, he said the embankments had prevented the river from regularly shifting its course over the swath of its floodplain and so had allowed for development. But as everyone discovered later, embankments essentially destroyed the very development they had allowed, leading to the crumbling houses I saw in Supaul in 2009—the remnant of the garden that Supaul had become for half a century. Nonetheless, in an e-mail after the Koshi flood, Sinha defended embankments, saying "The present failure is not the failure of technology, it is a failure due to managerial and human efforts."

Sinha discounted the problems of the people living between the embankments, people like those I saw near the Kamla and later the Koshi. He said local people and the organizations that work on their behalf have exaggerated the problems created by the yearly floods.

He pointed out that far too many people now live in the "protected" floodplain to think of letting the embankments erode and the people go back to living with floods. When people lived with the floods there was a small population in Bihar, and they resided away from the river. Now Bihar's population is approaching a hundred million; it has tripled since the embankments were built.

Even with a dam some embankments will still be necessary, Sinha says, to keep the water released from the dam inside the river's channel. And managing a dam on the Koshi will be challenging. Generating hydropower and storing water for irrigation call for keeping the reservoir full. Flood control requires keeping it empty in preparation for receiving floodwaters. A dam on the Koshi won't be much help to Bihar unless its priority is flood control. But to be attractive to Nepal, the dam's host country, the priority would be hydropower. Sinha says the dam could be managed to the benefit of both countries; but the reservoir would have to remain partially filled to ensure continuous power generation, so it would not be available to hold back the high discharge of the Koshi. Thus a dam would merely reduce the severity of floods, not prevent them. This incompatibility of interests is one of the reasons the project has been stalled for decades.

The Barakshetra gorge where the long-discussed Koshi dam would be built is about a mile wide. The reservoir would be narrow and long, stretching down the valley below the confluence—the *triveni* of the Arun, Tamur, and Sun Koshi—for about thirty miles. Most reservoirs are built for a single river. The reservoir on the Koshi would capture all three rivers—a tricky proposition, given their unpredictable and varying sediment loads.

Ajaya Dixit says he sees no way to avoid such a dam trapping 120 million cubic meters (150 million cubic yards) of sediment or more each year. He thinks the dam might have a life of thirty-five to forty years and then be useless because it would be filled with sediment and unable to hold any water. Building it would take fifteen to twenty years, even if nothing goes wrong, he says. So the dam would not offer flood protection in the short term or the long term, just a little benefit in the medium term. It would cost a lot of money, displace tens of thousands in Nepal, and there could still be floods from all the rain that falls below the dam.

Some of those who fear that a dam is destined to fail, just as the embankments have, suggest another solution. They want to study the possibility of directing the Koshi back into some of its old, natural channels. The strategy would still involve infrastructure, but would not be so at odds with the Koshi herself and where she wants to take her sediments. Floods from several river channels would theoretically not be so dramatic as recent floods because the water could spread across the land. But there are still the millions of people who live in the floodplain. The "social engineering" required to effect this solution would be daunting, given the limitations of the governments on both sides of the border.

As an engineer himself, Ajaya Dixit says he's not against infrastructure but believes many things can be done right now to help people in north Bihar. His organization, Nepal Water Conservation Foundation (NWCF), has done some cost-benefit analyses in another flood-prone area of Bihar along the Bagmati River, the sacred river that originates in the hills above Kathmandu. Down in the plains, after it receives monsoon rains from the hills and *tarai*, the Bagmati can become tempestuous and prone to flooding.

Researchers at NWCF have assessed the value of various interventions, such as embankments, raising hand pumps, and creating warning systems. Ajaya says people-centered, small-scale, decentralized activities have a higher rate of return than large-scale, government-initiated projects. And financially, they are less risky in a region where climate change is always the joker waiting in the deck.

Ajaya says there should be a robust warning and forecasting system. But first officials have to acknowledge the threat; it's not possible to have a warning system if you don't admit that embankments can breach, which is why the Koshi flood of 2008 was so much worse than it should have been. People and authorities were not prepared in any way because of the pretense that floods were under control.

Many other manageable changes will help, says Ajaya; for example, building houses on stilts, or at the very least, schools and hospitals. Like Eklavya Prasad, he says the people in the region must have other ways of making a living than agriculture alone. A series of small interventions, implemented slowly and steadily, like houses on stilts and alternative livelihoods, are the solution. There is not one big thing, like a dam, that is going to solve the problems of North Bihar—of that Ajaya is sure.

Neeraj Lab, a development consultant who specializes in natural and human-induced disasters in India, has been focusing on Bihar since 1990. He says he still doesn't understand how people manage to live in some areas of North Bihar, like those I saw. They survive on the barest minimum, perpetually in a state of mal-nourishment and ill health. "It's a slow death," he says.

The embankments never should have been built, he says, and believes that eventually the people of Bihar will have to learn to live without them. But in the meantime there is no blanket solution for North Bihar, only what he calls "credible micro-interventions," like the changes Ajaya and Eklavya recommend.

For the most part Neeraj thinks the groups working in North Bihar have not tackled structural issues, like empowering the people to help themselves. Their donors want accountability, good value for their investment, and that means easily measured and often short-term projects. The yearly cycle of disaster and disaster relief fits this description but feeds the structural problems. But there are exceptions, like what Eklavya and Chandrashekhar are doing in Supaul. One of Dinesh Mishra's good friends, Satyendra Prasad, who runs an organization called Nav Jagriti, showed me another.

Satyendra took me out to the district in which he works, Saran, a district surrounded by the Ghaghara, the Ganga, and the Gandak, another of the big rivers that come over the Nepal border. Though not as tempestuous as the Koshi, the Gandak has also been embanked and there is an Indian-owned and-managed barrage just over the border. There is also an irrigation canal, the Gandak Canal. It has brought water to some areas, but blocked drainage from others. That was the case in the area Satyendra took me to see, where his group plans to help restore the old drainage methods.

We walked through a village named Bhatha to see a narrow drainage ditch, created in the Mughal era, around 1600. Until twenty years ago this ditch provided good drainage for monsoon waters. The ditch is one of myriad small-scale water management structures that people in the Indian subcontinent built over the centuries to reroute, trap, or drain water.

An older couple in Bhatha, Mahendra Singh and his wife Demati Devi, said they used to grow both wheat and rice on their land, enough to feed their family for the whole year. Then the big irrigation canals were built, blocking the flow from the small canal. Eventually the canal filled with sand, could no longer drain monsoon rain, and part of the land Mahendra and Demati had farmed became waterlogged. The ditch had been six to ten feet deep, but when I visited it was less than two and filled with weeds and leaves. Mahendra and Demati Devi were ready to help clean out the canal so floodwaters will drain and they can grow crops again. If the canal is deep again, they said, it will drain floodwaters and also carry irrigation water in the dry season.

Satyendra took me to see another village project that was almost complete after a year's work. The villagers had cleared a wide, shallow drainage channel to carry water away in the rainy season. The drainage channel, about ten feet wide and two feet deep, will allow water to pass through to a pond a mile or so downhill and then on to the Gandak River about four miles away. Several thousand people from sixteen different villages helped dig the channel and cart away the sand and grass for their own use. They will be able to farm the nearby fields here for the first time in thirty years. In return for their work they received a little money and some clothing, along with constant reminders that it was their own land they were maintaining. For now they may be working for money, but Satyendra believed after the next monsoon they would see how the water drains away and they will never let the drainage become blocked again.

Empowering the village residents to help themselves was every bit as important as the drainage scheme, perhaps even more so. It's one way to break the cycle of recrimination that characterizes public life here.

On the way back to Patna after visiting Saran, we stopped on a bridge over the Gandak River. It was a moonless evening, the dark water sliding beneath us barely visible. The small swells in the river caught an occasional glimmer of light and reflected it for an instant. Only the sound of the river was constant: a sleek, animal murmur.

As we leaned over the railing of the bridge, an educated man who worked with Satyendra and was getting me safely back to Patna lamented that during the monsoon the Nepalis released water from dams across the border, causing floods here. When I pointed out there were no dams upriver in Nepal, he was surprised. Such propaganda had become so much a part of the fabric of common wisdom, even a relative expert in the water issues of Bihar could be mistaken about what lay just across the nearby international border.

People on both sides of the border have absorbed erroneous beliefs from incorrect TV and newspaper reports over the years, reports that fail to make clear there are no dams in Nepal on the shared rivers, just the barrage at the border. Still

Indians hear that Nepal released water upstream, which caused flooding in Bihar. Reports often inform people in Nepal that India wouldn't let water flow out through the "dam" so it backed up and flooded the Nepal *tarai*. And so it goes. The Bihar government blames the central government for not negotiating with Nepal, and they both blame Nepal's government. And many have blamed Nepal's people for their overcutting of trees in the hills, which is in fact not the problem. Trees were cut down, others replanted. The amount of sediment the Koshi brings down has not varied much over the years. But the finger-pointing is endless.

It's almost ironic how much Indians downriver want a dam in Nepal in hopes that it will protect them from vicissitudes of floods and droughts. "When we really have dams—think how much we will be blamed then!" says Ajaya Dixit.

On the way to the villages on the Kamla River that first day I spent in rural Bihar in 2007, I perhaps glimpsed something about how things work in this benighted state, and why change may be slow in coming, though not impossible.

We had traveled north out of Patna for more than six hours on more or less paved roads, then we finally left the bumpy pavement and proceeded by a dirt road to that first Kamla embankment I saw.

We drove down parallel to the embankment and turned left onto a one-lane combined rail-vehicle bridge to cross the river. Two thirds of the way across, our driver Chandra stopped. I was sitting in the back seat trying to stay alert, calm, and friendly despite the rising heat. I was somewhat dazed and disoriented already, unsure where we were going and how long the grueling ride would go on, let alone what our destination held.

Chandra got out. A small red car coming from the opposite end of the bridge blocked our way. The driver of the red car sat behind the wheel yelling at our driver, who yelled back, gesturing with his right hand as he leaned out the window, telling the other driver to back up a couple of hundred yards to the side he came from. Our driver was lean and handsome with dark, thick, wavy hair; his full beard framed full lips. He looked like someone from an Etruscan wall mural, but he was Bihari and spoke the modern version of the ancient language of Maithili. He moved with a kind of liquid grace; he smiled wryly, laughed easily, and wasted no energy even when he was angry.

Both drivers eventually got out of their vehicles and continued yelling and gesticulating. It was very hot inside the immobilized vehicle, so I started to get out for a breath of hot but less stifling air. Then I paused, not sure whether my white face would aid or impede progress. Ajaya Jha, mild mannered and soft-spoken, which many in Bihar are not, got out of the SUV at that point, saying "wait just one minute; this will be resolved."

In the back seat of the small red vehicle sat a pompous, looking older man in a long-sleeved white shirt and black vest with a *topi* on his head. Eventually he got out of the car along with another young companion who was sitting in the front with the driver. Ajaya looked on, uncomfortably. Our driver continued to look disgusted with the other driver, who kept insisting we back up. Their voices rose, gestures grew more forceful. I feared blows, but none came.

By this time a truck had stopped on the near side of the bridge, the side we were headed toward, awaiting the outcome of the struggle. Bicycles, rickshaws, and motorbikes had come up behind both vehicles. Those who could do so nonchalantly threaded their way through the gap between the vehicles and proceeded to the other side. A man in a brown uniform stopped to see what the altercation was about, but he had no authority and did not seem inclined to get involved.

Ten minutes after we had come to this abrupt halt the driver of the other vehicle continued to insist he started crossing the bridge first, though it was obvious that was not likely— even if we had driven twice as fast as he had. The officious man returned to the back seat of his vehicle, folded his arms over his chest, and sat there looking sour.

The pileup at both ends of the bridge increased and vehicles finally started honking, having so far shown great restraint in this typically horn-addicted country. People came to kibitz. Many agreed with our driver. Finally one young man on a motorbike with a young woman in a *sari* sitting sidesaddle behind him spoke to the other driver pleasantly but persuasively. He reasoned with him that regardless of who was at fault would he please back up so all these people could go about their business. The driver huffed a little more, then complied. A general sigh of relief passed through the crowd. Some face-saving had been provided.

Pushpa and I were sitting in the back discussing South Asian attitudes and the impediments to progress they may pose; Ajaya returned to his seat in front and our genial driver resumed his work. The latter eased forward toward the end of the bridge as the red car retreated. I asked Ajaya if he knew the man in the back seat of the other vehicle. He answered "public" followed by a word I did not quite catch.

Feeling woozy from the heat and beginning to relax from worrying about how long we might have been trapped on that bridge, I queried with a smile: "public mosquito?"—thinking that's what the phrase had sounded like.

Ajaya laughed heartily. He had said "public prosecutor," but from that moment the official and his ilk became "public mosquitoes" in our conversations over the next two days.

If the "public mosquitoes" are reluctant to change things, perhaps there are young men in Bihar who can find ways out of the state's economic and social impasse—like the one on the motorbike who reasoned calmly, and the farmer we met out in Basantpur village who tried to start a fish farm when his fields were flooded and buried under sand.

A strange creature makes its home in Bihar. As we were riding around the maze of embankments in North Bihar one cool morning before I was overwhelmed by the heat, I saw several dark shapes in the distance. They looked and walked rather like birds, but they seemed enormous, prehistoric. The sight startled me, but at that distance I had no reason to feel fear or repulsion, as I might have had I come upon such a creature suddenly at close range. It seemed stately and quiet.

As I watched, one took flight. It seemed to hang in the air, unhurried, the Boeing 777 of birds. I learned that it was an adjutant stork. There are two subspecies, greater and lesser. Most likely what I saw was the lesser, but even the lesser is a huge bird: it stands about four feet tall and has a seven-foot wingspan. The bird is black with a white breast, long legs, webbed feet, long orange neck, a big beak almost as long at its neck, and a bald head. The lesser adjutant likes to forage in marshy areas, like paddy fields, where it eats frogs and other amphibians and fish. Bihar has an abundance of such wetlands year-round, thanks to the waterlogged stretches outside the embankments.

Though the stork is big, ungainly, unattractive, and a little scary, it has sudden grace when it takes to the air. It seemed a good emblem of Bihar, an appropriate state bird for a land whose beauty is not easily grasped, a land that was once the resting place of the Buddha and the home of the great university at Nalanda. It was once and could be again one of the richest agricultural regions in the world, the Ganga running right through the heart of it, giving nourishment and embracing its floods.

# 13

# The Garland

The answer to nature's variability is not necessarily
trying to tame her.
—Ashok Khosla

Bundelkhand is a thirsty land. When I arrived there early in 2008, my skin—
already parched from the dry winter air of Kathmandu and Delhi—immediately
felt itchy. The cool air hit my sinuses with a prickly thud. They ached, and my eyes
smarted as moisture left them.

The land was an expanse of beige sand and rocks; beautiful, I thought, save for
a dryness so intense it made me feel a little anxious. Most of the trees were not
very tall, except for the water-thrifty "flame of the forest," with its dark green
dust-covered leaves, several inches wide. In the spring the leaves drop off and the
tree's bright orange blossoms, shaped rather like bird beaks, pop out to give the
tree its other English name, "parrot tree."

Bundelkhand is sometimes called the heart of India. It sits in the center of the
broad upper half of the subcontinent and its many ruins from the nation's Mughal
and Hindu past evoke the shifting suzerainty of pre-British India.

Most of the ancient kingdom of Bundelkhand is now in Madhya Pradesh, also
known as "MP," or "middle province." It's a large landlocked state south of Delhi;
Bhopal, the site of the devastating 1984 explosion at the Union Carbide pesticide
plant, is its capital. The remainder of Bundelkhand is in Uttar Pradesh, "UP," or
"northern province." Many would like to see Bundelkhand secede from both and
become a separate state. With a population of fifteen million, it would be a substantial state on its own. And some people believe this poor and undeveloped
region will have a better chance of progress if it is independent of both MP and UP
and their politics.

I stayed in Jhansi, a large district in the UP portion of Bundelkhand, at the
campus of a nonprofit endeavor called Development Alternatives. The group
works to help people in Bundelkhand manage water and develop small industries
as an alternative to agriculture. There was a simple guesthouse on the campus
with hot showers, which revived me and rehydrated my dry eyes and nose in the
evening. A warm dinner of chapatti and dal was delivered to my room. A cozy

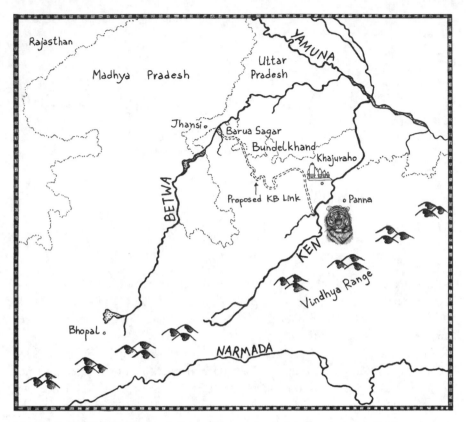

Map 6 Bundelkhand—Ken and Betwa Rivers. Created by Kanchan Burathoki

sleep followed, with the booms of a rock crusher, the horns on the road, the slamming of doors in the building all subsiding for a quiet stretch of hours.

The nearby town of Jhansi would be the capital if Bundelkhand ever became a separate state. The region's heroine is the warrior queen of Jhansi, Rani Lakshmi Bai, who died fighting for independence from the British in 1857.

This may never have been a lushly tropical land, but before intensive logging in the nineteenth century it was far more forested than it is today and had plenty of water for its people and their farms. It was in the midst of a particularly dry phase when I visited. Because of this drought there was growing support for constructing a long concrete canal to import water from a river about 150 miles to the east.

That canal would be just one of the links in a colossal chain envisioned by engineers and politicians who want to shunt the wealth of the monsoon around the Indian subcontinent from river to river, soaking up floods and hydrating parched places.

The "Garland Canal," connoting something natural woven by the hand of man into a lovely artifact, was the name given to one of the early schemes for linking India's rivers. The first such concept cropped up in the mid–nineteenth century

A farmer going to town in drought-stricken Bundelkhand

Ruins near the Betwa River in Bundelkhand

as the British Raj sought to extend its control over India's productivity. Later the "Garland Canal" sketched a sweeping series of channels uniting major rivers from Nepal to South India, moving water from one region to another over the vast subcontinent. That approach was discarded some decades ago as not feasible, but in its place eventually emerged a similar idea with a less elegant name: Interlinking of Rivers, or "ILR." This also embraced the concept of moving water from wet areas of the country to dry ones by linking the subcontinent's rivers through a grid of dams and concrete canals. Proponents even likened it to an electricity grid.

The theory behind the proposal is that some river basins in India have more water than they need and are thus "surplus basins." These basins, likely prone to floods, are located mostly in the north and east. Other basins are called "deficit basins." If engineers just link them up and move extra monsoon waters from the surplus basins to the deficit ones, India would have the right amount of water everywhere and would no longer suffer from drought and flood. A lot of concrete canals would be needed for the thirty proposed links, in addition to the many dams and reservoirs required to consolidate water in a "wet" river basin before moving it into a "dry" one. The estimated cost in 2002 dollars was 120 billion.

When I first heard of this plan I was incredulous. Such a stupendous reengineering of nature as was proposed even in the less ambitious versions of the ILR was audacious, certainly, and also anachronistic. Many engineers in the world were questioning the cost of big dams in proportion to what they achieve and the damage they do to river ecosystems and the people who live in them.

In one sense, the interlinking of rivers in the Ganges basin was what I came to the subcontinent to study. Long before Indian engineers dreamed up their plan to make cross-continental channels of concrete, the rivers of the subcontinent had already been linked by tectonic, geological, and hydrological forces. This vast and magnificent network, stretching from an array of glaciers in the Himalaya down to the Ganga and on to the Bay of Bengal, already linked glacial rivers to nonglacial ones via the Ganga. And for thousands of years that river system nourished countless creatures and a succession of civilizations. People found many ways to adapt to what the rivers brought, just as they adapted to what the monsoons brought, whether that was surplus or scarcity.

Just a few years ago there was great excitement in Delhi and some of India's states over the ILR concept. With the backing of the government and even the nation's supreme court, the plan seemed unstoppable. But soon the idea came under intense scrutiny and received so much criticism that government officials began to downplay the scheme. They said it was being investigated and would perhaps be implemented piece by piece, not as a whole. Yet the ideas continued to generate a great deal of discussion in India and suspicion in its neighbors.

After my initial reaction, I tried to find some balance. Who am I, I asked myself—an American who has been the unconscious heir for decades of profligate use of the world's resources—to criticize what India does to exploit its own resources and keep a billion people supplied with food and water? Clearly it's none of my business. But it is very much the business of India's water intelligentsia. So I discussed the subject with a number of those luminaries during several trips to Delhi.

Respected natural resources writer and former newspaper editor B. G. Verghese supported the government's aim to create "national water security." One way to accomplish that was to look seriously at the concept of river-linking, which he pointed out was not a new idea. Verghese tried to calm what he saw as unnecessarily hysterical opposition to river-linking by pointing out that inter-basin transfers of water were common throughout the world. And that's all the so-called river linking plan was—a set of possible inter-basin transfers.

Verghese said there was no difference between these transfers and what India had been doing for fifty years by building dams and irrigation canals. Some such transfers were going to be necessary because water scarcity and drought would result in migration of people if something were not done to prevent it. "Rather than have people move, you move water. This doesn't mean you're excluding other means," he continued. "It doesn't ignore groundwater management, watershed management, rainwater harvesting, demand management, or recycling." Demand management is basically conservation, which some people in India practice chiefly because of lack of supply. But none of these methods had been applied very exten sively in India at that point. Verghese was sure the government would pursue all such methods to attain water security.

Verghese adamantly supported the government's intentions in terms of river linking, but not all of its methods. He deplored the government's lack of candor with neighboring countries and its clumsy public relations at home. I managed to meet with some government officials in the water resources ministry during early trips to Delhi. They were pleasant but did not say very much about the policy beyond the standard explanation of surplus and deficit basins. They kept referring me to their website, on which I had found only sketchy information. "If there is storage, there will be water for everyone," said one, referring to the many dams and reservoirs for which the plan called.

The ILR, however limited the information on the government's website, continues to generate its own reality. Logging on to that website you might still discover a flashy front page with a map of India, sketched quickly in gold lines by an unseen hand against a purple background. From its heart emanate the names of the proposed river links, the letters growing larger as they loom out at the viewer, changing from white to turquoise: Gandak-Ganga, Yamuna-Rajasthan, Farakka-Sundarbans, Koshi-Mechi, Ken-Betwa, and twenty-five more.

B. G. Verghese estimated the entire river-linking proposal would take forty years to build if the government tackled the whole thing all at once. Even forty years seemed an optimistic estimate for such a massive undertaking. Yet when India's supreme court weighed in on the proposal in 2002, the chief justice urged the government to get going and get it done in a dozen years. In the face of opposition, the government backpedaled on a portion of the scheme, saying for the time being it would only pursue links that did not involve any Himalayan rivers, and thus would not require agreements with Nepal or Bangladesh.

But damming and diverting only on those rivers that began and ended in India eliminated much of the sought-after water from the north and east and thus much of the rationale for the scheme. The government may have decided to wait for a more propitious time before pursuing any transboundary river links, but negotiating interstate links was not going to be easy either. India's individual states have jurisdiction over water, so river linking is never going to be something the central government can just proceed with on its own. Some Indian states have been in court for years trying to adjudicate water disputes. River linking would only work if states agreed, because most of India's rivers cross state boundaries.

Gopal Krishna, environmental researcher and activist, says that without the Himalayan rivers, "the entire exercise is meaningless because it's the Brahmaputra which is surplus." The Brahmaputra flows out of western Tibet, where it is called the Yarlung Tsangpo; it takes the name Brahmaputra and flows through remote northeast India before crossing into Bangladesh, thanks to the vagaries of the borders drawn during the partition that created East and West Pakistan and India. India's river-linking plan called for diverting Brahmaputra water to the Ganga and then on to South India. This possibility continues to alarm Bangladesh. But perhaps before any such Indian plan materializes, China might divert the Brahmaputra and send water north, to the dismay of both India and Bangladesh. When China took over Tibet more than half a century ago, it may have had its eye on the Yarlung Tsangpo, one of the triad of great rivers in South Asia, along with the Ganga and the Indus.

Gopal Krishna says a single south Indian state, Tamil Nadu, is the source of much of the support for ILR. The state is desperately in need of water, having used up its own resources. Even though it is largely alone among the states in supporting river linking, it has political clout. He thinks government officials support river linking because it's a good tool for garnering votes and making people believe their representatives are doing something for them.

"We are following the same development paradigm as the U.S.," says Krishna, referring to the big river valley projects of the twentieth century that altered the American landscape. "And as long as we do that we will have the same consequences. Even though the days of big dams are over in the rest of world, India is in a time warp."

A. D. Mohile is an engineer and former chairman of India's Central Water Commission. He was director general of the National Water Development Agency, where he worked on river linking before he retired. He says it's not quite correct to call river linking a plan; it's an assembly of plans.

Though some of those who have expressed doubts about the efficacy of the ILR have championed other big dam projects, Mohile thinks many of the critics of river linking oppose it because they simply don't like the idea of development. "It's imperative that for the good of humans some water is stored," he contends. He sees it as the only way to produce enough food and electrical power as India's population surges forward, as well as the only way to help alleviate rural poverty for the majority of India's people who still have no source of income other than agriculture.

Mohile says there are no easy solutions to these problems. "They require a scientific temper and analysis. People should not jump to conclusions without deep study of the issues." Mohile himself spoke very methodically, without emotion, carefully phrasing long answers to my questions.

Ramaswamy Iyer, one of the most respected water experts in India, is equally methodical in his approach to the subject but has come to different conclusions. He is a retired civil servant who held various government positions in his career, culminating as secretary in the Ministry of Water Resources. He has since written several books analyzing India's water problems from legal and economic angles as well as environmental ones, and has outlined thoughtful approaches to solving them.

As he sees it, the river-linking plan got a burst of energy during the elections of 2002. The prime minister of the ruling party, Atal Bihari Vajpayee, announced his support for a river-linking project to bring water to needy areas. When India's Supreme Court put its weight behind the idea, recommending that the government get busy and complete the massive project as fast as possible, the Congress party did not have much choice but to support it. Water officials liked the idea of such a big project, which also appealed to nationalistic pride at a time when archrival China was busy assembling the Three Gorges Project.

Ramaswamy, originally from South India, is thin, alert, serious, and courtly in manner. His apparent caution in style belies thinking that is seen as radical in government circles. He is now shunned by India's hydrocracy. "In government, they all think the answer is in big projects, so I am very unpopular in my old ministry, more or less ostracized. It's all in the game, one takes it," he observed wryly.

"My view is that *if* there is a crisis, it's not one of availability but of management," he continued. "What I am arguing is to reverse and give highest priority to local water harvesting and management all over the country, keeping big projects as the last resort to be taken up where there is no other option or it's the best one. And to put severe restraints on groundwater extraction. I must say this is a maverick view."

Bihar is a good example of a state that doesn't want to join the scheme, even if Nepal could be convinced to do so. The comprehensive river-linking plan advocates storing water behind a dam on the Saptakoshi in Nepal, and then bringing it through Bihar to the Ganga. Bihar's government showed some interest in river linking, but not in any plans that would take water outside the state's borders. None of India's states, even those that sometimes suffer floods during the monsoon, are likely to say they have water surpluses.

"We shouldn't be talking about interlinking rivers. We should be solving the water problems of India," says Dr. Ashok Khosla, who founded Development

Alternatives, the group that works on sustainable development in the Jhansi district of Bundelkhand.

"Do the right thing and you don't need to link rivers," he says, referring to the same set of water management techniques Ramaswamy Iyer wants to see exploited to their fullest before any rivers are linked. With a Harvard PhD in physics, Ashok established the Indian government's first environment department and also directed the United Nations Environment Program. Two years after the government set about planning how to accomplish the river-linking scheme, it had divulged few details. Ashok helped form a committee of experts to allow for some exchange of ideas on river linking. By including strong proponents from government along with anti-dam activists on the committee, he hoped to stimulate public debate of the scheme.

"It's a cycle. Every thirty years somebody comes along and says there's a lot of drought in this area and a lot of water in that area, why don't we join them up? Pretty logical," says Ashok. "It has come up over the past 150 years on a regular basis. In my lifetime it's come up three times. It will be a bubble and then they'll find out it doesn't make any sense—that the cost is greater than the benefits. It slows down and then nothing is done and it becomes dormant. Like a virus, it just stays there in the system and then blows up again."

I wanted to see an area in India that could benefit from one of the proposed interbasin transfers. So I went to Jhansi after talking to the experts in Delhi in January of 2008. Ashok Khosla had said I could stay at the campus of his organization, Development Alternatives. Everyone agreed that the plan to link the Ken River (in the Madhya Pradesh part of Bundelkhand) to the Betwa River (in the Uttar Pradesh part of Bundelkhand) would be one of the first links to be tackled, if and when any of the river links were built. The two states had formally agreed to pursue the plan in 2005.

In Jhansi I found wide support for the idea, and it was not difficult to see why. If my skin and sinuses were smarting from the dryness, how were crops faring? And the animals? And what difficulties were falling to the women and children who are the ones who always fetch and carry water for household uses? And to members of still shunned castes who are often barred from the wells near their homes?

Santosh Prajapati, who worked for Development Alternatives, was assigned to be my guide. A slender young man in a black leather jacket, sneakers, and dark brown slacks that looked a little too big for him, he had a sparse mustache on his upper lip and straight black hair. His face was both kind and serious, sometimes wearing a big grin, sometimes a wistful smile. At twenty-three he was already married and awaiting his first child.

Santosh was polite, welcoming, and attentive. The day I arrived by train from Delhi, he took me to see some of the local tourist attractions first. I was equally interested in seeing the Betwa River, the proposed recipient of water from the donor to the east, the Ken River. Both rivers flow north to meet the Yamuna before that river joins the Ganga. In the midst of the drought, the Betwa was low. The water did not reach the knees of a smiling, muscular teenager standing in the riverbed in briefs to bathe. The wide bed was mainly an expanse of sandstone boulders, beige with a slightly pink cast like the nearby stone temples and fort.

Santosh had lived here all his life. He saw increased poverty and suffering for lack of water and believed the so-called "Ken-Betwa link" was a good project. He used the terms *deficit* and *surplus* that the government engineers use, though there is no such concept in hydrology.

The next morning Santosh took me to visit some villages east of Jhansi. Our first stop was Barua Sagar, a dam and reservoir built several centuries ago by the Chandela rulers who held sway in Bundelkhand before the Mughals. This man-made lake, still serviceable, would be the end point of the canal proposed to bring water from the Ken River.

The lake's dark waters mirrored the arc of the old dam itself. Its inside curve, from a distance, appeared sculpted with an abstract linear design like rows of triangles, staggered and stacked one on top of the other to form diamond shapes from the rim down to the water's edge. This design was really an array of stairways carved into the sides of the thick walls. The multiple triangular shapes were practical as well as attractive, allowing access to the water's edge from many points at the top of the dam down to the water, whether the lake was full or low, as it was now. One could easily fetch water, do laundry, or bathe. The water in the lake was so low that none was pouring over the dam and into the three stone spillways on the other side.

Santosh and I walked out on the top of the dam. A local man named Hargobind Raikwar told us that the recent monsoon had brought some rain, but before that the lake was almost dried up. Behind us the land—green with a new crop of wheat and mustard—sloped down toward the water. That land would be under water if the lake were full. Hargobind told us half of the 200,000 people in the area around Barua Sagar are farmers.

We also met Mathura Prasad atop the dam. He wore a dark scarf around his head in the morning chill; a loose tan sleeveless sweater hung from his thin shoulders over a flannel shirt. We could see a temple on a nearby knoll. The temple owned the acre of land that Mathura farmed as a sharecropper. He could keep half the wheat he produced, or five sacks, but could feed his family for only six months with so little grain. He said he went to Jhansi or Delhi to do construction work after he harvested his crop.

Mathura said if more water were available he could farm all year, bringing in at least two crops, and make a living from his own eight acres. He had not farmed them for the past seven years because of the drought. His land was not irrigated, so he had to depend on rainfall. Mathura believed a canal bringing water from the Ken River would allow him to irrigate. Even as recently as five years ago, he told us, there was enough water here for farmers to grow rice—five thousand sacks of it. Even though we were standing next to a reservoir, he said he did not know anything about ponds for harvesting rainwater, but would be happy to use that or any other means to save water.

At our next stop, near Ghughua village, women in bright saris carried aluminum jugs on their heads as they walked down a dirt path from the road where we left our rented vehicle. They had wrapped the ends of their saris around their heads; other pots or empty plastic containers the size of gasoline cans dangled from their hands or rested under their arms. One woman in dark blue modestly covered her entire face with the end of her sari as she went to fetch water.

We met Hari Om Kushwaha as we walked down the path. He told us he farmed seven acres here, which yielded a hundred bags of wheat—fifty for him, fifty for the owner of the land. But out of his fifty he had to pay expenses, so he got only twenty-five to thirty bags for his family. That amount would feed him, his wife, and three children for seven months at most. Hari sometimes borrowed money from his landlord and had to pay him back with interest from the next crop.

There was too little water in the reservoir to irrigate directly from it, so Hari pumped water from a tube well. The power for the pump cost about three hundred rupees each month, or $6.50—which for him was expensive. He said he had to pump water for three days to irrigate the seven acres. After two days he would have to pump and irrigate for another three days. One crop could dry up the well. If there was water in the reservoir, the well also had water, so indirectly they were using the stored water. Hari told us the reservoir used to hold enough water to irrigate seventy thousand acres when good rains came.

Tube wells were not always a good alternative to the reservoir, Hari said, because this is such a rocky area. Either the tube wells can't be sunk at all or they can't be sunk very deep. Now that the water table is sinking, few wells can reach it. He says once wells could reach water at twenty feet, but now they must be sunk ninety or a hundred feet.

Like Mathura Prasad, who sharecropped the temple's acre, Hari also had land of his own. But he had not farmed his own two acres for five years because his well had gone dry. The land was his father's, which the latter had divided among three sons. He and his brothers all had to find other work. He worked locally while his brothers went to Delhi. He said a national rural employment act guaranteed jobs for farmers like him, but corruption kept its benefits from reaching people. Some of the farmers could get work but had to give half the hundred rupees they might earn to the labor boss.

The well Hari used to irrigate the land he sharecropped is also the source of drinking water for the whole village of Ghughua, which explained the parade of women I saw as we walked down the path. When the well runs dry, the government tankers come to give the villagers free water.

A small, whitewashed shed at the end of the dirt path housed the pump. Fields of green wheat and yellow flowering mustard surrounded it. Bright red letters painted on the side of the shed announced "NWDA" and "KB LINK"—National Water Development Agency and Ken-Betwa Link—heralding the central government's plans for the inter-basin transfer. The National Water Development Agency drew up preliminary plans for the Ken-Betwa link in 1995, aiming to irrigate up to 250 million acres. Hari believed water would come down the new canal in six or seven years. That's what government authorities had told him, and he felt confident the water would come.

We walked back to the car and drove out to a village called Mudara. A group of men gathered around us as we got out and Santosh talked to them for a few minutes. "They all want the KB link now," he told me.

Munna Ahirwar, a Dalit, had five acres of his own but had grown no crops on them for the last four years. His well had also dried up. The day was getting warmer; Munna was in his shirtsleeves, a light cream-colored scarf hung loosely around his neck. His pants, too big for his slight body, were a golden sandy color like the color of the dry earth of his land, which stretched behind him. He had a slight beard and mustache on his young face and his ragged hair had the look of a home cut. He bent his head downward sadly, but his mouth formed the beginnings of a gentle smile.

Munna said he used to grow wheat. Now he has to go to Delhi to work in construction, helping to build new factories and hotels. He lived at the construction

Women going to fetch water from a well near Jhansi

sites while he was working; his family stayed here in Mudara. He said he needed rain or the canal or something. According to his understanding the canal would be finished as early as 2011.

A lanky, darkly bronzed man with a full head of white hair came down the strip of asphalt in a cart, driving a team of black oxen. The plastic frames of his big square glasses, taped at the nosepiece, matched his frosted hair. Just down the road behind him was a dry canal, the men told us. It was constructed in the 1950s before the villagers here started using their own wells. But the canal dried up twenty years ago, as did the monsoon-fed pond from which it was supposed to bring water.

Matadin Sahu rode up on his bike as the oxcart passed. Bald with a short grizzled beard, he wore clothes that showed the instinct for color all South Asian women seem to have. Maybe his wife bought his clothes, I thought, admiring the light blue long-sleeved shirt, blue fleece vest, and pumpkin checked scarf with a blue stripe that he wore around his neck. Matadin seemed a little better off than some of his neighbors, though he was not rich. Looking out at this dry land, he told us, made him feel afraid.

People here were friendly, but Santosh also felt fear. He said the roads would not be safe if it did not rain soon: people would turn to dacoitry—banditry. "Bundelkhand will become like Bihar," he said.

Back on the main road we continued to the east. Old-fashioned scarecrows dotted the few green fields. Low boxlike markers painted red popped up at intervals across the brown earth, marking the path the canal would take: "KB Link, 216/906, NWD." The red markers, each about a foot cubed, were spaced 300 feet apart. They made the link seem real, something that was sure to happen someday, once environmental impact studies were completed and approved. The land for the canal would have to be purchased, and land here, close to the road, would be more expensive. Some farmers would likely be happy to sell their barren land to the government.

We stopped to talk to a trio of men near the road. They told us the wheat crop used to need only one or two irrigations; but now, because the soil is so dry and the groundwater table so low, it takes seven or eight irrigations to bring in the crop. Only a quarter of the arable land is being farmed now, one said, because of the shortage of water. While many others were leaving for work, the men in this group said they had been selling their cattle and goats to get by.

As we were starting to think about lunch, we saw a group of men and boys working near a traditional deep dug well by the side of the road. I wanted to stop and look at the well. The inside walls of the circular cistern were brick. Lower down the bricks were smaller and older; bricks near the top looked as if they had been formed in a modern assembly line process. The bottom of the deep well, about a dozen feet across, was almost dry. A wheel at the top, which would have secured a rope and bucket, sat useless. Small plants grew on the inside wall of the well. But a pump brought water up from a nearby tube well; some of the men were clearing out a narrow channel by hand to send the water to nearby fields. On a low wall lay neat piles of freshly picked, long white radishes with bushy green stems.

A preadolescent boy in a scanty pair of shorts and sleeveless undershirt held a black baby goat. He posed obligingly when I wanted to snap a photo. Later he and his older brother followed us to the car, looking shy and curious. I asked Santosh if it would be all right for me to give them a little money. He assented, a little dubiously.

I got in the car and found a fifty-rupee note in my bag—a little over a dollar at the exchange rate but worth more in buying power here, maybe enough for a schoolbook. They had told us they both went to school, though sometimes irregularly. As I extended my hand to give them the money, both of them grabbed roughly at my outstretched hand and wrist. Startled, I quickly pulled my arm back into the car, money still in hand. Santosh scolded them and told the driver to leave. Maybe if I had held out two equal amounts, my clumsy attempt at charity would have ended better for all of us.

Before I left Jhansi, Santosh took me to his home to meet his family. His heavily pregnant young wife made us tea, then sat smiling as she watched us sip. Outside in the large courtyard in front of the modest brick house, preparations were underway for his brother's wedding. White lines and floral designs painted on the hard-packed red earth were the first decorations readied for the festivities.

The idea of bringing water to this dry farmland is no whim, even though the scheme would benefit engineers and contractors as much as or more than farmers. This land and these people need water or the people will continue to leave. The question is—is a dam and a long concrete canal that will cost millions of dollars the best way to help Jhansi's farmers in the long run? And what will that do to the Ken River and people who depend on it?

Just before going to Jhansi I attended a seminar at Jawaharlal Nehru University (JNU) in Delhi. One of the student political groups on campus had arranged for a debate on the river-linking plan. The current water minister, Saif Ud Din Soz, arrived forty minutes late. Slender with upright posture, he sailed into the room to take his seat, surrounded by a gaggle of young men.

The debate began with a question posed by JNU professor of science policy Rohan D'Souza: "Is a river a resource or an endowment?" Ecologists say the latter, that a river is a set of relationships: mountains to wetlands to floodplains to mangrove forests. Engineers tend to transform rivers into quantities. If rivers are reduced to numbers, to the amount of water in them, it is easier to create technical interventions. Those interventions end by disturbing or even destroying the relationships that make the river a river, and doing the same to the creatures—birds, fish, deer, tigers, people—that may have thrived in them and on their banks.

Rohan D'Souza spoke quickly, leaning forward in his seat. Balding prematurely, with a broad forehead, Rohan peered over the glasses perched on his nose. He told the group of students seated around a large circular table that those who control the definition of water scarcity control the political process. And those who control the political process can control the definition of water scarcity. "Supply-side hydrology," he said, is sustained by an "iron triangle of engineers, contractors and politicians," none of whom are the people who live on the banks of the river.

Local communities see water scarcity as transitory and have traditionally had means of coping with it. Only engineers see permanent scarcity, which lends support for big dams over smaller interventions and adaptations to variations in water. But "variations in water are like the beating heart of the river system. When you alter the variability of a river," Rohan concluded, "you are virtually flatlining the river."

The water minister did not meet the professor's energetic arguments head-on. He summarized the standard rationale for river linking, saying its goal was "more crop and income per drop of water." He said there were sixteen feasibility reports ready, and the Ken-Betwa link was at the top of the list of the five best possibilities. A detailed report would be finished in a few months. If the Ken-Betwa link were a success, India would decide whether to proceed with the other four links on the short list. During most of his rambling speech he talked about climate change, lamenting that JNU and other universities were not focusing enough on that, and saying that the West was trying to hoodwink India to slow its development.

The meeting broke up after various students had weighed in, largely in support of the minister. I had to choose which of the two speakers to run after and meet. I chose the professor, thinking I probably *should* try to talk with the minister. But I was no longer meeting deadlines and satisfying editors, so I chose the interesting speaker over the one who currently had power.

Rohan did not have time to chat at that point, but I met with him on a later trip to Delhi. In the meantime I had read his book *Drowned and Dammed* about embankment-induced ecological disasters in Orissa. He was happy to hear I was trying to understand some of India's river issues. He said India could use an institution like Cal-Fed, a stakeholder group in California that had debated, in public, how to use California's rivers. It was not an efficient process, and had been such an unwieldy organization it eventually folded, but it had created a forum for vastly differing viewpoints—farmers, environmentalists, fishermen, urban government officials—to rub elbows and debate each other in person instead of through accusations in the newspapers.

Engineers gained power over water resources during the past century, Rohan believes, because of the great shift from rural living to urban living, and the resulting "modern quest for getting you to open a tap and brush your teeth."

Not long before our conversation, Congress party leader Rahul Gandhi and India's new environment minister, Jairam Ramesh, called river linking a potential disaster. But Rohan does not believe this means the idea has been killed. The environment ministry is, according to some observers, the weakest. Its voice has frequently been drowned out when it was time to launch ambitious development projects in India. Like Ashok Khosla, Rohan thinks river linking is more likely shelved just for the time being and will probably reemerge as it has before, perhaps under a different name.

At that point something along the lines of sixty billion rupees—well over a billion U.S. dollars—had been spent just on the many reports generated by those studying the thirty complicated links. Himanshu Thakkar, an engineer who founded the South Asia Network on Dams, Rivers & People to keep a bright light focused on the government's development of rivers, says that the studies alone are so expensive that it does not make sense to continue with the effort. But, according to Himanshu's monitoring of government data, studies were continuing, and the budget for them increased each year.

The Ken-Betwa link in particular made no sense, said Himanshu, because both rivers are in spate at the same time and dry at the same time. A massive investment in the project seemed pointless if in any given year there could be a drought and no water to send from the Ken to the Betwa. The expensive concrete canal would run dry.

Development Alternatives was working on a different way to supply water, helping to construct check dams in Bundelkhand. Check dams don't stop water from flowing down streams and small rivers; they slow the water so the land can absorb it and recharge the groundwater. The check dams also allow irrigation without displacing people or radically changing a river's flows.

Most likely there were once thousands of reservoirs in this region similar to the one at the Barua Sagar dam I saw that morning in Jhansi with Santosh. Some of them were completely lost. They dried up, and people started living on the land or cultivating it. But many could be revived—cleaned out, de-silted, and readied to receive the next monsoon. They offered a huge potential water supply, one that would cost less and do more good consistently. But it did not offer the apparent convenience of the single, big solution. It would involve many small solutions, help from many local people, and local management strategies.

Ashok Khosla told me that the balance had not yet shifted to sustainable methods like these. "If you're looking for your next trip to India to see these changes, you won't. It will be fifteen to twenty years before these things happen." Twenty years sounded like a hopeful scenario to me.

In recent decades, just these sorts of changes have come about in the neighboring state of Rajasthan—the land of kings and camels, fortresses and frescoed palaces that attracts hundreds of thousands of tourists each year. In the far western reaches of this large state stretch fabled deserts of sand dunes; but less than two hundred miles west of Jhansi as the crow flies, the land looks very much like that in Bundelkhand. In the river-linking scenario, Himalayan water was to be shifted to Rajasthan via the Yamuna; but farmers in eastern Rajasthan were far too desperate to wait for that. They took water into their own hands.

A young woman named Devayani Kulkarni took me to see what they had done. She works for Tarun Bharat Sangh, an organization that helps people in Rajasthan manage water resources. Its leader Rajendra Singh, often called the "water man of India," won the Magsaysay Award, the "Asian Nobel," in 2001.

Long before that recognition, he and a few other young social workers sought to help the desperate villagers of eastern Rajasthan. Much as in Bundelkhand, traditional rainwater harvesting systems had been abandoned in favor of state-sponsored pumping of groundwater. By the 1980s animals were starving, people were leaving their land for work in the cities, community life had deteriorated.

But old people remembered the old ways. With support from the young social workers, in 1986 they directed villagers in Gopalpura village to start restoring a *johad*—a check dam to hold the monsoon rain, provide water for animals and irrigation, and recharge nearby wells. Devayani took me to see this johad after we traveled a few hours from the capital of Rajasthan, Jaipur. Green fields surrounded by trees burst from the reddish-brown semi-desert near the pond.

We took a walk into the foothills of the Aravallis, an old eroded mountain range that long predates the Himalaya. We passed a *baoli*, a stepwell, said to be five hundred years old; it had dried up during the drought but was full of water now. As we walked up over a low hill, the higher hills of the Aravallis framed the drainage. I saw no other signs of water harvesting structures. A girl was singing, gathering fodder for animals. Taking fodder or grazing was allowed here, but no agriculture, Devayani told me. This land was held in common by villagers. I saw a cactus with many arms; it looked like a miniature organ pipe cactus, the kind that grows in Arizona. There were few trees, only low bushes, which Devayani told me were used for making a tonic for cough and cold. She said the animals also ate it and liked it, especially when they were sick.

The first check dam surprised me. I would not necessarily have known I was looking at a man-made structure. I peered over the edge of the johad into the depression the local villagers had carved out. They labored for six months, digging the soil, piling it up, adding water, then compressing the soil to create an earthen dam about twelve feet high. It blocked the outlet between two small hills and looked, at first glance, like a continuation of them. Low vegetation grew on the dam and in the dry pond.

Winding further up into the hills, we came to the edge of another johad. This one still had some water. Below us in the distance a few dozen goats were moving toward the pond, herded by a villager. Pointing up past the next small hill, Devayani said there was a sanctuary, a forest villagers protected for the sake of the animals that lived there.

Villagers in this region have built many such structures—though at first government officials tried to stop them, saying they were illegal, poorly designed, and in any case belonged to the state. The government eventually backed down because the local communities were focused and organized. They planted trees to restore forests stripped decades before for their timber. Rivers that had dried up are perennial again, not just storm drains during the monsoon. Fish have returned. Wells are full. The people can grow up to three crops a year using water from their old, deep dug wells for irrigation. No bore wells are needed here. The result is healthy land. The villages I saw had an air of solidity and order: not wealth but security, calm, continuity.

Devayani told me about a "water parliament" that had helped all this happen. Every six months, she told me, the parliament meets to discuss problems that may have arisen, such as people who have been cutting green wood. Seventy-two villages send two or three representatives each, the group deliberates, the representatives take the decision back to the village and see that rules are enforced, according to traditional village practices of persuasion or punishment. Through this parliament, the *Arvari satsad*, villagers have agreed not to sell their land to outsiders and to stick with traditional crops that use little water. There are to be no new industries here, and the parliament has been working to persuade marble quarry owners to fill in empty pits so they won't suck away the groundwater the johads capture.

Ashok Khosla told me the way to start better using the 80 percent of India's water that goes to agriculture would be to stop growing water- and chemical-intensive crops. That would be the best way to get "more crop per drop," which the water minister at the JNU discussion said was the benefit of river linking. "Match crops to water availability," says Ashok. "This is not being done right now—it's purely a political issue."

The economic arguments in support of the Ken-Betwa link are based on growing rice in Bundelkhand, says Kanchan Chopra, an economist at Delhi University and the director of its Institute of Economic Growth. "Paddy shouldn't be grown in that soil," she maintains. "So if you take that away, the financial justification for the project falls apart."

Chopra disagrees with those who see linking rivers as the only way for India to achieve water and food security. She advocates managing water resources locally as more cost effective. "Smaller systems can be managed locally better than large mega projects," she says. "Governance needs to be pushed down to the local level because that has local values. And water is not just an economic good; it is also a social good. In some situations we need to use economic instruments for better managing water—like prices and so on. But when it's the basis for livelihoods, then we have to think in broader terms, in terms of human welfare."

Manoj Mishra, the retired forest department officer who launched a campaign to save the Yamuna River, worked in Madhya Pradesh during his career in forestry. He said he had seen the pain of drought-stricken farmers. "Nature is an engineer in her own right," he mused when we spoke in Delhi. "And when nature has decided something, we think that we have no authority of any kind to tinker with what nature has done or has not done."

Even though he is a staunch advocate of rivers as "endowments," not "resources," Manoj said it was hard to make a definitive judgment on the Ken-Betwa link. "Water is crucial for people's livelihood and prosperity. This area, Bundelkhand, has seen very severe droughts in the last few years, and in any case the area doesn't get too much rainfall. Their traditional water harvest systems have all fallen by the wayside, so the river is their primary water source. The Ken has got a lot of good water and not much pressure on it."

"We see all such efforts as technological fixes, engineering dreams," he said. "It's a very serious matter and should be avoided to the extent possible—until and unless it becomes a question of life and death."

Two years after my first visit to Bundelkhand, I had a chance to return. I wanted to see the famous temples of Khajuraho and the yearly festival of Indian classical dance held in their shadow.

A newly constructed railway spur allowed overnight trains straight from Delhi to Khajuraho, which had not previously been so accessible. That was partly why I had not gone to see the temples on my first trip to Bundelkhand. Yet in artistry and magnificence, Khajuraho's temples rank with the Taj Mahal and the temples of the Ajanta and Ellora caves. Khajuraho's temples are famous for their erotic sculpture, and some of the figures are indeed acrobatically sexual. But the majority of this wealth of graceful, sophisticated sculpture is not overtly erotic, though it often celebrates female beauty and loving couples.

Traveling almost due south from Delhi, I left winter behind. In Khajuraho, near the northern border of Madhya Pradesh, nights were chilly and mornings fresh, but afternoons under the bright sun were already hot in early February. The temperature rose quickly after dawn, and I discarded the turtleneck, vest, and socks I wore in the chill in favor of sandals and light shirts.

The simple hotel where I stayed was built around a garden with trees that offered shade. Outside my third-floor room were tables and chairs on what served as a terrace. It was really an extended landing, a wide, open hallway. Like so much of South Asia, the hotel seemed to be under construction. Men carried sacks of concrete to the roof, presumably to build another set of rooms. But the wing I was in hadn't been quite finished yet. Rebar poked up at the edge of the "terrace" and there was no railing, not even a low wall where the floor ended and open air began. An American guest with an East Coast accent said "not code," smiling a little uneasily as he climbed the railing-less stairway.

From this terrace I could see piles of plastic garbage, one of them five feet deep, in the backyard next to us. Below me were fountains, red hibiscus, marigolds, and a vine with delicate purple flowers shaped like orchids. The back of the hotel was fenced off with brick and green netting to hide the view outside for people sitting in the garden. In the plastic-strewn field behind the hotel, bullocks foraged. Three children squatted briefly, then pulled up their pants. The two girls and a boy each carried a little can of water. A cell phone tower loomed above us.

My second morning in Khajuraho, as I sat at a table in the open breezeway surveying this scene and waiting for my pot of tea, a pleasant-looking gentleman in his fifties emerged from a room and queried, "ça va?" I responded in French, pleased not to be immediately identifiable as an American. The short exchange ended when his friend came out onto the landing, sized me up, and said "good morning." My brief masquerade as a European over, we all laughed and proceeded in English to find out what had brought us here.

Johan D'hulster, who greeted me first, is a Belgian farmer who has been taking his winter breaks in Khajuraho every year since 2005. He's been getting to know its farmers and trying to help them recover some of their old methods of farming. His friend Peter, who had been buying Johan's organic produce for twenty years, accompanied him on this visit.

Johan never had any particular desire to travel, but he had seen pictures of India in his brother's books when he was young. He had a kind of vision that he would end his career there after many years of farming. He studied economics and philosophy and then started an organic farm that he and his wife worked as they raised their children. He joined the Belgian chapter of an Indian heritage foundation working to restore some derelict gardens near Khajuraho and then made his first trip to India. "Everything in my life came together," Johan said. "India, farming- and philosophy."

I had come to Khajuraho mainly for the temples and dance festival and was happy to have time for yoga and reading away from claustrophobic and frenetic Delhi. But I was even happier to meet someone who understood my obsession with water and could help me understand some of the history of the land surrounding us.

The array of magnificent temples here, built a thousand years ago, could not have been crafted by people with empty stomachs. This had to have been a far richer, more productive land then. Guidebooks mention how remote Khajuraho is, puzzling over the mystery of why such beauty cropped up in this out-of-the-way place. That fortunately preserved it from destruction when some idol-destroying invaders took over much of what is now India. But why was it here to begin with? Clearly it had to have been a good place to live. The land sustained a healthy and—judging by the faces on the sculptures— happy population.

Johan agreed. And he believed that rediscovering ancient practices of composting and mixing crops could revivify soil throughout India. Returning to a style of agriculture practiced before Western methods were adopted during the "Green Revolution" would more than feed the increasing population of India, he believed.

Around the time I had visited Jhansi and talked to the farmers who were just scratching by, waiting for help from outside, Johan had been here in Khajuraho, meeting the few farmers who were still managing to feed their families during the worst of the recent drought. The "lost gardens of Khajuraho," the Indian-Belgian project he had joined, held a vestige of the old style of land use. Now those gardens, which had thrived several centuries ago, were either covered in weeds or broken up into small patches worked by a single family to eke out a living.

Johan realized that he would have to learn about local agriculture before he could help restore these gardens to functioning farms and orchards. So he rode around on the back of a motorbike with an Indian colleague from the lost gardens project, looking for farmers who were still practicing a more traditional system of farming. There weren't many, but they were not hard to find. After three years of no rain theirs were the only green fields, scattered oases dotting the brown landscape.

One of the farmers Johan got to know was Bagawandas from the Kusawaha caste. Five families, all members of this traditional farmer caste, were farming the same way and doing well during the drought. Johan took me out to visit Bagawandas one morning. We walked away from the noisy tourist street and down into the old part of Khajuraho village, out past a dried-up lake—one of the many large ponds that once stored the monsoon and replenished groundwater.

The family home was a rectangle of whitewashed mud brick surrounding a wide, sunny inner courtyard. A beamed, thatched roof created a perimeter of shelter on four sides. Underneath it were a few enclosed rooms-but much of the house was open to the air. The women cooked with wood and kerosene-but they had electricity for light and Bagawandas had a mobile phone. His wife had made a delicious breakfast "soup" with carrots from their garden: shredded carrots, milk, sugar, and almonds cooked together. They served us bowls of this with some store-bought sweets and chai before we took a walk around the farm.

Bagawandas exuded energy and enthusiasm and tried hard to tell us what he wanted us to know despite his limited English vocabulary. He was dark and thin, quick-moving, a broad grin showing strong white teeth. Johan was taller and moved more deliberately, slender but better fed, white hair, pink pate. He smiled gently at his young friend, who beamed in return.

What had kept this land greener while the rest of it turned brown? The family had kept many of their trees, Johan told me, which provided shade to protect soil and groundwater from the fierce sun. They kept the soil covered with some crop all year-round to shield and replenish it. And they did not use chemical fertilizers. They also used their own seeds—local vegetable and grain varieties that tolerated local conditions better—not hybrids from foreign companies.

Though Bagawandas was doing pretty well, Johan lamented the slow loss of local knowledge about traditional methods. Johan says the farmers see too much of what others are doing, using fertilizers and hybrid seeds. Recovering the old systems would make the farmers more prosperous, Johan believes, but it won't be easy. Johan wants to help Bagawandas and other traditional farmers in the area make their soil richer and more resilient by instituting crop rotation and returning to composting. Johan told me composting started in India long ago, then came to Europe.

The cow is central to the traditional Indian composting system, and that's why it's holy, Johan believes. The manure of a healthy cow is the medicine that keeps a family's soil healthy and productive, yielding sweet hay for cows and good food for people. The manure from the well-fed cow is the basis of the compost that goes back into the soil and the cycle goes on, creating a closed system.

Traditional agriculture in India fell into disuse partly because pumping water out of the ground became a cheap and easy alternative to more labor-intensive methods, thanks to government support during the "green revolution" that began in the 1970s. With more water, more land could be cultivated, so trees were cut down. Then came foreign varieties of seeds—some of which demanded more

Johan D'hulster and Bagawandas at his farm near Khajuraho

water—and chemical fertilizers. Johan outlined for me what happens when chemical fertilizers are used. At first the soil productivity goes up. Then it declines, and the cost of more fertilizer and added water to combat the salinity the artificial fertilizer induces eventually cuts into the added profits.

Ecological balance can't be regained with composting alone, because a third of the land should be forest and that's no longer the case. A mixed landscape with many trees helps keep the water table high so that wells dug by hand can reach the water. Reclaim the dozens of local ponds that once stored the monsoon, plant trees, bring the groundwater table up, and this will lead to more normal rainfall as well, Johan said. It was the deterioration of all these elements, he and his Indian colleagues believe, that slowly led to the dried-out, drought-stricken situation Bundelkhand finds itself in today. And it has led people in Jhansi to believe that a dam and "borrowed" water from the Ken River is the only solution.

We visited another farm, started five years earlier with help from Vandana Shiva's organization Navdanya. Shiva successfully challenged the patenting of Indian varieties of grain and trees by Western companies. Two families were working together on this farm and having some difficulty getting along; but the fields, which looked more like orchards, were green. Looking around at the thriving young mango, papaya, and guava trees with vegetables growing in the brown soil beneath them, Johan said: "This is the past and the future."

I started to look at the irritating tourist touts in the three blocks between our small hotel and the entrance to the temples and dance venue in a new light. Perhaps some of them were once farmers, now caught up in the game of harvesting tourist money because their families had lost the productivity of their land.

In archways leading into Khajuraho's temples I saw *makaras*. These are the same mythical creatures that are carved on the stone spouts of Kathmandu, from whose mouths the clean water from underground aquifers flowed. Hundreds of miles separate Kathmandu and Khajuraho, which are not all that easy to traverse even now, and would have been covered only on foot hundreds of years ago. But the cultures were connected, over the millennia and over the miles, in their reverence for water and their husbanding of it.

I walked around some of the "lost gardens" one day. Some of them, comprising ten acres or more, had beautiful temples of a more recent vintage that had begun to crumble. And there were expertly crafted and elaborate stepwells, in some cases several in each garden. These were surrounded by overgrowth and not maintained.

The stepwells, built several centuries ago about the same time as the Barua Sagar I saw in Jhansi, were barely functioning because the groundwater level had dropped so low. But some still had water in them. One of these was about thirty feet across and three times that deep. With the water level so low, it was easy to see almost all of the steep stairways. Like the stairways at Barua Sagar dam, they created diamond patterns on the interior walls of the wells. There were also archways, shrines, and balconies in those walls. It was like a temple. Another stepwell was square, even larger, and equally elaborate. Many of the wells were small, round, or octagonal, but still showed careful craftsmanship.

The style of architecture was very different from the Malla stepwells of Kathmandu, but I felt something similar as I descended the stairs into the coolness of the larger ones. I could not quite put my finger on it. Admiration, awe, reverence, gratitude? For certain I felt delight in this marriage of the functional and beautiful, along with sadness that they have been forgotten, though they still seemed to be struggling to do their jobs despite the neglect.

I thought I could see India's dilemma here in Khajuraho as I had seen it in Bihar, Delhi, Uttarakhand. How is development going to proceed? Will India continue to follow the "West," an industrialized, mechanized, highly engineered model, which in some pockets the "West" itself is abandoning? Is the best way a return to some of the ancient, indigenous methods of land and water use? Or should India use some combination of the two, if indeed they can be combined?

"Sixty percent of India's population is agrarian and most of those farmers have less than one hectare," Johan said. "If you lift them up and give them a good price for their crop, they will be OK. Peasants are so invisible, there is no other way for them than to return to their essential farming." By this Johan did not mean to keep India's peasants barefoot and down on the farm—or that farming families should not enjoy the fruits of "development": better educations, better health

A *baoli*, or stepwell, near Khajuraho

care, opportunities for other employment if they choose. But the path India seems to be on does not look as if it will necessarily deliver those goods to people like Bagawandas anyway.

So why not make the land healthy again, restore its water table, and give India's peasantry a better life through indigenous methods? The thought of even more people moving to cities like Delhi and Mumbai, camping on the sidewalks, is horrifying. But that will be the outcome if the land can't sustain them. Water alone, through engineering, is presented by river-linking advocates as all that's missing for farmers to prosper. The main arguments against returning to older methods of water storage focus on the enormously greater population on the land now, and the loss of the community structures that maintained the old systems. Johan and people like him acknowledge it won't be easy or simple, but would cost less and last longer.

The virus of river linking—the highly engineered and concrete-bound approach to water management—will probably keep "blowing up," as Ashok Khosla said, unless and until India's discarded traditional methods of capturing and storing water are revived. Such methods would have to be applied throughout the nation, not just in isolated pockets. And they would have to be combined with conservation, urban rainwater harvesting, and a host of other methods, including protecting the rivers and the groundwater that are now lost to pollution. Even if rivers are linked in some cases, those links alone won't give India "water security."

As the year of the tiger began, early in 2010, I saw a sign in Delhi at the clean and shining toll entrance to a new bridge crossing the smelly Yamuna River. It pictured an adorable dozing tiger cub with the caption "Just 1411 left. You can make a difference." India's butchered wild tigers, poached largely by Indians, are carried to China to make health tonics and tiger "viagra." It seemed likely the year of the tiger was going to stimulate demand for the products made from tiger bones and organs.

I soon learned that the figure of 1,411 was outdated. The count was probably down to 1,000 at most by early 2010. Joanna Van Gruisen told me this during my visit to Khajuraho. Joanna was a wildlife photographer and her husband Raghu Chundawat was a wildlife biologist and one of India's best-known tiger experts. Together they had spent ten years studying the tigers in the nearby Panna Tiger Reserve.

The Ken River, the "donor" river in the proposed Ken-Betwa link, runs through the Panna reserve, just east of Khajuraho. Starting in 1995, Raghu and Joanna studied, tracked, and came to love Panna's big tigers, some of the largest types of the big cat ever seen. The females were the size of most males; Panna's males were so big the standard radio collars would not fit them and special ones had to be made.

When Joanna and Raghu arrived they counted fewer than twenty tigers in the reserve. In the next six years the population doubled as almost all the cubs born survived. Between 1998 and 2002 Panna had the best recovery rate in the subcontinent, they told me. Raghu says tigers are very resilient. Given the chance, which means enough habitat, and peace and quiet without human disturbance, they can breed—well, like cats do.

Then the park management changed. The cooperative working relationship with the Forest Department, which Raghu and Joanna had established with the previous director of the park, deteriorated badly. In 2002 and 2003, tiger after tiger disappeared. Some were caught in the snares local villagers set to catch other animals for food and were inadvertently killed. But others were deliberately being poached for their parts. Interviews done by BBC and local reporters with admitted poachers in subsequent years verified these suspicions. Poachers were very focused on their work, and knew how to elude the reserve's sleeping staff in the middle of the night.

Within three years all but one or two male tigers were gone. The tiger reserve's authorities did not acknowledge the loss until 2009, even insisting there was no poaching going on. They had already begun importing some females from another park whose population was stable. But by then even Panna's males were gone.

Joanna and Raghu told me this story as they were working on a new endeavor. Thwarted in their long, patient efforts to protect Panna's tigers, they bought some land on the west side of the Ken River—only a few miles from the tigers' onetime forest home—and started building a unique eco-lodge. They hope to share some of the region's beauty with visitors. They also want to create training and jobs for local workers in plumbing, electrical work, and carpentry to stem some of the outflow of drought-stricken farmers to city construction jobs. They plan to continue research, shifting from the big cats to the effects of climate change on the region's agriculture.

Joanna went through a pile of fabric samples for curtains while she reminisced about the time she and Raghu lived in a hut on the other side of the river, just inside the Panna reserve. "I could sleep to the sound of snoring elephants," she said, holding up a sample. It was a local sari cotton—a sheer butter yellow with a cream-striped border.

The cottages have thick mud walls and thatched roofs, but inside there is understated elegance: cool polished concrete floors, warm earth tones on the walls, tile and stone bathrooms with views of the arid hills dotted with trees. The furniture is carved from local woods. Raghu said they are not using the local teak, because it was hard to find any that had been harvested legally. Instead they were using neem and other plentiful local woods, and even pieces of fallen wood in striking shapes, which they had used for tabletops and headboards. He added that it wasn't easy to find a carpenter who would build with these local woods because they are harder to work than teak. Joanna told me about their "air-conditioning" system. A pipe from underground brings cool air up from Bundelkhand's rocky belly. A single fan will blow the underground air into all the rooms. They'll use solar power for lights and a few electric appliances.

It's easy to imagine that at one time tigers lived here where the cottages were getting their finishing touches. Even if the horrors of poaching were by some miracle to be halted, the tiger's habitat has diminished to a fraction of what it once was throughout the subcontinent. The species continues to succumb daily to bad management and human encroachment. Perhaps guests at the new lodge, exposed to the owners' knowledge of wildlife, can help spread awareness of how much of this habitat has been lost. But it is too late for Panna's big tigers; their unique gene pool is gone forever.

Joanna and Raghu were busy talking to carpenters, making decisions about the doorways for the red brick kitchen, so I strolled down to the banks of the Ken River. I walked across an old stone causeway that once allowed vehicles to cross the river in the dry season. Now it was unused and crumbling. The river channel was about half a mile across. Upstream most of the rocks in the riverbed were submerged because the causeway acted somewhat like a weir and made the water slightly deeper. But downstream the boulders were so dense and exposed it was possible to get across the river by leapfrogging. A lone boy walked out on the rocks, playing, or maybe going to bathe in the river.

I had somehow imagined that this "donor" river would be filled with water and the land around it lush and wet. But the Ken looked much like the Betwa, the river that was supposed to receive its apparent surplus. The difference between this area and that 150 miles to the west seemed to have more to do with the fact that there were fewer people here, less development, no large cities like Jhansi, and fewer diversions of the river for agriculture. There is more forest in this part of Bundelkhand. The trees of the Panna reserve that I could see to the east were protected, for now, better than the tigers had been.

All around Khajuraho I saw green, but not because this side of the region is actually wetter. It's because the whole region got its best rain in twenty years during the past monsoon. Bundelkhand sometimes gets a lot of rain in the monsoon. To the south I could see a new bridge; occasionally a car crossed it. It was some thirty feet above the water, but I heard that one year a flood had reached the bridge and washed over it. Above on the riverbank I could see the thatched roofs of the guest cottages, blending into the landscape like brown hillocks. Raghu said water could rise that high, twenty seven feet above the current level in the river, and flood the ground around the cottages. I sat there, watching the herons and cormorants and kingfishers, trying to imagine so much more water in this shallow river.

The fate of Panna's tigers would be just another, if extraordinarily sad, chapter in the ongoing tragedy of the eradication of India's tigers to make Chinese people feel healthy. But the final twist to this story is that the dam included in the proposed Ken-Betwa link would inundate a third of the Panna reserve.

Is Panna still a tiger reserve with no tigers? Is it important now to protect it and its remaining habitat from inundation? There's an abundance of wildlife in the park other than the tiger, but if the keystone species has been eradicated, will other species matter? Perhaps water could not be allowed to trump tigers, but maybe it would trump leopards and deer, monkeys and elephants.

Late in 2009, when Jairam Ramesh, India's current environment minister, said the river-linking plan would be a disaster, he voiced particular doubts about flooding any portion of the Panna reserve. But the Congress party's heir apparent, Rahul Gandhi—who agreed about the potential disasters of river linking—had also made relieving Bundelkhand's distress his pet project. And at that point, progress in the region was still being pegged to linking the region's largest rivers, the Ken and the Betwa.

# 14

# Susu

Gods and Goddesses watched from their heavenly thrones and
drew near to celebrate the birth of the river. "Now show me the
way, Bhagiratha" cried the daughter of heaven. And a great
procession began to wind through the land of India: in front
went Bhagiratha in his robes of penance, and after him came the
river with her myriads of fishes, turtles, frogs and leaping
dolphins—all creatures that live in the rushing mighty stream.
—From the story of Ganga's descent to earth

"You'll never get a dolphin with a digital camera," Sushant Dey said, as we floated with the current on the Ganga. "They're only on the surface for a second. By the time the shutter clicks, it's already gone."

Early one morning in April 2007 I walked down to the bottom of a wide concrete stairway just outside Bhagalpur in the Indian state of Bihar to board an old fishing boat. Where the stairs met the river, the prow of the twenty-foot-long boat rested on the riverbank. A narrow plank, its ends positioned on the bank and the edge of the prow, allowed me to board. Sushant Dey and his brother Subhasis were taking me out to look for dolphins before the heat built up, when the dolphins might still be looking for food.

Dolphins jump out of the water when they're hunting, which they typically do early in the morning and again in the evening. The boatman tried repeatedly to start the wooden boat's old diesel engine. Finally it coughed and caught; we chugged upriver a short way. He turned off the engine and we floated. After a few minutes, I heard a swish of water. A slick muscular body slipped back into the river before I could get a good look. A few minutes passed: another swish. I was looking in the wrong place and missed him.

In spite of Sushant's warning that my effort would be in vain, I tried again and again to catch a dolphin, pointing my camera to a likely spot on the opaque gray-green water where the animal might surface after I had missed a breach. Then I missed again.

The Ganga flowed smoothly. It was about a half mile wide now, in the dry season. In the monsoon it grows to three miles wide and can be twenty-five feet higher in some places. The boatman took us to places where the dolphins were known to rest in the deep waters. I got half a dozen good glimpses as a dolphin surfaced briefly to breathe: an arc of dark gray, a shiny comma. I never saw the full length of the creature's strikingly narrow snout; the eight-foot-long adults, or the smaller juveniles, moved too quickly.

"You can stay here a whole day, but you can never tell where they will surface. There's a big dolphin over there," said Sushant. "We'll stay here a few moments."

"Do they ever come up to the boat?" I asked.

"No, you don't have anything interesting. They can't see, so it doesn't make any difference to them."

These river dolphins are ancient, among the oldest creatures on the planet. The charismatic and less endangered ocean dolphins descended from them. The river dolphins are virtually blind, having no crystalline lens in their eyes. In the muddy Ganga and its silt-laden tributaries, they don't really need eyes. They can perhaps detect light, but not shape or color. They rely solely on echolocation to find their food as they skim along the bottom of the river sideways, touching it with one of their flippers.

*Platanista gangetica gangetica.* The Ganges river dolphin, widely known now as *susu*—an onomatopoeic name that resembles the sound the dolphin makes when it breaches and breathes. I had come to see the only sanctuary for these endangered animals, located in eastern Bihar. In 1991 the Government of India designated the Vikramshila Gangetic Dolphin Sanctuary, which covers a thirty-mile stretch of the Ganga near the city of Bhagalpur. As early as 1972 the dolphin was included under India's wildlife protection act.

Ironically, more *susu* have been found dead in the sanctuary than elsewhere on the river. But this is not for lack of heroic effort on the part of a handful of men I met in Bihar who have dedicated their lives to the dolphins. They are up against deadly odds.

One of the biggest obstacles to preserving the dolphins seems to be how the laws that should protect them are being enforced, pitting local fishermen against the sanctuary. We stopped to talk to one of these fishermen shortly after we started our search for breaching dolphins. Shankar Sahni was out on a small fishing skiff and came up alongside our larger boat. He climbed aboard to talk to us while his son remained below on the skiff. Wiry, his skin a deep bronze from the sun, he had a trim, grizzled beard and mustache, and short wavy hair. He squatted in the boat in his white cotton shirt and *lungi*, elbows on knees, long fingers laced, as he talked to Sushant.

Shankar Sahni had fished this stretch of the Ganga for fifty years. "Shankar does not support the sanctuary because it's not good for fishermen," Sushant said. "He understands that the sanctuary is meant to save dolphins, but he is not pleased with the sanctuary or the wildlife act because the forest department officials come regularly and take his nets and disturb his work."

Shankar wanted the government to allow the fishermen to fish. "They should not kick them out, that's all he can say," said Sushant. "The department took away his boat and some of his nets last year and is always telling him they'll give it back, but they haven't done so."

The little fishing skiff alongside was likely a borrowed boat, or perhaps Shankar had two boats. The forest department took boats and nets from a few fishermen like him. "It doesn't happen to a lot of fishermen," said Sushant. "There is no written order saying they can do that. There are hundreds of fishermen, so why take from only a few?" Sushant speculates that the department harasses a few fishermen to look like they're doing something. "It creates an environment of fear and panic. It doesn't help the sanctuary at all."

Sushant's quieter brother, Subhasis, added that the department tends to harass fishermen within the city limits to demonstrate they are doing their work. Ten miles upstream there are plenty of poachers, but the officials will not go up that far.

Fisherman Shankar Sahni wants the government agents to go arrest the dolphin killers instead of keeping men like him from doing their work. He said he is always careful and does not catch dolphins by accident. Things are getting worse in general for fishermen. The river is no longer deep, the current has decreased, and there are fewer fish in the river. There is an infestation of what he calls "anti-social elements."

Shankar returned to his own boat and we went back to looking for dolphins. "There's that big dolphin," exclaimed Sushant. The cetacean slipped back under the glinting surface of the water.

Fisherman Shankar Sahni on the Ganga near Bhagalpur

The susu has a nine-month gestation period like humans and the females give birth each year. They don't have natural predators, but some big fish will eat neonates. Dolphins regularly become entangled in fishing nets.

I looked away from the water for a moment to rest my eyes from the glare. I heard the quiet splash as a dolphin surfaced just then. "Oh! Look away for two seconds and he's gone," I moaned.

"At times when you are very close to them their breath smells like kerosene because of worm infestation in their stomachs," Sushant said. I immediately thought of the sewage and chemicals from industry and agriculture that spill into the Ganga. But Sushant said the worms are natural for the dolphins, not a sign of disease.

Sunil Kumar Chaudhary is trying to make the Vikramshila sanctuary live up to its name with the help of his young disciples, the Dey brothers. A professor of botany at Bhagalpur University, he became interested in dolphins and has devoted much of his life to them since the sanctuary was created.

The sanctuary is named for Vikramshila University, one of the great ancient Buddhist centers of learning in North India, along with Nalanda University. Founded in the eighth century, Vikramshila's ruins are in Bhagalpur district, about thirty miles outside the city.

Sunil Chaudhary met the Dey brothers when they joined a nature club he organized at Bhagalpur University. Sushant said his father had wanted to be a wildlife photographer so he often took his sons into the countryside in West Bengal, their family home, to watch the animals.

"Dad used to take us to the jungles when we were kids so we were in touch with wildlife. It's a childhood hobby which kind of got serious when we grew up. We loved the wildlife, jungles, and rivers. I am actually a physics major; my brother majored in zoology. But it doesn't matter what you major in when you want to save something. The passion for conservation is all that's important." The brothers have been working with Sunil for more than a dozen years.

Before my trip to Bihar, an environmentalist in Delhi gave me Sunil Chaudhary's e-mail address. I wrote him and said I might come to Bhagalpur after I visited Dinesh Mishra in Patna. Bhagalpur was a university town; I imagined a green, quiet place, students reading under trees.

Sunil wrote back: "Come, it's not as bad as people say." He knew only too well Bihar's reputation. Once there, I couldn't entirely agree with him; a lot of things in Bihar tended to live up to the state's bad press. And given my fantasies of Bhagalpur's quiet, it was worse. My hotel was at the major four-way intersection in town, where every passing truck had to stop and start up again. From early in the morning throughout the day they pulled up, brakes squealing, diesel engines rumbling loudly, and pulled away again.

But like Mishra and his friends in Patna, Sunil and Sushant did their best to make me and Pushpa, my twenty-five-year-old Nepali assistant, feel welcome.

Pushpa and I took the train from Patna early one morning and arrived in Bhagalpur pretty much on schedule that afternoon. Our hotel had seen better days, but I got a large room with a settee and table and chairs so I could have Sunil and his colleagues in for tea and talk the day we arrived. The room seemed fine. Later I discovered I had a resident family of mice, along with a family of lizards, perhaps geckos, of which I had grown quite fond when I lived in Hawaii for a couple of years. They kept the mosquitoes in check.

My shower didn't work, but the one in Pushpa's room did, so I borrowed his bathroom the next day. And my room was quiet enough late at night. I fell asleep quickly as the overhead fan whirred.

Pushpa and I laughed a lot. By now this stranger, hired to help me negotiate Bihar, had become my pal. In Patna some people had called me Charley, and Pushpa's name had morphed into one more common in India—Pushkar. I modified that to Pushcart. Like two teenagers we dissolved in gales of laughter, calling each other Charley and Pushcart. Despite our kind and attentive hosts in both cities, our encounters with the travel infrastructure in north India had ranged from bearable to disconcerting. But none of the famed *dacoits* of Bihar ever bothered us. We just kept laughing, partly from the strains of travel in a strange land, partly because we had seen many sad sights in Bihar and they must have created a psychic undertow for which the laughter surfaced as antidote.

After a couple of hours on the river that morning, we wearied of trying to find the elusive dolphins. The sun was rising toward the zenith; it was getting hotter. Pushpa took off his shirt and plunged into mother Ganga, surfacing less elegantly than the dolphins, who seemed to have finished feeding until evening.

Sushant Dey said there were some other fishermen nearby if I wanted to talk to them. The young boatman headed toward a modern concrete bridge downriver, then steered toward the riverbank.

The murky, slow-moving river lost its silver-gray sheen as the sun rose higher and hotter. The water was now the color of pea soup, which Sushant said was its natural color, not entirely the result of pollution.

Still, raw sewage is dumped here just as it is all along the Ganga. A sewage treatment plant for Bhagalpur was built about twenty years ago as part of a massive "clean up the Ganga" campaign. There was enough money to build it but not to start it up, so apparently it sat idle, with one watchman, ever after. Just as well, since it was built upstream from the city instead of down.

As we eased diagonally toward the bank I saw a litter-strewn stream trickling down into the river. Less than two hundred feet below this sewage outlet, children splashed in the water. A group of adults bathed at the river's edge.

We climbed out of the boat and walked up a steep slope of hard-packed dry earth. Under the fat concrete pilings of a modern bridge a little shopping area had developed at the bottom of a road. A constant stream of cars and motorbikes brought people down to the river to get water to take home for puja. A pretty young woman in an indigo sari arrived on a scooter with her father; a serious look on her face, she carried an array of empty plastic cooking oil bottles to fill with water. A car full of people, an entire family, arrived soon after. A funeral ghat was nearby, just to the west.

An old fisherman named Suresh Mahaldar ran a little café in the shade of a corrugated tin roof, serving tea and frying gram cakes in a wok over an open fire. His wife used chickpeas to make the batter, which he then squeezed through a cup with a perforated bottom into the frying pan. The noodle-like strands of batter clumped together to create a crunchy pancake. Piles of them sat golden and inviting near the earthen stove.

Suresh, about seventy years old, had fished in the Ganga for fifty years. Like the fisherman we met earlier, he said the river used to be deeper and the current stronger. After the Farakka Barrage was built downriver, the Ganga's current diminished. Suresh lamented the lack of fish in the river. A whole day's effort yielded only one or two kilo of fish, so it was no longer possible to support a family through fishing. Insecticides were killing the fish, as well as the insects the fish eat, he thought. Men had to become laborers or do something else to earn money, which is why he opened his café.

Suresh still went out fishing at night about seven, coming back home at four in the morning, after which he worked in his café. He slept in his boat for two or three hours a night. The other men fished at night as well, he told us, then went to work building houses in the morning.

The Forest Department staff often stopped Suresh from fishing. Then there was the "mafia," what the fisherman we spoke with earlier that morning had hinted at when he complained of "antisocial elements." When the catch was better, in October and November, and in May and June, the "mafia" came to steal their money or beat them up and take away their catch.

"When we go out to fish in the night they surround us. They say 'You have to pay 2,000 rupees ($40) or tomorrow we will take your nets and boat.' If we don't pay we can lose our property or our lives. When there's increase in the catch they always come. If the catch is normal they come once a month." The fishermen are too afraid of these criminals to go to the police, and in any case the police know what is happening. The fishermen think the criminals bribe the police. This "mafia" sometimes barricaded the coves in the river to trap the fish.

Suresh values the dolphins. They have always helped the fishermen by herding the fish for them. They look beautiful when they surface and make him feel the river is still alive. So for him there is no conflict between dolphins and fishermen. The thugs who barricade the coves and hoard the fish that are the livelihood of fishermen and the food for the dolphins, they are the problem. If this were not happening, Suresh said, the number of dolphins in the river would increase.

Local fishermen here would have to stop fishing if things continued this way, Suresh feared. They would all have to become migratory laborers. He sat thin and erect in his white tee shirt and lungi. Two of his sons, also traditional fishermen,

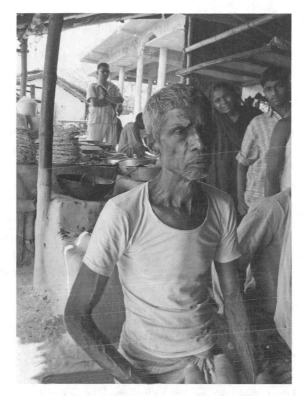

Fisherman Suresh Mahaldar at his café near Bhagalpur

flanked him on a long bench In their forties, they wore long-sleeved shirts with scarves around their necks. All the men's faces were serious, unsmiling, and they looked very tired. One of the sons said some of their friends had already sold their boats and gone to the Punjab to pull rickshaws or work as laborers. A group of young men crowded behind them to listen as Sushant and I, sitting on a bench on the opposite side of the table, asked questions. I noticed how much softer and better fed we looked than the sinewy fishermen.

One of the sons took me to meet his wife. Their home was a five-minute walk up from the river in a village where vegetables grew and dung cakes for fuel dried in the sun. The houses, made of bamboo and thatch, were open on two sides.

Milam Devi wore a blue and turquoise sari and invited me to sit on one of the several wide wooden platforms that served as seating, dining table, or bed, depending on the time of day. She told us she supplemented the family income by making *papadam*, spicy crisp flatbread made of chickpea flour. In eight hours she could make three hundred and earn about thirty rupees, less than a dollar. She had six children, five girls and one boy, but now her tubes were tied. She and her husband sent their son to school in hopes he could find a better way to make a living.

Her parents arranged her marriage. Being the wife of a fisherman is a very hard life, she said. The rest of the family crowded around as Sushant tried to ask questions for me. Sometimes her husband and other men answered before she had a chance to respond.

Most boys don't go to school. They help their parents, then they leave for work in the cities. Literacy rates in this area, according to one estimate, are at best 10 percent. For women the rate is barely above zero.

When Professor Ravindra Kumar Sinha was eight years old he participated in the last rites for his grandmother. After the cremation by the riverside, the family scattered the ashes in the Ganga. During the rites, Sinha saw black things moving in the water. He wanted to go bathe in the river, but family members told him not to, warning him the black creatures would drag him away.

Now a zoology professor at Patna University, Sinha leads the Indian government's dolphin conservation project. I talked with him in Patna before Pushpa and I took the train to Bhagalpur.

For years after his grandmother's funeral Sinha watched the same creatures swim in the river, but never heard of anyone being dragged by them. He wondered why people were so afraid. Over the years he kept asking questions. He found out about the dolphin from fishermen; they told him that, like humans, the dolphins give birth to a single baby and suckle milk. "Old fishermen loved the dolphins; for them more dolphins meant more fish. They called them 'flower of ganga' or 'cow of ganga' because it nurses its young with milk."

Sinha heard that steamers that once plied the Ganga always enjoyed an escort party of dolphins. And in school he learned that the first known laws protecting wildlife were in his home state of Bihar. In 300 BCE, King Ashoka, who followed the Buddha's philosophy after recoiling from the horrors of the many wars he had won, had decreed the dolphin should never be killed for any reason. This was the beginning of Sinha's lifelong fascination with the elusive susu.

When he was young, Sinha said he used to see ten or twenty dolphins each time he went down to the riverbank in Patna. "The population has declined to an unimaginable level now." A total of only two thousand susu are left in India, about half of them in Bihar. As shallow and slow moving as the Ganga has become in this state, it still offers better conditions than the neighboring state of Uttar Pradesh, once the home to many of the cetaceans. Less than thirty years ago, according to some estimates, there were up to five thousand susu in the Ganga. In the nineteenth century there were many dolphins in the Yamuna River, even in the dry season.

When he started his research, Sinha could find little information about the susu and thus had to start from scratch. He discovered that even wildlife specialists thought the dolphins were fish. The International Union for Conservation of Nature (IUCN) declared the Ganges River dolphin an endangered species in 1996. The organization has since tried to support projects to monitor them throughout their range, especially along the Ganga and the Gandak River, which Nepal and India share. There are also a few susu in Nepal in the Koshi and Karnali rivers.

An ominous warning came in 2006, when a related species, the Yangtze River dolphin, was declared extinct. China's Yangtze, another muddy river, had deteriorated in the same way as India's Ganga. Now that the Yangtze dolphin is gone, only three river species are left: the susu and two close cousins in the Amazon and Indus river systems.

Sinha says the biggest threats to the susu are accidental catching and poaching. The dolphins often become entangled in fishermen's nets, and some people kill them for their blubber. Their oil is used in folk medicine and also to attract fish. Now there is a synthetic substitute officials give free to fishermen, but some people say it doesn't work as well as the dolphin oil.

Dams and barrages damage the dolphins' habitat and reduce the size of breeding populations, decreasing genetic diversity and the species' resiliency. Water extracted for irrigation means there is often not even a minimum flow in the dry season, a source of stress for the dolphin. That combined with heavy loads of pollution, especially organo-chlorine pesticides, can make the water toxic for the animal. Studies have found a million times more DDT than is safe in their blubber.

Dolphins like to congregate in parts of the river where there are areas of both shallow and deep water. They fish in shallow water, rest in the deep. Embankments and barrages have tended to streamline this complex habitat. Sinha says he has tried to persuade government engineers to help preserve the dolphin by comparing their river habitat to arteries and veins. The dams and barrages on the river can be like blockages. He wants them to build bypass gates with shallow water so the animals can migrate up- and downstream during the monsoon when the river is in spate. Fish ladders I had seen on West Coast rivers, intended to allow salmon to surmount high dams as they swim upriver to their spawning grounds, have sometimes helped keep a species alive.

Sinha says dolphins like to congregate near places humans frequent—ghats and ferry guards. He thinks fish come to these spots looking for food and that attracts the dolphins. During festivals when millions of people come to the Ganga, dolphins come closer and closer.

In a faint echo of King Ashoka's prohibition on dolphin killing, the current chief minister of Bihar, Nitish Kumar, declared saving the susu a priority. At his urging, the Government of India declared the Gangetic dolphin India's National Aquatic Animal. This gives Sinha a little hope.

The morning after my frustrating search for dolphins, Sunil Chaudhary took me to the outskirts of the city to meet with the head of the Bhagalpur forest office. The forest officer was late, so Sunil and I sat on the verandah at the forest headquarters, surrounded by trees. The honking horns of the noisy city receded into the background as Sunil identified the calls of cuckoos and sunbirds.

We sipped tea and talked about all the many threats to the Ganga. "The dolphin is part of the natural heritage of India," Sunil began. "Ganga is a lifeline to India and the dolphin is its flagship species. It is symbolic of the physical and spiritual health of river Ganga. Dolphins are the apex of the food chain. If they are healthy the river is healthy, and millions of people who depend on the river for their livelihood are happy, so conserving dolphins and the welfare of the people are two sides of the same coin."

I asked Sunil where the fishermen I had met on the river fit in this picture. "No conservation project can work without the people. Local fishermen's families have fished in this river for centuries. Now the wildlife act doesn't permit them to fish. They are allowed to do subsistence fishing, but it's very difficult for the government to distinguish subsistence from commercial fishing. We are trying to lobby for the fishermen.

"The fishermen know they are not competing with the dolphins. They have coexisted with them for thousands of years. They know that where there are dolphins there will be fish, so they know the importance of conserving them."

Sunil told me the fish the men try to catch and those the dolphins like to eat are different, so they are not competing. "It is people from the outside who have created the clash between them. The wildlife act and the sanctuary have helped create conflict. There was no management plan for this sanctuary. There were some good gestures in the plan, which was made in haste. But there is no real recognition of stakeholders.

"The wildlife authorities' approach is irrational," Sunil continued. "They seize the nets and boats the fishermen need for their livelihood. And there is no communication between government agencies and local stakeholders. So we are trying to link them up. Excluding the fishermen will not help. It will only make them oppose the sanctuary. It will not help the river or the dolphin."

Sunil would like local communities to be part of "each and every decision about the plan. It should be the people's management plan, not the government's. The government should give them fishing rights over a limited segment of the river, and the fishermen should help decide this and the type of nets. They must participate and be responsible. Government agencies should monitor the fishing and they can withdraw fishing rights for infringements. But excluding the people— that is not going to succeed."

While the Forest Department oversees all wildlife areas in Bihar, Sunil wants to educate the police about their role in the effort. Higher-level officials support measures to protect the wildlife, but "the lower-level government agents treat us as though we're against them, because the work we're doing in the sanctuary exposes them, shows they're not working, and makes them angry."

Sunil explained the evolution of the "mafia" that has made life hell for the fishermen. Historically a fifty-mile stretch of the Ganga in this region was under the control of water lords, "panidars" who oppressed the fishermen by collecting taxes and requiring them to lease a segment of the river in order to fish. In the 1980s the Ganga Mukti Andolan, or Free Ganga Movement, sought to boost local people's livelihood. People in the movement later realized that pollution was also a threat to the fishery, so pollution abatement was included in the movement's efforts. The movement lasted for eight years.

In 1990 the Bihar government repealed the act that gave the water lords jurisdiction. The river was free—for a while—and the fishermen were very happy. But there was essentially a power vacuum in a land with weak laws and law enforcement. Soon the stretch of the river near Bhagalpur came into the hands of many antisocial elements—the river "mafia." The mafia threatened the fishermen and extorted money from them, making the situation even worse than it had been under the water lords.

"We believe fifty percent of the problems in the sanctuary would be gone if the police could control this mafia," said Sunil. "The police should be patrolling the area, but they're not doing so. The Forest Department will say they have no fuel for operating patrols. Police say this is their last priority, because there are so many problems in the city and villages.

"If law enforcement did its job, the fishermen would feel secure and would not practice detrimental fishing. The extortion demanded by the mafia forces them to do extra fishing, beyond subsistence, in order to pay extortion money and feed their families. They know it's detrimental, but they feel they have no choice.

"We have to work very patiently," Sunil continued. "We have completed the initial phase. When we started very few people knew this was a sanctuary and that the dolphins are important. Through street theater and education campaigns we have taught people that there is a sanctuary, that the dolphins are here, and why they are important. At one time there was directed killing, intentional killing of dolphins for their oil. But now this is rare." Dolphins get entangled in nets and die, Sunil said, but few are killed outright. When his group started its work in 1998 there were ninety-five to a hundred dolphins in the sanctuary. In 2007 there were 150. "In ten years we have done better than stabilize an endangered species' population here in the sanctuary. The population has increased by 50 percent."

"So the threat to the dolphin is basically the essential economic conditions of Bihar?" I asked.

"Yes, exactly," said Sunil. "If socioeconomic conditions improve, the problems for the dolphins will resolve. But the dolphin doesn't have that long. The dolphin can't wait for the economy of Bihar to improve. The species could go extinct before then, so we have to work faster."

"And the conditions are alarming," Sunil continued. "The river has narrowed, the magnitude of water has lessened, and pollution has increased. If river Ganga dies, the dolphin dies. If river Ganga dies, India dies."

The birds in the trees all around us kept singing. We both sank into silence.

The crunch of tires on gravel grew louder as a jeep sped up to the headquarters. Ajaya Nayak, the conservator of the forest in this region, who is also called B. C. Nayak, got out of the land cruiser and gave his assistant orders to bring fruit and refreshments.

Sunil demonstrated the diplomacy he uses to establish rapport with forest department officials. He told Nayak that he and his staff are the caretakers while Sunil and his team are always willing to help them whenever they ask. Sunil said Nayak listens to local people's problems, but the problems can't be solved in one day.

Nayak told us he was short-staffed. He had only three foresters who tried their best to protect the animals, but maintaining the sanctuary was difficult: "It's a thirty-mile stretch of river and we have only two motorized boats based in Bhagalpur. Once every two weeks staff go out on patrol, near Bhagalpur only. So we have difficulty finding offenders." Nayak said his staff could not patrol the whole sanctuary because they lacked money for fuel and boat maintenance. "We patrol just to generate some fear among the antisocials."

The forest officials were not patrolling the whole sanctuary, nor did the police know what activities were taking place there. In the small area they did patrol, people who were fishing illegally somehow got the message that a patrol was coming. As soon as the patrol left, they would restore their nets. The mafia employed laborers to do illegal fishing for them, Sunil and Nayak told me. The forest department does not have guns.

Nayak had been in the Forest Service for more than twenty years and here in Bhagalpur district for only two years. He admitted he didn't know what he was getting into. He had no knowledge about dolphins and didn't know why they were important. He didn't even know he would have both a forest and a river as well as dolphins to oversee when he started this job.

The sanctuary did not have a separate staff, Sunil explained; Forest Department staff oversee the river in addition to their other work. But supervising the forest and the dolphin sanctuary are completely different efforts. Sunil asked the Bihar government for one staff member just for sanctuary administration, more funds for patrolling, and guns for the forest officers.

These measures should help, said Sunil, because right now forest guards have no protection. There is no insurance on their lives. Lacking motivation to take any chances fighting offenders, they keep themselves safe because the needs of their families are more important. Sunil and the Forest Department were also recruiting a group of local volunteers to report illegal activities and injured dolphins to the authorities.

I asked Nayak about the fishermen's complaints that his staff had taken away their boats and nets. He said he didn't know how many fishermen had been arrested, maybe ten seizures in the past two years. None of the cases have gone to trial yet, so the boats have not been returned. In spite of the difficulties and the lackluster record of his staff, Nayak said he liked Bhagalpur and wanted to try to help the dolphins.

He admitted that the real perpetrators were not being caught. Most of the illegal fishing went on in the middle of the night, but the forest patrol did not go out until after daybreak, which is the same thing I had heard about the forest officials in the Panna Tiger Reserve. By that time the culprits had left and taken their catch with them. The moment they see the patrol boat coming they leave their nets and flee, said Nayak.

Disagreement over the size of mesh in the fishermen's nets was often a pretext for seizing boats and nets, Sunil told me. The fishermen were using nets with a mesh smaller than the legally allowed size. Sunil had been lobbying to get the legal size reduced to something more realistic so that fishermen could catch some fish for themselves while the fish the dolphins like could go through the mesh.

The dolphins get caught in these nets and sometimes die, because they go to them looking for fish. The general shortage of fish in the river drives them to the nets.

One evening Sunil invited Pushpa and me to have tea with his family; it was virtually a meal, abundant with fruit and sweets. Afterward Sushant Dey took us down to the dolphin team's office, a cubicle-size room on the ground floor of the building where Sunil had an apartment. The team is chiefly Sushant and his brother Subhasis, along with a former fisherman named Karelal Mandal. He was a fisherman for thirty-five years, but has been working for the sanctuary for more than a decade. He shares the Dey brothers' interest in conservation, and it's a job in these lean times.

On the computer, Sushant showed videos documenting the kind of work the team does. When fishermen report dead dolphins to the brothers, they first go to investigate, then contact the wildlife officials who are responsible for disposing of the bodies. Because it's an endangered species, no one can keep the carcass—even, apparently, for scientific study.

The forestry staff defers to Sunil Chaudhary and his team to do autopsies, even though it is their responsibility. The fishermen contact the team instead of the authorities for fear of being blamed for the death. Once when the team reported a dolphin kill in the sanctuary the authorities tried to file a case against Sunil and the Deys, saying they had killed the dolphin. Sunil was called to the high court, where he showed one of the group's videos. The court dismissed the case and admonished the forest officials.

There are no funds for research on the dolphin carcasses even if the team could keep them. The Deys and Karelal must learn as much as they can about the animal's health and how it died by performing a crude autopsy, often on the floor of a dark and dirty warehouse at the compound where the forest officials live and work.

The drama of an endangered species unfolds as the grainy amateur video flickers on the computer monitor. A noisy fan dispels some of the April heat in the tiny room. Smoke from a mosquito coil winds lazily in the air. The power goes on and off several times while I peer at the screen intently. The building's noisy backup generator kicks in after a few minutes.

In one scene Sushant takes dolphin "juice" away from some fishermen. This is part of their patient campaign to persuade the fishermen who still take dolphins for their blubber to abandon such harmful practices.

Another shows an illegal style of fishing—a rowboat with a diesel motor that has lines with multiple hooks. Again the brothers try to persuade the fishermen to abandon the practice, for their own sake and the dolphins'.

As the video plays, Sushant points out a dead dolphin entangled in a net, an old male. He and Subhasis take measurements and pictures and inform authorities they should pick up the carcass.

The next scene shows yet another dolphin caught in a net. Sushant and Subhasis have taken a forest ranger to see a dead pregnant female. They hoist her body onto the bed of a small pickup truck.

The forest department veterinarian is supposed to do the postmortem but demurs, so Karelal cuts through the thick blubber with a small curved knife. The baby dolphin pops out in its fetal sack, like a live birth. The baby was fully formed, so this snaring, whether accidental or intentional, took two members of the endangered species.

Karelal uses the recalcitrant veterinarian's dull instruments to continue the autopsy, unskillful butchery in a good cause. Because of the dull knives and the thick blubber, it's a slow, unpleasant process. After this incident, Sushant told me, he went to the market and bought some surgical instruments for future autopsies. He said he would have tried to bottle-raise the baby dolphin himself if only he had arrived in time to open the mother and rescue the baby.

The men look at the contents of the dolphin's stomach, to learn what fish she ate and to see the endemic worms that create the dolphin's turpentine breath. The pictures flicker on. One shows a dolphin with a fishnet in its stomach. Another shows a decomposed dolphin, weighed down by bricks. Poachers were waiting to bring it ashore and take the blubber, some fishermen related to the dolphin team.

Sushant mentions a particularly horrifying but common practice. Some of the men who fish illegally pour pesticide into the river's side channels to kill the fish and make them easy to catch. They net only 40 percent of the fish that die. The rest wash away, the poisoned fish enter the food chain, and the river's load of dangerous effluent increases. This practice is common in other parts of South Asia. It seems to be an unfortunate, postindustrial version of an age-old practice of using plant poisons to stun fish and make them easier to catch.

Sushant told me the river mafia threatened his brother and Karelal several years earlier. Through backdoor channels, some fishermen managed to let the team know about a planned attack on them. The river bandits called off the attack, apparently convinced the repercussions of harming the dolphin team would be too serious. However, the team has twice been fired on by the mafia, Sunil said.

The latest census of dolphins in the sanctuary showed 174. Sushant said that 230 would be optimum. He said they hadn't seen a carcass for several months at that point, though they had heard reports of bodies upstream.

Karelal, the former fisherman, believes that longtime fishermen are part of the ecosystem too and must be included in any attempts to save the dolphins. The government should try to restrict nontraditional fishermen because they create the biggest problems in the sanctuary. Anyone can just buy a net and drop it in the river and call himself a fisherman. One of Karelal's sons still earns his living by traditional fishing. If only this kind of fisherman were allowed to fish and the government controlled the real criminals, this would save the livelihood of traditional fishermen and help protect dolphins.

Sushant agreed. "Farmers just go and buy nets in the market and become fishermen."

"Because they're not making enough money as farmers?" I ask.

"Not making enough is not the point," says Sushant ruefully. "Making more is the point. Sometimes farmers with land on the river won't allow the traditional fishermen to fish near their land. Nontraditional fishermen use any technique to kill fish; it doesn't matter to them. But the traditional fishermen have their own techniques. Every season they catch a different kind of fish using different types of net. But the nontraditional ones use poison, illegal nets—any method to catch fish."

Sushant had no patience with forest conservator Nayak and his staff or their excuses about why they could not improve this situation. Sunil cut him some slack, because he knows Nayak gets so little support from the government to do his work,  but would be happy if the government took over and did a good job policing the sanctuary. If the authorities could make Vikramshila a real haven for the dolphin, outlawing their slaughter with the firmness of the ancient King Ashoka, Sunil and his colleagues would withdraw.

Some observers are not optimistic. They think another ten years could wipe out a species that has survived for one hundred million years. The susu was already in these rivers before the Himalaya were born.

# 15

# Beyond Barrages and Boundaries

> When the people of Bangladesh cry, the sound
> does not reach across the border.
> —Mustafa Kamal Majumder, Editor, *New Nation*

A low dam girdles the Ganga about sixty miles beyond Bhagalpur. More than a mile and a half across, the structure is the longest barrage in the world. It has 109 gates, almost twice as many as the Koshi barrage I traveled over near the Nepal-India border. Its name, Farakka, is anathema to people throughout Bangladesh. In India mainly fishermen on the Ganga know much about it. The barrage, which sits just eleven miles from the international border that separates the tiny nation from its big neighbor, has poisoned relations between the two governments for forty years.

The story of Farakka is one of the thorniest river disputes on the subcontinent. Whole books have been written about it on both sides of the border as well as by international commentators, not to mention the technical treatises it has engendered. The barrage did not accomplish the task for which it was built and has harmed people in both India and Bangladesh. Farakka offers a warning about how *not* to handle transboundary rivers to prevent complex subcontinental water-sharing problems from becoming crises in the future.

Borders fragment the river system in the Ganges basin, creating unique transboundary water management challenges. To visualize the Indian subcontinent's river-sharing problems, imagine a slice of pizza. Take a bite out of the middle of the bumpy top crust. That's Nepal. Then take a small bite out of the right, or eastern edge, just below the crust. That's Bangladesh. The rest of the slice is India. These three nations share the greater Ganges basin. The river spills into the Bay of Bengal in Bangladesh after flowing across the wide top part of India. Many of the river's major tributaries come from Nepal. The smaller slice of pizza to the west would include Pakistan and the Indus River, but that's another complicated story.

Now move the piece of pizza to North America and pretend the United States is the majority of the slice. There's a tiny "Canada" stretched along a portion of the border to the north. It's small both in terms of population and size, but it has hydropower potential in its mountains and a wealth of clean, clear water. Many of

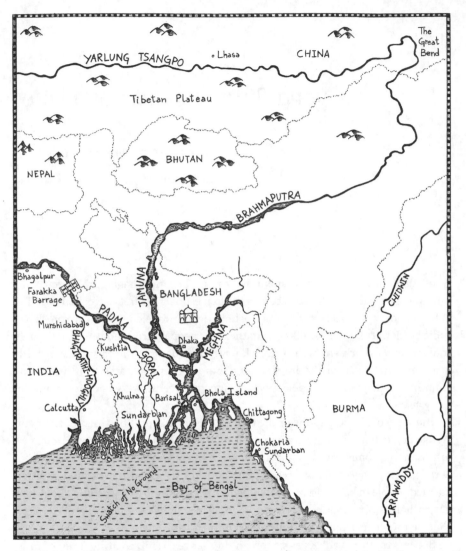

Map 7 Bangladesh and Lower Ganges–Brahmaputra Basin. Created by Kanchan Burathoki

the rivers that its southern neighbor relies on flow from the smaller country. Imagine "Mexico" to the southeast, in place of Florida perhaps, and partly engulfed by its enormous neighbor. Almost all the rivers that arise in "Canada's" mountains flow down through the agricultural heartland of the "United States," then coalesce and flow through "Mexico" to the ocean. Like Florida, this "Mexico" is low lying and vulnerable to rising seas. This picture roughly sketches part of the power dynamic that governs South Asian water resources.

India, much like the United States, is sometimes so bent on its own imperatives that it becomes oblivious to its size and power, trampling other countries' sensitivities. Or at least the other, mostly smaller, nations bordering India perceive this to be the case. Thus, much as the United States would do if its water relationship with Canada and Mexico were like India's with its near neighbors, India has been trying to establish some control over rivers coming out of Nepal and flowing into Bangladesh.

Vis-à-vis Nepal, India is the "lower riparian," the country downriver; upstream nations like Nepal can typically claim a certain advantage. Vast as Nepal's water resources may be, the country has relatively little infrastructure. Nepal's political instability and lack of funds hamper intelligent development of the resource. So the small nation has depended on its big, stable neighbor to the south for some development. India has the resources to build infrastructure and would like to capture and store Nepal's water, but so far the latter has resisted the control that extensive development would entail.

In relationship to Bangladesh, India is the "upper riparian." For much of the 1,500-mile length of the Ganga, India is a fairly dry upper riparian, as well as a heavily populated one. The country feels entitled to hold on to the water flowing within its borders. The Farakka Barrage, just west of the border with Bangladesh, allows India to grab some of the Ganga's water before it can flow into Bangladesh. The water is diverted southwest through India, to Calcutta.

Until about three hundred years ago, the mouth of the Ganga was probably near Calcutta, now called Kolkata. But the Ganga began shifting her course centuries ago, as is the wont of rivers in this region. About twenty-five miles downstream of the Farakka barrage, the river splits into two. The channel on the left flows out of the Indian state of West Bengal into Bangladesh; the channel on the right continues on down to Calcutta. The left channel is called the Padma in Bangladesh; the right channel is the Bhagirathi.

As some still see it, the Ganga spills into the Bay of Bengal through the Bhagirathi; but to most people, the main stream of the river is now the Padma. The Padma meets the sea after joining with the great Brahmaputra—the "son of Brahma," Hindu god of creation, progenitor of mankind—as that river sweeps down from the north following its long journey across Tibet as the Yarlung Tsangpo. In Bangladesh the Brahmaputra is called Jamuna. Like the aspects of Hindu deities, these rivers take new names as they travel and shift.

For orthodox Hindus, the true mouth of the sacred Ganga remains the mouth of the Bhagirathi, also known as the Hooghly in its lower reaches. This is the Ganga distributary that carries the sacred waters to the sea near Calcutta. Distributaries are a sort of mirror image of tributaries. Tributaries carry water into the main stem of a river; distributaries take water from the main stem to the sea.

Up in Uttarakhand, the Bhagirathi is the tributary that descends from the Gangotri glacier and later takes the name Ganga, so there was a kind of poetic symmetry in the Ganga's emptying through another Bhagirathi, named for the mythical king who was instrumental in bringing Ganga down from the heavens. But for hydrologists, there is really no true mouth for a river system as vast as the Ganga's. As there are many sources of these great rivers, there are also many outlets.

I suspect there could be an element of religious chauvinism in the Indian feeling of ownership of the holy Ganga, more than three quarters of which flows through India. The real beginning and end of the Ganga should be in Hindu India, according to such thinking; it would not be right for mother Ganga to join the ocean in a Muslim land. But it might be clear by now that the idea of owning any of these rivers misses the truth of them. In the words of Manoj Mishra, that is just an "engineer's dream." Who owns the Ganga as she finds her way to the sea? India? Bangladesh? The goddess herself? A Bihari or Bengali farmer might give us the best answer. I wish I had asked.

Whether or not religion has anything to do with this sense of ownership, India as a sovereign power clearly believes it is not only entitled to make decisions about the water that flows in the Ganga, but that it has a duty to apportion water resources according to its citizens' needs. Hence, soon after independence, the new nation began to make plans to divert the river's waters south from what had become its main stem, the Padma, into what was likely its main stem a couple of centuries before, the Bhagirathi.

The reason given for diverting the Ganga's waters was that the riverbed near the port of Calcutta had an excess of silt. If the plan worked—and the engineers who conceived it were certain it would—silt would be flushed from the lower reaches of the Hooghly River so that large ships could navigate up to the port of Calcutta for the sake of Indian commerce. Theoretically, more water in the Bhagirathi-Hooghly would disperse the excess of silt, move it out to sea, and deepen the channel for ships.

Farakka, about 150 miles north of Calcutta, was conceived as a way to save the port from further deterioration, though there were already experts who thought Calcutta port not worth saving. When Bangladesh was still East Pakistan, soon after the subcontinent's independence and partition, the engineers of the young India began planning a barrage that would slow the Ganga, allowing water to flow into a canal and thence into the Bhagirathi-Hooghly.

The plan might have been in the works even before then, while the future leaders of India were jockeying for territory in the process of partition. The idea of such a diversion had been kicked around during British rule, but it didn't go anywhere. At the time of partition, the potential site for such a barrage was placed in Indian territory. In Britain's hasty drawing of lines that created India and Pakistan in 1947, the majority Muslim district of Murshidabad was given to India, despite protests from Pakistan, and India got control of the headwaters of the Hooghly. Farakka is located in Murshidabad, now part of the Indian state of West Bengal.

The chief engineer of West Bengal, Kapil Bhattacharya, opposed Farakka. He said it would not work because there was no way to divert enough water in the dry season to send the quantity required down the Bhagirathi to Calcutta. He blamed the silting up of Calcutta port on a TVA-inspired project in the Damodar Valley, which had reduced flows down the Bhagirathi. Pakistan, understandably, embraced this argument. The Indian government branded Bhattacharya a traitor and he was fired from his job.

When government officials in Pakistan learned of the plan to build a barrage at Farakka, apparently from an article in a newspaper in West Bengal in 1951, they worriedly inquired of the Indian government whether this was true. The Indians demurred, saying it was just an idea, only in the planning stages. Correspondence between the two governments ensued, but for the next ten years, India more or less continued to keep its plans vague. Finally, in 1961, Indian officials admitted to Pakistani officials that construction had begun. Diplomatic meetings followed; construction continued.

Then came the 1970 war of independence. India sent soldiers to help East Pakistan, now Bangladesh, gain its independence from West Pakistan—a country with a different language and culture with a thousand miles of India separating it from its eastern sibling. After this war of independence, the relationship between India and Bangladesh was at its "high-water mark," according to Mohammed Maniruzzaman Miah, former professor and vice chancellor at Dhaka University and a water resources expert. Trained in France and Britain as well as in Bangladesh, Miah was also an ambassador and later a member of a national anticorruption commission.

"We wanted to believe what India said," Miah recalls. According to his account of the sequence of events, India told Bangladesh there would merely be a test run at the barrage for about a month during the dry season, and after that the two countries would come to an agreement about sharing the water. During this friendly period, prime ministers agreed there would be joint studies on issues of common concern like flood control, irrigation, and river basin development. A Joint Rivers Commission was set up.

"But that was a ploy," says Miah, who was born in the majority Muslim district of Murshidabad. His family relocated to Bangladesh a few years after partition. "They told us it was just for test running. Then they said since you have allowed us to run it, it can go on." The "test" was supposed to end May 31, 1975, but water diversion continued beyond that date and has never stopped. Still, in the very fact of the Joint Rivers Commission, Bangladesh may have acknowledged India's right to Farakka; and India may have conceded Bangladesh's right to share the water. Whether that share would ever be enough is a different question.

The year after the diversions began, as might be expected, less water flowed into the Padma. "In 1976, during the dry season," says Maniruzzaman Miah, "the water was a trickle down Farakka—some 6,000 cusecs (cubic feet per second). Down from 60,000 cusecs before. People were alarmed."

Hundreds of thousands of Bangladeshis marched forty miles toward Farakka that May to protest the sudden reduction in water flowing into their rivers. Soon after, Bangladesh brought the issue before the UN General Assembly, but most of the body's member states were disinclined to take sides in the dispute. Bangladesh had to resume trying to negotiate with India. Late the following year the two countries signed an agreement.

"In my opinion it was the best bargain we could have," says Miah. The agreement included a formula for sharing the water and assured Bangladesh of a minimum flow during the driest time of the year. But the problem wasn't really solved for either country. Both wanted more water than was available in the river. The five-year agreement expired. During the next twenty years, meetings continued as the two countries haggled over the number of "cusecs" flowing below Farakka during the dry months.

An interim agreement in 1985 was superseded by a treaty in 1996 that Maniruzzaman Miah faults because it doesn't guarantee Bangladesh a minimum amount of water. It provides instead for emergency meetings between the two nations should the amount of water in the river fall so far below historical levels there is not enough water to parcel out according to the agreement.

"The whole agreement comes to this," Miah continued. "Whatever amount of water arrives at Farakka would be shared between the two countries. Now, upstream of Farakka, the Ganges travels about 1,500 miles. There are many water works, diversion canals, small barrages. So they can manipulate the water anywhere and very little comes down. Some people might say that because water passes through West Bengal, their territory, they won't bring it down to a catastrophic level. But our experience doesn't show that. They have other sources, upstream dams here and there.

"The issue is complicated by global warming. There will be more water in monsoon season and less and less in the dry season. It appears that even nature is going against us," Miah said with resignation. "Our country will become desert in 50, 60, 100 years. I won't live to see this desert."

Today Bangladeshis attribute a host of problems to the continued diversion of the Ganga into the Bhagirathi-Hooghly. Chief among the problems is the matter of the Gorai River dying. Some say it's already dead.

The Gorai, a major distributary of the Padma, was once a perennial river full of fish and the source of irrigation water for the western districts of Bangladesh. But a mound of sand blocks the mouth of the Gorai, cutting off the flow of water from the Padma. For miles downstream there is a swath of silt, punctuated by isolated pools growing stagnant as the dry season progresses.

A Bangladeshi named Dr. Anwarul Karim took me to see the mouth of the Gorai in February 2007. Short, with a little extra girth that showed his fondness for the delicious Bengali sweets he urged me to try, Dr. Karim dressed more formally than most professional men in the tropical climate of Bangladesh. The days I met him he wore a dark suit and tie with a blue shirt. Professor and treasurer at the Islamic University in Kushtia at that time, his gray hair was thinning, and his goatee already white, though he was not old.

Karim was a busy man, full of energy, though somehow he seemed fragile to me. He coughed often. I worried that the air pollution of Dhaka had irritated his respiratory system. Most of all he was sad about his wife, whom he loved intensely, crediting her with having made it possible for him to pursue his academic career and many other interests—in Bengali folk music and literature, as well as a crusade to help villagers cope with arsenic contamination of groundwater.

His wife was very sick, weak, and jaundiced. Doctors had intervened with the wrong treatment when she fell ill some years before, ruining her kidneys. Now she would never recover her health; she had to have regular dialysis and round-the-clock care. She could not return to their home in Kushtia, to her garden and flowers, but had to stay at her daughter's small apartment in oppressive Dhaka.

In his effort to be a good host, Karim, who went by his last name, talked continuously. I began to feel he was trying somehow to ward off the flood of calamities that fall on his people with this flood of words. For the most part he fell silent only when he was praying or listening to his friends make music.

Karim gave me a ride to his district, located a half day's drive northwest from Dhaka, so that I could get a sense of Bangladesh outside the vast metropolis. To help me understand the effects of Farakka, late one day he took me to the edge of the Padma near his hometown, Kushtia. The Padma here was very wide—much wider than the dry-season rivers I saw up on the Gangetic plain—even though the driest season of the year was approaching. It looked more like an ocean than a river, its banks more like beaches. Here as the river moves into Bangladesh is the beginning of the final gathering of the waters before the rivers give themselves to the sea.

Some workmen were gathered on the low dunes. They seemed to be harvesting sand for construction; neat piles of it sat close to the shore. The men had a boat, and Karim asked them if they would give us a ride closer to the mouth of the Gorai.

We chugged down the river just as the hour for late-afternoon Muslim prayers arrived. Karim spread a kerchief on the rough deck of the boat and kneeled, facing west toward the setting sun, toward Mecca. He lowered his head to the kerchief. A local official who had joined us spread his prayer cloth on the boat's rough boards. The rest of the men on the boat did not.

I was accustomed to such reverence, because I often visit Hindu temples and Buddhist monasteries. But here I was surprised. I tried not to stare. I thought of their prayer as private, and as a gentile (or whatever I am) I felt I was an intruder who was not necessarily invited to look on. But I was moved by Karim's reverence, here on the boat on the way to the Gorai. I had been with him a couple of days earlier in Dhaka about the same time of day. He went to a nearby mosque, then returned to continue our conversation. Here on the boat, the prayers seemed almost directed to the river.

The two men prayed briefly, less than five minutes. Then they got up and started telling me about the river as we neared its mouth. They told me the water from the Padma could not pass the mouth of the Gorai, for it was now completely blocked by silt. As a result there was not enough water flowing down to the southwestern districts of Bangladesh, which had hurt industry and agriculture there.

"The Gorai is finished," said Karim. A few years earlier the Netherlands had funded an effort to dredge the mouth of the river to allow water to flow in it again. Work continued for a couple of years, but when a new Bangladeshi government came in, officials stopped the effort. In any case, dredging would never be a permanent solution for the siltation, because in a few years the river's mouth would just be blocked again. All that Himalayan silt from the upper reaches of the watershed comes down from the mountains—through Nepal, Uttar Pradesh, Bihar, West Bengal—to rest here.

Karim favored a more permanent solution, another barrage to force water into the Gorai. This barrage would be placed downstream of the mouth of the Gorai so that water in the Padma would slow down, pool, and flow into the Gorai. For lack of funding, it seems, this idea hasn't been implemented. "The time will come," Karim predicted, "when the women of Bangladesh will sell their ornaments to construct this barrage, the conditions on the Gorai will be so bad."

In a few minutes we reached the mouth of the Gorai. I saw only an expanse of sand. Trying to get up high enough to see over the pale ridge of sand on the shore, I stepped onto the narrow prow of the wooden fishing boat. The men urged me to be careful; but I wanted to see for myself what was on the other side of that ridge.

The orange sun was setting behind us; the balmy winter air was hazy. The flat surface of the water, the sand in the distance, the pale sky, blended into each other. I saw only water and then sand, sand, and more sand. Except for the wide river around us, whose current gently rocked the boat, it looked like a desert.

I saw nothing that resembled the mouth of a river. "For four kilometers, from here on down, it's all silt," Karim said. "Then there are ponds." During the monsoon the water rises high enough to flow over the mounds of sand. As the monsoon subsides, ponds remain in the river channel. Downriver, tributaries enter the Gorai and it becomes a river again, but not the robust one that used to carry water into the southwest districts and beyond, to the Sundarban—the mangrove forests on the coast.

Without the regular pulses of freshwater that once flowed down the Gorai, the delicate balance of fresh water from the rivers and saline water from the Bay of Bengal has been upset. Increased salinity in the swath of mangrove forest along the coast has altered the mix of trees and changed the ecosystem, according to some experts. Saline water has now pushed far inland, rendering precious agricultural land far less productive, in some cases useful only for shrimp farms.

Yet the diversion of water at Farakka has not yielded enough water to flush out the silt deposited by the Bhagirathi-Hooghly. And so the Gorai continues to die and still there is not enough water in the Hooghly to rescue Calcutta's deep seaport.

Whenever I think of the Gorai I think of Karim's pain, and the twin griefs he told me about: the drying Gorai and his dying wife. The river may have been shifting and silting up anyway, as a number of experts maintain. But its demise was hastened by an intervention: a barrage that failed to accomplish its central purpose for India and did damage to Bangladesh. There was a disturbance in the system;

the engineers made it worse, like the doctors whose clumsy treatments of Karim's wife made her illness a slow death.

Bangladeshis aren't the only ones who would like to wish Farakka away. Farmers in West Bengal are afflicted by a surplus of water. The backup of water behind the Farakka barrage creates waterlogged areas in that state; and the barrage keeps silt from flowing downriver, so the riverbed has risen. Riverbank erosion has increased. Villages and farmland have been washed away, leading to landlessness and marginalization. Thousands of Indian farming families have been displaced, creating more rural refugees looking for work in the big cities.

An anadromous fish called hilsa, the "salmon" of the lower Ganga, is threatened. The numbers of this fish, highly prized in both countries, have dropped sharply because the hilsa can't swim upriver past the barrage to the places where they once spawned.

Some river experts think the Ganga will eventually "capture" the Bhagirathi. The course of the Ganga has been drifting west since the construction of Farakka. Now at a point below the feeder canal that takes water from the Ganga to the Bhagirathi, the channels of the two rivers are only half a mile apart, having been about two miles distant fifteen years ago. The Bhagirathi's bed is higher than the Ganga-Padma's, so all the water would flow into the lower channel and go to Bangladesh, defeating the purpose of the canal. To avoid such a capture, India may try some additional engineering, though it's not yet clear what that might be.

If, like the embankments in Bihar, the Farakka barrage is an engineered solution that went awry, leading to unintended consequences, why not open its gates and let the river flow freely at all times of the year? At this point, even such a move—unlikely on India's part—would not solve the problem. By now there are so many irrigation canals sprouting from the main stem of the Ganga upriver from Farakka that the remaining water would not restore the Gorai to its pre-1975 flows. And districts below Farakka in India have become dependent on irrigation water from the Bhagirathi in the dry season. As India sees it, the water is somewhat wasted on Bangladesh, a nation that suffers from floods even more than others in South Asia, while India is crying for water. That need, combined with being the upper riparian, justifies keeping the water.

"Even today, if you talk to Bangladeshis, they say it's a problem," says Ramaswamy Iyer—former secretary in the water ministry, widely praised for his subtle understanding of water resources though now often seen as a maverick by India's water establishment.

"It's not. It's a solved problem. There's a treaty."

"Then why are they still so unhappy?" I asked him during one of our conversations in Delhi.

"There are two different questions here," Ramaswamy continued with his customary careful, measured analysis, a skill honed over years working within the

Indian bureaucracy. Ramaswamy worked behind the scenes during negotiations that led to the 1996 treaty, which still governs the sharing of the Ganges at Farakka.

"One question is whether Farakka is a good or bad project," Ramaswamy continued. "It was the best advice that Nehru got, in any case. He asked, 'Will this cause difficulty with East Pakistan?' He was told at the highest level of engineering that there would be no problem. But then Pakistan started raising objections. In the meantime Bangladesh was created.

"I'm not saying it's a sound idea. It was the idea of engineers at that time. If you ask Pakistan for concurrence they will not agree, so that means you will not be able to build it. Therefore, Indian engineers said 'start building.' It was ready for commissioning by '67 or '68. But it was not commissioned. Talks were going on between India and Pakistan."

Both nations saw the waters of the Ganga as essential to their economies. India stalled on talks and proceeded to build what it felt it needed.

The other question, according to Ramaswamy, is whether the water diversions are both legal and justifiable. He says Indian negotiators were inclined to think that the early agreements in the 1970s and 1980s gave Bangladesh too much water, an attitude he does not defend. "One thing is clear—an agreement was needed. It was not a question of giving water but of recognizing their rights. I tried to put it across to the Government, 'Look they have a right.' Internally, of course. I couldn't say this in public. The tendency then was to assume that Bangladesh has a whole lot of other rivers. There's a huge country and population here and that part of India was entirely dependent on the Ganges. They had the Meghna, the Brahmaputra, and so on to choose from. We didn't.

"They had a strong grievance that this was a unilateral diversion and that it caused grave injury. Undoubtedly it caused some injury, any diversion does. It caused some difficulty in India too. They tended to overstate; we tended to understate."

Mustafa Kamal Majumder, the editor of *New Nation*, one of Dhaka's daily newspapers, disputes the Indian belief that Bangladesh has more water than it needs. "It's generally assumed that there is more than enough water because there are so many rivers. The three great rivers of the eastern Himalaya—the Padma, the Brahmaputra, and the Meghna—meet in Bangladesh. Why should people be crying about scarcity? But scarcity has to be explained on the basis of what you normally get and don't get anymore because of some action.

"To be a journalist here is to cover water, floods, embankments," Majumder says of the expertise he developed over thirty-five years. He adds that the Bangladeshi perspective is seldom understood. While well-informed Indians believe the water is just going to waste when it's allowed to flow to the ocean, Majumder counters "in nature there is nothing in excess.

"What is carried by rivers from the Himalayas is very important to the sustenance of certain species in the Bay of Bengal," he continues. "Bangladesh owes its origin to the rivers. They have built the delta. The ecosystem is based on a one hundred percent water system. If you divert the water, it is not Bangladesh. It is something else. If you don't have fresh water flow from upstream then there is salt water coming up from the ocean.

"Bangladesh has a huge flow of water, but you have to compare Bangladesh to Bangladesh. You cannot compare it to Rajasthan." The agreements with India "basically say give some water to Bangladesh. They do not say that Bangladesh has a natural right to this water." Majumder says a third of the millions of farmers, fishermen, and boatmen in Bangladesh have been affected by the reduction in the amount of water. Some of them have become environmental refugees because they can no longer make a living. "India says they are all coming to India. We say no—they are coming to the cities in Bangladesh. In any case, the people who will leave our country will not be the highly skilled ones."

Ainun Nishat agrees. He was Bangladesh's country representative in the International Union for Conservation of Nature (IUCN) for some years as well as a Bangladeshi representative on the Joint Rivers Commission. A former engineer, now vice chancellor at BRAC University in Dhaka, he says Farakka has seriously harmed Bangladesh, hampering the development of its economy—especially in the southwest region, the portion watered by the Gorai and other Padma distributaries. "The Sundarbans are dying. It is not global climate change. It is salinity due to lack of sweetwater. I am still convinced the main problem is Farakka, because sea level rise is not yet that great."

Agricultural productivity has not increased in the southwest the way it has in other parts of Bangladesh because the soil has become saline, the land unproductive. Industry has not developed in the region. "All the industry is in Dhaka and Chittagong, not Khulna and Jessore, which was a major industrial belt in the 1950s. But after the 1970s the region collapsed because of the rise in salinity. Freshwater for an industrial plant would have to be brought from up north, so the economy collapsed."

Nishat cites Government of India statistics that there may be twenty million Bangladeshis working in India. "Most of these people are from the districts affected by the Ganges—Kushtia, Jessore, Khulna." And they are working at the very bottom of the economy, he says, "as street cleaners or prostitutes." He says there has been poor documentation of the damage done to Bangladesh's economy and environment. He would like to correct that, document the effects, and ascertain the minimum flow required to reverse them. Then negotiate with India.

Though Nishat doesn't minimize the damage done by Farakka, he is realistic about having to work with India. Bangladesh has no choice, because it cannot survive alone economically. "We have neither military power nor economic power. But our economic and environmental refugees can cause harm, so it's better to help Bangladesh" to keep its people at home and productive.

Anti-Indian feeling in Bangladesh is "bad for our economy." And economic growth in Bangladesh would put many Indian worries to rest, he says. "But there is no urgency on the part of India to do anything. Urgency is only on the part of Bangladesh."

B. G. Verghese, the former newspaper editor and water resources expert who supports the Indian river-linking plan, says "the whole Farakka thing is a great tragedy. We are the bigger country, we have many more options than Bangladesh. It's very frightening for them in their situation. They've got to do business with India. We should be mindful of our size. We can innocently trample on people without knowing."

Still, Verghese believes Bangladesh "got a good deal more water than they bargained for. They've got the treaty and the water and the problem remains. They're unable to use the water; it goes to the sea.

"The problem wasn't Farakka that dried up Bangladesh," Verghese continued. "It's the whole Ganges system. The rivers are moving eastward." Verghese cited records in Pakistan showing that sometimes the mouth of the Gorai was blocked before Farakka. Water might stop flowing down the river as early as November or December, well before the driest time of the year. Other sources concur.

"Farakka barrage may have contributed something, ten percent I imagine. Ninety percent of the problem was there. You've got the water, but you can't use it because that Gorai hump is there. The Ganges carries exponentially larger quantities of silt than any other river in the whole world."

Verghese is not sentimental about Calcutta's port. He thinks India is wasting water on a dying port and should go build a new one. "Water from Farakka would still be needed for irrigation. But we could save a fair amount of water. We should trade this off with Bangladesh."

"I don't think it did either side very much good," says Ben Crow, professor of sociology at the University of California, Santa Cruz, who has studied the conflict over Farakka from its origins to the time of the 1996 treaty. The problems at the port of Calcutta were not solved, yet much of the damage to Bangladesh may not be the fault of Farakka. "I don't think the diminution of flow in the Ganges is due to Farakka but to upstream irrigation. I don't think there's any certainty about the causes of all that. My guess is that it was upstream withdrawals, but Farakka takes out a significant chunk. I don't think we understand the rivers well enough to be sure."

Crow thinks it's safe to say that Farakka exacerbated what was already happening in the watershed. The rivers were changing course and silting up, and upstream withdrawals of water reduced the amount that reached Bangladesh. Farakka compounded these problems. "Indians should never have been so cavalier as to say there would be no problems downstream."

The growing demand for irrigation water in both India and Bangladesh would have led to conflict eventually, Crow believes. Farakka precipitated this conflict several decades earlier than might have otherwise been the case. Most of the back-and-forth over Farakka happened when the two (later the three) nations carved from the British Raj were young. They were sorting out their places in the world and in South Asia. The fact that the tussle focused on Farakka, not on broader issues of watershed management, may have limited later discussions to arguments over cusecs during the dry season instead of opening up an exploration of the two nations' irrigation needs, a discussion so badly needed now.

"India is an easy target, and certainly it is at fault to some extent," says Crow. "But the tendency to replay the scene of the Farakka loss, hoping that repetition will lead to a different outcome, has not helped Bangladesh."

As serious as are the issues of river pollution and urban water mismanagement I have often dwelt on, there is a greater resource crisis on the subcontinent. The amount of water that agriculture requires and wastes looms as a larger threat for the future. This is why Farakka seems unsolved even though there is a treaty; there simply is not enough in the Ganges River system for current needs, using water the way it's being used.

Agriculture often sucks up three quarters of available fresh water, whether in South Asia or the United States. This was the central problem in California during the years of the "Cal Fed" forum I mentioned earlier. Interest groups haggled for years over how to allocate water to urban and agricultural interests, and leave enough for "nature"—for environmental "uses," which ranged from the survival of native salmon species to whether the state's fishermen could stay in business.

California agribusiness wanted to keep its lion's share, estimated at up to 90 percent at that time. Environmentalists said nature was getting short shrift because of dams and agricultural diversions. Serious water conservation efforts in California (drip irrigation, changing cropping) could save water substantially, but agriculture still required far more water than cities did.

In South Asia, water lost to pollution hurts both cities and farmers. One argument against river linking noted that some links would merely move one basin's cruddy, toxic water into another—which would certainly be true if the rivers continue to be fouled at current rates and the river links are eventually built. But no amount of Ganga cleanup can provide enough clean water to satisfy future increased demands for food as agriculture is currently practiced.

South Asia has known famines in recent memory. Fear of famine stalks governments here, which is why food security takes precedence over almost any other water-related crisis. India won't let go of what water it has; Bangladesh can't give up trying to get back what it has lost.

Many water resource specialists would say the solution is conservation, which would include rainwater harvesting, groundwater recharge, and changing crops in some areas. But even the most efficient use of agricultural water might not provide enough water soon enough to avoid the highly engineered alternative: dam a "surplus" river—one that is flush with "excess" water—and divert it. Rainwater harvesting and sewage treatment will put water back in the system, but agricultural overdrafts and waste would squander it. The complexity of coordinating so many different efforts makes big engineering solutions look good. The Indian government, with support from international water specialists, can argue that it must get started on them.

Dams in the Himalayan foothills of Nepal and of India itself are still seen as cost-effective, large-scale rainwater harvesting systems to hold back some of the water that rushes to the sea during the monsoon. If the best solution to the on-going crisis of too little water for both India and Bangladesh is to grab water from a Himalayan river by building another dam, then the next question is: which dam?

Bangladeshis have long advocated dams in Nepal on rivers like the Koshi to store monsoon water; theoretically the water could then flow down the Ganga throughout the dry season, putting enough water in the river for both Indian and Bangladeshi irrigation needs. India would also like dams in Nepal—for flood control, hydropower, and water storage. But progress on negotiations with Nepal has been slow, so India is looking elsewhere for now.

A canal taking water from the Brahmaputra would be a more tangible goal for India because the barrage could be placed in India. The structure that would link the Brahmaputra to the Ganga just above Farakka could also remain in India. The link would require a long, curving canal through the "chicken neck." This strip of land is another of the oddities of partition; it's a narrow piece of India sandwiched between Nepal and Bangladesh, connecting India's far northeast region to the main part of the country.

A shorter and cheaper canal might cross through Bangladesh, if the two countries could agree about this method of augmenting the water in the Ganga-Padma. But to Bangladesh, diversions from the Brahmaputra look like a new way to steal even more of the water the country needs. Were India to dam the Brahmaputra, there would then be a second river on which India could choose, unilaterally, to send less water to Bangladesh. The consequence of losing Brahmaputra water might mean seawater pulsing farther inland in the coastal region—an outcome Bangladesh could not risk.

The proposed Brahmaputra-Farakka link is part of India's river-linking scheme. India wants more water to send to its southern states, not for the failed flushing of the Bhagirathi-Hooghly. Diverting water from the Brahmaputra is proposed as a benefit to Bangladesh because it could help stem floods in the monsoon as well as put more water in the Padma in the dry season, keeping seawater intrusions in check. The Brahmaputra is typically the river most responsible for flooding Bangladesh; but it's also been the only salvation for southwest Bangladesh in the dry season, especially after the Gorai dried up. Bangladesh has opposed India's river-linking plans vehemently and it's hard to imagine Bangladesh tolerating such a risk after the experience of Farakka.

At this point, it all would seem to depend on how much water India could guarantee Bangladesh. And that would depend on the relations between the two countries and what direction transboundary negotiations over water in this region take in the future.

Building dams in Nepal to help Bangladesh would require a breakthrough in regional planning and cooperation. Bangladesh sees the Ganges as an international river and has long wanted decisions on this and other shared subcontinental rivers to be a three-way discussion with Nepal. India, seeing the Ganga as an Indian river, has resisted the idea for decades. Bilateral agreements are hard enough to work out. Besides, India has admitted it doesn't want to have to cope with Bangladesh and Nepal ganging up. But India's long-standing aversion to regional cooperation may have to change soon, because India is not the upper riparian on the crucial Brahmaputra. China is.

Transboundary river issues have become a field of study, an academic discipline. Emerging theories in this field draw on laws governing other natural resources, and on the success or failure of past river-sharing treaties. A body of precedents is emerging. Some water-sharing agreements from other parts of the world have worked better than others. India has a treaty with Pakistan over the Indus—an agreement widely seen as a model, though Pakistan is increasingly unhappy with it. Agreements can become less satisfactory over time because of changes in the watersheds caused by population growth and new diversions or hydropower. Global warming makes the future behavior of these rivers even more uncertain. Existing treaties were based on old assumptions about flows and weather.

Intelligent choices about river management for the future in this complex basin would seem impossible without regional cooperation. India's upwardly mobile population will continue to put stress on water resources. But the majority of the 600 million people in the Ganges basin live on a few dollars a day and have few resources to fall back on when crops fail, floods come, or bad water makes them sick.

Fine minds in South Asia are seeking solutions for the looming water and river crisis in this region. Some outsiders would like to help. Many feel genuine concern about the rivers and people of South Asia. In addition, the consequences of war, famine, and "natural" disaster in this part of the world are scary prospects for the "first world." Perhaps South Asia could use help from outsiders, perhaps not; but governments in the rest of the world are starting to pay attention.

One sign of this is an effort led by the World Bank called the South Asia Water Initiative, funded by Great Britain, Australia, and Norway. The bank has organized forums in different parts of the region and urged the various nations to share information. The bank has so far focused on the Ganges basin, but yearly meetings are drawing participants from Afghanistan, Pakistan, Bhutan, and even China—countries with rivers that arise in the Himalaya. The bank has set up a grant program to encourage research projects that reach beyond borders and disciplines. It is being coordinated by ICIMOD, the Himalayan think tank where scientists from the seven-country region—like Arun Shrestha, with whom I trekked up to various glacial lakes in Nepal—work together.

Ben Crow and a colleague proposed in 2009 that ICIMOD might even play a larger role in the region, beginning to serve in some capacity as a water regulatory institution, monitoring the health of river basins and overseeing participating nations' adherence to international agreements.

"There are unique challenges here and they've only got each other to learn from," says Claudia Sadoff, leader of the World Bank's South Asia Water Initiative. Soft-spoken, petite, and blond, she may seem an unlikely linchpin for the bank's planning effort in this massive watershed. But Claudia is no less passionate than she is knowledgeable about the Ganges basin and wants to see the best possible solutions chosen for this largest, most populated river basin in the world. "Where you allocate your water resources," she notes, "is where you promote growth."

An economist and mother of three vivacious children, Claudia says the World Bank is no longer the "army of engineers" it was when she started working there in her late twenties. Now sociologists, environmental scientists, and communications specialists join teams of engineers and economists. Still, bank teams tend to work on the assumption that infrastructure of some sort will be central to water management and that most traditional systems no longer work in South Asia because they are overwhelmed by large populations and industrial development.

"You can't throw out any option. You want to put as few structures as possible. You want to make the most use out of every development you create, because they all change the ecosystem and disrupt a society," Claudia says. "If you're going to disrupt a site you have to be sure you're going to get the most out of the infrastructure." She says small dams are not necessarily better than larger ones. "Sometimes it makes more sense to have fewer larger, not more smaller, ones. The question is never large or small but what works, and how can you minimize impacts?

"We started with the conventional wisdom, which is that what you need is large dams in Nepal to stop flooding and provide water for agriculture—that there was a way to regulate the system all the way into Bangladesh." That was the accepted wisdom for all three Ganges basin governments. The World Bank team wanted to encourage discussion about that accepted wisdom, to let a little fresh air into the discussion after "so much contentious history over treaties."

Claudia says research on the watershed is continuing; but based on what the team has seen so far, the "traditional wisdom" about dams in Nepal to save India and Bangladesh from floods is not accurate. "If you can't reduce floods in Bihar or saline intrusion in the Sundarban with infrastructure in Nepal, then that simplifies the discussion. Maybe you don't have to negotiate everything across boundaries." If Bangladesh were convinced that dams in Nepal would not help them, whether with flooding or low flows, then it could simplify discussions between India and Nepal over developing hydropower. This would be good news to India, though maybe not to Bangladesh.

It would also lend weight to India's argument that diverting the Brahmaputra water to the Ganga could benefit both India and Bangladesh, by putting more water in the Padma and the Gorai in the dry season. But the bank's own analysis indicates that water stored in Nepal could be of some benefit downstream, so Bangladesh will not be in a hurry to concede the issue. The bank seems ready to referee discussions if asked.

In the meantime, after recent government upheavals in Bangladesh settled down, meetings of the Joint Rivers Commission resumed. Cordial discussions have lately focused on less volatile issues. India and Bangladesh share fifty-four rivers, so there are plenty of other questions for the two countries to sort out. The sharing of the Ganga was the big topic of the past half century; the sharing of the Brahmaputra may be the topic for the next half century, but right now there seems to be a productive lull.

One eventuality might make fierce allies of Bangladesh and India. If it were to become clear that China's plans for the Yarlung Tsangpo would disrupt the Brahmaputra's flows, both countries would have grievances. A shared cause like this could create another "high-water mark" in their relations—as Maniruzzaman Miah described their relations just after Bangladeshi independence.

The Yarlung Tsangpo, a Tibetan river, starts near Mt. Kailash and flows east across the Tibetan plateau. North of Bangladesh it turns sharply south and west, then winds its way through the Himalaya. As the river turns, it drops through a spectacular gorge, sacred to Tibetan Buddhists and so remote it has been only partially explored, and that only by the most intrepid adventurers. The river drops ten thousand feet in this gorge, called the Great Bend of the Tsangpo. The hydropower potential of such a drop is estimated in gigawatts. Chinese engineers would like to tap that power.

India is monitoring the river from satellites. As of 2011 only a 540-megawatt hydropower plant called Zangmu, well upriver of the gorge, was under construction. Indian officials announced they are satisfied with Chinese assurances that this plant is "run of the river." The water taken out at the diversion dam will come back to the river; the flows in the Brahmaputra after the river reaches India will remain about the same.

But on the drawing board are other dams. A whole string of diversion dams in the same region as Zangmu is proposed. And a monster dam at the Great Bend of the Tsangpo called Motuo would supposedly have a capacity of 38,000 megawatts, about double that of China's massive Three Gorges Dam. Another proposed dam at the Great Bend called Daduqia is even larger. There is also talk of diverting water not only from the Yangtze but from the Yarlung Tsangpo into the Yellow River. Some of these ideas may prove so unfeasible, environmentally disastrous, or costly that they will never happen—or even be attempted. Still, it just might be a good time to get the idea of serious regional, cross-border dialogue off the ground.

Claudia Sadoff says World Bank advisors now believe the Ganges system is just too big to run as one entity. This makes a lot of sense; the vastly different dynamics of the few parts of the system I have visited and described here lend weight to that understanding. Yet it seems a little ironic as well. Nature was doing just fine for quite a while running the system as a whole. The greater Ganges river system has been one entity since the upthrust of the Himalaya. Only relatively recently have national boundaries and burgeoning populations, combined with the almost wholesale abandonment of traditional South Asian water management, thrown things out of whack.

Bangladesh is literally the child of the Himalaya. The soil that made Bangladesh started its journey from the highest mountains on earth. The distance from Everest to the Sundarban is less than a thousand miles, about the length of California, yet that distance spans a descent of more than five miles from the world's highest mountains to the sea. The sediments deposited at the bottom of the watershed are several miles thick.

Some bits of soil I walk on in Bangladesh have perhaps come down the Dudh Koshi, stripped from the sides of the river canyon where Dig Tsho burst, or traveled down the Bagmati from the gentle hills around Kathmandu. Silt came down the Rolwaling Khola to the Sun Koshi, or down the Melamchi to the Indrawati. The Ganga delivered it to form this country—once part of Bengal, then part of the British Raj, then part of Pakistan.

Over the centuries the ashes of millions of cremated Hindus have contributed to what is now mainly Muslim Bangladesh. The ashes have floated down from the ghats all along the Ganga, mingling with sediments from the largely Buddhist high Himalaya in Nepal, deposited finally in the delta, where the soil and ash have nourished crops, maybe later to be shoved into the bay by a monsoon flood.

This mingling of earth and humanity belies the fretful borders and uneasy relationships on the subcontinent. Both the rivers and the earth itself say that these national borders are transitory. But right now, in this particular century of South Asia's long history, the borders are very real.

# 16

# Poisoned Blessings

> Do the unbelievers not see that the heavens and the earth were
> joined together, before we clove them asunder? We made every
> living thing from water. Will they not then believe?
> —Quran 21:30

"Any water-related stress you can imagine, we have. Abundance, shortage, pollution. We have them all," a young woman named Afifa Raihana told me on my first trip to Dhaka, the capital of Bangladesh. Afifa was working for the World Bank at that time, coordinating environmental initiatives, having earlier worked as a journalist. The list of water-related problems in Bangladesh is long and sometimes contradictory: waterlogging as well as desertification, floods along with shortages. Bangladesh sees frequent cyclones and storm surges; it copes with salinity and sedimentation of riverbeds. Industrial chemicals, agricultural runoff, and urban sewage pollute the nation's ponds and rivers.

The problems sometimes stem from the sheer abundance of water in this near-liquid land. In the monsoon, a quarter of the land is regularly inundated. When rivers flood, two-thirds of the land may be covered by water—drowning people and their animals, displacing families, destroying crops. This is the bottom of the Ganges watershed; any water and sediment that has not been held back upstream comes to rest here or washes into the Bay of Bengal. On occasion the abundance is a curse, but usually it is a blessing. Maniruzzaman Miah told me that drought is a far greater threat here than floods, which are essential for growing rice and jute and for keeping the water table high. "Rain and the need for rain. That is what Bangladesh is all about. Floods are part of the ecosystem."

The oddly shaped country that is now Bangladesh was once part of a prosperous realm stretching from Bihar to the Bay of Bengal. Bengal, which was partitioned in 1948—half to India, half to Pakistan—was known as the best-educated, most literate, most cultured part of India. It was a grain basket, a seat of Buddhist learning in ancient times, and later had a well-developed textile industry until England's East India Company strangled it to promote English-made textiles. Britain essentially launched control of the subcontinent from the east. Calcutta was the seat of British power until 1911, when the capital was moved to Delhi.

Even then Calcutta remained the queen of shipping, which is why India tried to save its port by building the Farakka barrage.

Bengal was key to Britain's war effort, as well as a focal point in the struggle for independence from Britain. Then, a few years before partition, a drought caused a famine. Five million Bengalis died. The region was crippled, and Bengal became associated with pictures of starving people.

Bangladesh has become the poster child for the nations most vulnerable to rising seas now that global warming is a hot topic. The low-lying nation is vulnerable as well to a whole set of water-related problems due to planetary warming: increased evaporation from water bodies, soil, and plants; stress on animals and people from rising temperatures; more intense cyclones and bigger storm surges; longer dry periods and fewer but wetter rainy days. On top of these stresses, this small country has the highest population density of any nation in the world. About the size of the state of Iowa, Bangladesh has a population of 150 million.

But the river delta that is Bangladesh also has some of the richest alluvial soil in the world. And there's a largely overlooked story of human resilience, of aid money put to good use—not merely manipulated by corrupt politicians. Farmers have overcome repeated famines and are now able to feed the nation so that it no longer relies on imports. Remittances from expatriate Bangladeshis and profits from the industry of successful entrepreneurs could in time move this country out of the list of least developed nations. It's also a place of great charm and hospitality.

Yet it's precarious, this success. Once one crisis is managed, another one seems to crop up for Bangladeshis. One such crisis arose in the 1990s, took center stage as Bangladeshi authorities and foreigners scrambled to cope with it, then—unsolved—ceded its place on the crisis list to global warming. That crisis is known simply by the name of the chemical element that caused it: arsenic. A chemical associated in the West with an Agatha Christie mystery or with an obsolete treatment for syphilis, this toxic substance abounds in much of Bangladesh's groundwater. The very Himalayan silt, transported by rivers for millennia to create the thick layer of alluvial soil that makes up most of Bangladesh, carries the element down to the plains, where chemical processes release it into groundwater. But no one knew anything about this until 1993.

It all began innocently enough. Surface water in Bangladesh, as in the rest of South Asia, had become heavily polluted with human and animal waste, causing diarrhea, typhoid, and cholera. To give people a source of safe drinking water, UNICEF, USAID, and other Western aid organizations funded the drilling of wells to bring groundwater to the surface. Called "tube wells" because a metal tube several inches in diameter is bored into the ground to reach the aquifer, these wells allow water to be pumped up to the surface from a level uncontaminated by the organisms that cause waterborne diseases.

Water can be pumped up from one hundred feet or more by means of a simple, inexpensive hand pump. These wells tap relatively shallow aquifers, but are much deeper than wells that can be dug by hand and that are so susceptible to contamination from surface pollution.

The tube wells were drilled throughout rural Bangladesh beginning in the 1970s. People's health improved. Then, in 1993, the first signs of trouble appeared. Community health workers in northwest Bangladesh saw patients with symptoms of arsenicosis. Ingested in small amounts over time, arsenic can be very dangerous. In the early stages of the disease, rusty brown spots may appear on the skin. In a more serious case of poisoning, these may become bumps and welts. In severe cases, caused by long exposure to arsenic, a person would feel extreme fatigue and be unable to work. Eventually cancer could develop or the victim's vital organs could fail.

Tests on groundwater in this region, which borders the Indian state of West Bengal, showed arsenic. The government did some surveys in other parts of the country using simple test kits. Based on early results, the problem seemed to be widespread throughout rural Bangladesh, but generally not every tube well in a particular area was found to be contaminated. Still, pockets of contaminated groundwater seemed to be present here and there throughout most of Bangladesh. Rural Bangladeshis had unknowingly traded one invisible harm for another.

It was a scary time. The World Health Organization called Bangladeshis' exposure to arsenic the largest mass poisoning of a population in history. According to early estimates, eighty million people might be at risk—more than half the population of the country, most of whom still live in rural areas. This figure, based on the number of tube wells in use—upward of ten million, multiplied by the number of people typically taking water from each—has since been revised downward because arsenic is not present in all areas. Possibly water in only a quarter of the tube wells in Bangladesh is contaminated, which is still a formidable number. And in some heavily contaminated areas, up to 90 percent of the wells might contain arsenic.

Using water from deep in the ground is a relatively recent development in South Asia, so the arsenic problem caught people by surprise. In time arsenic was discovered in groundwater in the flatlands of Nepal, right at the foot of the Himalaya, and in Bihar. It had already been discovered in West Bengal. Thus Nepal, Bihar, West Bengal, and finally Bangladesh, the most contaminated, are coping with this problem—the dimensions of which are still not clear.

After the discovery that groundwater was contaminated with varying degrees of arsenic, the Bangladesh government, aided by various international organizations—some the very ones that had promoted the tube wells—tried to get a handle on the problem. In the face of the staggering number of potential victims, one of the first things to do was simply identify which tube wells had arsenic in them and which did not. After testing them, the government started marking the wells with red and green paint to indicate which were safe, which unsafe.

Several thousand wells were eventually marked as part of a program funded by the UN. This method protected many people from contaminated water, but it was not a guarantee. Sometimes people simply ignored the colors and continued to use the pumps marked with red. If a well dried up, villagers might move a green hand pump to a new location where water was contaminated. And in any case wells painted green might become contaminated later. Nothing was permanently safe.

The painting was roundly criticized as a stupid idea. Dr. Iftikhar Hussain says he suggested painting the tube wells red and green. A Bangladeshi public health professor, now dean at the new Atish Dipankar University of Science and Technology in Dhaka, he was the project director on arsenic under the minister of health during the late 1990s as the size of the arsenic problem was becoming clear. He now readily acknowledges that painting the pumps was a bad idea. He says Bangladeshis are not tuned in to the use of these color signals as people in the West are (something that I can attest to as a pedestrian hoping red and green traffic lights will keep me safe crossing streets in South Asia).

Painting a red X on a contaminated tube well did not deter people who needed water because that need is an immediate problem, while the signs of arsenicosis develop slowly. Exposure and effect are not at all the same thing, Iftikhar told me. Later he recommended painting a skull and crossbones on the contaminated wells, and that worked better, especially with younger users. It instilled real fear. But it did not necessarily tell them where to find an alternative source of drinking water.

Around this time Afifa Raihana was working on a master's degree in public health from Oxford University. Her research focused on communities suffering from arsenic exposure; Iftikhar Hussain was directing her research. Afifa saw suffering—people who could not work, kids ostracized because of signs of arsenicosis, wives rejected for the same reason. The signs of arsenic poisoning were seen as a curse from God. "People blame themselves. They think they should be able to fix the problem, but they don't know what to do."

Arsenic is both a slow killer and one with effects that vary greatly. People sicken at various rates, depending on their own genes and whether or not they have nutritious food to eat. "Some people are exposed for years and don't get sick. No one knew how to measure how many people were sick, or how many died."

"People die of malnutrition every day here. It's hard to know if it's arsenic that's killing them," said Afifa. The suffering she saw in the course of her research wore her down so much she decided to switch fields. "I can't live my whole life thinking about this issue," she told me in 2007.

After our visit to the mouth of the Gorai River that February day in early 2007, Anwarul Karim took me into the nearby countryside to visit some friends of his. As the sun set and darkness gathered, we bumped down a one-lane dirt road—a green tunnel of dense foliage, palms, and banana plants. Near the road from time to time sprouted tall, slender brown structures. These were curing huts for tobacco leaves, assembled to cure after their harvest in the dry season.

The darkness grew profound, magically interspersed with glimmers of light from candles or oil lamps. We stopped and threaded our way through the foliage in the darkness down a little path; a man and woman came out to greet us inside a compound. Inside an adjacent house a room full of people noisily greeted Karim. The hubbub of voices continued while we were seated, and given snacks, tea, and homemade sweets, including a divine rice pudding simmered for hours with milk fresh from the resident cow. A white-bearded man recited a poem; a younger man with a *dotara* sang a folk song. The group fell silent and listened.

Karim took me to this village, Ramkrishnapur, near his hometown Kushtia, to help me understand how some villagers had coped after they discovered arsenic in their water. In 1999, with support from the government and the World Bank, he had worked with people in Ramkrishnapur to survey all the residents and determine who had symptoms of arsenicosis, then to help people understand the risks of arsenic in water.

"Some NGO people came to the village and sealed tube wells—marked them as contaminated," said Karim. "But people didn't understand what was happening." The village of two hundred families had about two hundred tube wells, most privately owned, a few belonging to the government. Three quarters of the wells were tested; forty-three were found to be contaminated with arsenic. They were marked with red and sealed.

Karim found that about a quarter of the families in the village had one or more members with signs of arsenicosis. Two women, whose families showed signs of arsenic poisoning, had apparently died from the disease. He also found from his brief study what later researchers were to find: that people drinking from the same wells did not consistently show signs of poisoning, and that people with signs of arsenic poisoning who switched to water from safe tube wells got better.

Many of the villagers believed the disease was a curse from Allah or caused by evil spirits. People with the disease were shunned, barred from village ceremonies and work.

The depth of the wells in the village offered no clue to which ones might be contaminated, Karim discovered. Wells of the same depth did not always contain arsenic; and if two wells of the same depth were located near each other, one might be contaminated, the other not. In the face of such uncertainty, the villagers, with Karim's help, returned to using surface water.

Karim introduced me to a young man named Monju, a composer of folk plays and songs. Monju had written a play about the arsenic problem in his village and performed it in Ramkrishnapur and other villages in the Kushtia region to teach people about the dangers of contaminated wells. The songs and plays also reminded villagers how to protect water in their ponds and how to store rainwater.

A farmer with orchards of mango, guava, lichee, and jackfruit, Monju stayed in the village instead of going to Dhaka with the rest of his family. His auntie made the food we ate that evening. Monju had also worked with CARE Bangladesh on a program for strengthening good local governance. "It's hard to write a play about governance," I teased him. "It must be about love." The group nodded in agreement. "He writes love stories. Without love, no governance."

Other Bangladeshi villagers may have chosen, like those in Ramkrishnapur, to take better care of their surface water, thereby avoiding complete dependence on groundwater from tube wells. But there was no official effort to reclaim ponds and surface water from pollution in order to make the water in shallow dug wells safe for drinking. The government's chief strategy in the early stages, supported by international aid groups, was filters to remove arsenic from contaminated water. For the decade following the discovery of arsenic, a great deal of government energy and foreign donor money went into the search for such filtering methods.

Karim took me to visit a plant in Kushtia where one of these filters was being assembled. Stacks of empty green plastic buckets stood under an overhanging roof, ready to be used. Red plastic buckets with lids were lined up on the ground, their filters already assembled inside. Apparently red and green had not been completely discarded as signals.

Within the small factory, piles of sand and chipped brick were heaped on the ground. The sand and brick were used along with charcoal and an adsorbent called a CIM, a "composite iron matrix." These were layered inside the buckets to create a filter that could successfully remove arsenic from contaminated water. The CIM was the patented part of the filter, whose chief ingredient was iron. It was placed in the red bucket, into which the arsenic-contaminated tube well water would be poured. The cost of this fairly simple device, known as the Sono filter, was about $35.

At a home in the nearby village of Shastipur, a young woman named Champa showed me the family's two-bucket Sono assembly. It stood on a covered concrete-floored porch just outside the bamboo and plywood house. The five-gallon buckets were stacked in a simple metal frame, the red bucket on top, the green bucket below. A small spigot and hose allowed the water to pass from the top red bucket to the lower green bucket, where it was again filtered through sand and charcoal and chipped brick to remove impurities.

On the ground below the green bucket sat a tin jug into which clean water could drip. Champa said the family had been using the filter for five years and drank water only after filtering it. She said someone came to the house, tested their water, and gave her the filter. Though they drank only this water, tube well water was still used for crops. Champa's family grew pumpkin, cauliflower, and other vegetables. It's still not clear the degree to which different plants absorb arsenic and thus become as toxic as contaminated water.

Champa said her mother had suffered arsenic poisoning but was now cured. Outside under the trees, near a narrow canal, her barefoot mother cheerfully showed me the tube well and pumped a little water up so it would drip from the iron spigot. Brown spots, darker than her brown skin, covered her forehead and were sprinkled across her nose and cheeks. Except for these brown spots, signs of previous arsenic exposure, she looked healthy. According to Champa, her mother was the only family member who had any signs of arsenic poisoning.

At that point about thirty thousand Sono filters had been distributed to households in arsenic-affected regions. The filter was said to work for five years. But that estimate did not take into account the varying concentrations of arsenic in groundwater—which ranged from 50 parts per billion (ppb) to 200 ppb and in some cases much more. And if the size of the family exceeded the average, the filter might wear out sooner.

The only way to know whether a filter is still working is to test the water with sophisticated equipment. Not all villagers would know when or how to seek such a test even if the government had provided enough good labs, like one I saw not far from Kushtia. The cost for testing was then three hundred taka, about four dollars. That seems like little money but is not an insignificant amount in a poor family's weekly budget.

The main criticism I heard of the Sono filter was that it dripped too slowly. Rural families are typically large in Bangladesh, and they need a lot of drinking water in the tropical heat. Tired, thirsty farm families might abandon use of the filter unless someone took charge of filtering a reserve supply. If you were thirsty, and no one in your family seemed sick from arsenic, the Sono might start to look like just some silly foreigner's idea. A contaminated tube well, on the other hand, would provide immediate relief.

Apparently this is the reason Sono buckets were often found abandoned or used for other household chores.

The arsenic crisis might seem to have an easy solution: simply stop using groundwater—which at the time arsenic was discovered had been a source of drinking and irrigation water in Bangladesh for only a couple of decades, just a generation. Keep the ponds and rivers cleaner to begin with, then remove any remaining biological contaminants with filtration or chlorine or boiling. Those pollutants are easier to get out of the water than is arsenic.

But could we easily, say, give up mobile phones and go back to only landlines? Cell phones have been in widespread use for less than two decades. Perhaps we find out they really are making us sick, so that using them is not an unalloyed benefit. Going backward probably would not be easy here either. People would resist, or simply ignore warnings, especially when the danger posed by the convenience is unclear.

In somewhat the same way, rural Bangladeshis became accustomed within a generation to being able to pump water from a source close to their homes, instead of carrying water vessels to and from ponds. Women could go out for water at night if they needed it, just by stepping outside their own doors. And a well is a status symbol. Having one just outside the house is not only easier compared to the old way of gathering water, it confers prestige.

Filtering devices began to seem the best solution to the crisis, given this reluctance to going backward, combined with confusion over which water was contaminated and how dangerous it was. Added to those problems were the difficulties in alerting a large population to the hazards of arsenic. Besides, keeping surface water clean remained an enormous unsolved challenge.

Afifa Raihana had hoped rainwater harvesting would be a good way to avoid contaminated groundwater and surface water alike. Then she discovered the difficulties of setting up rain-water gathering systems on thatched roofs and the problem of how to keep stored water clean in the warmth of the tropics. That, combined with people's dislike of the taste of rainwater, made her change her mind.

Then too, tube wells represented a substantial investment. Filtering the water allowed villagers to continue using their tube wells, the life of which is generally at least five years, thus preserving the investment in the well and keeping a convenient source of household water. But finding the best filter for widespread use was a challenge, one that has still not been met.

Iftikhar Hussain told me he had tried to promote a filtering technology some years before I met him. A product called Arsen-X had been invented by a Bangladeshi working at a university in the United States. It was designed to hang from the tube well itself. Contaminated water was supposed to pass through the filter as fast as it came out of the ground, which would likely have appealed to villagers far more than the slow Sono. In 2000 this filter would have cost about $150—much more than the Sono, but for many rural families still an affordable investment.

A promising product may have gotten lost in the government-donor shuffle, however. The U.S. manufacturer wanted to sell the device in bulk to the government, which said it didn't have the money. The World Bank, which was essentially running the country's arsenic project at that time, had the money but was shifting its funding priorities toward surveys at that point. Iftikhar himself thought the device could be marketed filter by filter to individual families, but the manufacturer wanted bulk sales.

The device was successfully field tested in 1999 and 2000 and shown to reduce arsenic contamination to zero, but then the government instituted a new testing protocol for arsenic filters. Eventually Arsen-X fell out of competition for reasons that remain unclear. By the time I started looking into these issues, memories had faded, staffs had changed. Iftikhar said only that "many interests are there. We are not being supported." He didn't want to blame anyone else for the failure.

Iftikhar was optimistic about protecting Bangladeshis from death and disease caused by arsenic. "Even stopping bad water gives you a fifty percent cure," he says. So the disease itself was reversible in its early stages, as long as it had not damaged a person's organs. Mild cases may produce no permanent damage if the person drinks only safe water and can have nourishing food and vitamins—a very big if for some rural residents of Bangladesh. Curiously, some people don't seem to succumb despite long exposure, which makes it difficult to convince people who are short of water that their water may be dangerous. Research might even someday yield a cure for later stages of the disease. But the main problem was safe alternatives to tube well water and keeping people from getting sick in the first place.

One alternative was new and deeper wells that would get beyond the stratum where arsenic contamination was prevalent. At about two hundred feet there is typically little or no arsenic. Iftikhar Hussain thinks that is just a way of selling more expensive pumps. Boring the deeper wells requires heavy equipment; a deep tube well might cost ten times as much as the shallower wells. And there was no guarantee the deeper well would remain uncontaminated; arsenic-contaminated water might seep down from above after the deep well was drilled.

Iftikhar believed the best solution was to abandon tube wells. "We have abundant surface water, so the easy solution is to purify it. Use the money for arsenic removal to clean surface water and protect it. Retrain people to use surface water, which less than twenty years ago they were using." He felt people could once again embrace the methods their families had used a generation before. It was all a matter of communication. "People still listen. Change is possible."

The key, he believes, is for people to participate in the solution. Each village has a mosque, a graveyard, a school. Villages maintain these in their own way. Such organizations could create systems to protect water. Perhaps religious leaders could be trained. "We can have small solutions. We don't need large scale. Our people take initiative. We just have to ignite them."

In 2007 Iftikhar thought arsenic would "become a monster again." Three years later he said: "It's an issue. I can't say major now. It is a part of the environment. We are always concerned with massive loss of life all at once. A cyclone is more important than arsenic in this sense. Floods. Emergencies."

Still, he did not believe the arsenic problem had been adequately addressed. It had simply become an "old problem." Funding had become available for newer problems, like HIV. People were chasing that money, "even though there is no HIV problem here. There are 110 patients in the whole country. NGOs are doing research projects on HIV because we have borders and HIV is on the other side." There were billboards in Dhaka and in the villages with messages about HIV, not about arsenic. There was money for HIV billboards, not for arsenic anymore.

Iftikhar said he didn't blame the NGOs for switching to the issue of the day. "They have to live." The problem was complicated. A lot depended on government priorities. NGOs were subcontracted to carry out certain projects. They do so, they spend the money. But "water projects should not work this way. Just distributing the work and the money won't yield results." Iftikhar felt accountability was lacking and there was inadequate supervision from the government and the donors as well.

Dhaka is the seat of power in Bangladesh. It has the dubious distinction of being the fastest-growing city in Asia. A hundred and fifty miles upriver from the Bay of Bengal, it is sixty feet above sea level, but still vulnerable to flooding in the monsoon. A sprawling mess of a city, it nonetheless works somehow, and it retains a certain sweetness and humanity that many other Asian cities are losing.

Dhaka's culture was first Buddhist, later Hindu; then it became a Mughal city until the British took over. It was overshadowed by Calcutta until it became the capital of East Pakistan, then of Bangladesh. The people of East Pakistan began their struggle for independence when West Pakistan tried to impose Urdu on this Bengali-speaking people. The two languages are related, as siblings of Sanskrit. But asking the people of East Bengal to give up their language was like asking Spaniards to switch to Italian or French in order to remain in the European Union—and to substitute someone else's literature and history for their own. The idea was doomed from the start. It might have made more sense to leave East Pakistan as part of India, or perhaps better still for it to be part of an independent Bengal. But it was too late for that; instead, they fought.

Fierce protection of its language is now embedded in the country's very name. Bangla is the language of Bengal. And *desh* is the word for country in many South Asian languages. Bangladesh—the land of Bangla. Bengali or Bangla is spoken not only by 150 million people in Bangladesh but by another 150 million in India.

The Bengali Language Movement, or Bhasha Andolon, was launched in the early 1950s, soon after partition and independence, in protest of Pakistan's legislating Urdu as the national language. In 1952 protesters at Dhaka University were shot and killed. February 21 has been commemorated ever since in Bangladesh and other parts of the world as International Mother Language Day. It's a public holiday in Bangladesh. Thousands of flowers are placed at a monument commemorating the spot at Dhaka University where the protesters died. Schoolchildren march in parades all over the nation.

A month-long book fair in Dhaka coincides with celebration of the language martyrs. I visited the fair one Friday evening in 2007. Friday is the Muslim prayer day and thus a holiday, already the weekend, so the throngs at the popular book fair were even greater than on other days of the week. I stood in a long line, several blocks long, with a young man who worked at my hotel and who had offered to show me around Dhaka. After a while he started leading me up the line, asking if we could get in ahead since I was a visitor. Staying in the queue might have taken several hours, so I let him do it, even though I felt awkward, thinking of all those times I had been dismayed as I stood in lines and South Asians shoved in front of me. Foreigners weren't much in evidence at the event. Several times, Bangladeshis graciously let me get in front of them.

The crowd was overwhelming. As a Bangladeshi woman from the American Center commented wryly, "and you thought you had seen crowds before." If you haven't seen one in Bangladesh, you have never really seen a crowd. Several acres of a park were packed with people roaming from booth to booth where publishers and bookshops displayed their wares. The throng was so dense that at one point my young companion had to steer me to the edge of the milling mass. An eddy of humanity in one of the aisles doubled back on itself and the ensuing shoving threatened to knock people down. My guide pulled me out quickly, worried that something untoward might happen to his foreign charge. For the most part, the crowds were orderly.

I was amazed by this love of language and books. When later someone told me that minimal book buying goes on—it's mainly an event for socializing, for people to hear music, for boys and girls to look at each other while their parents talk, and for more book browsing than buying—I was still immensely impressed. The closest thing I could think of in comparison to the book fair was state fairs or rock concerts, whose excesses would not be allowed in a Muslim land. But for an English major, the idea that books and language were an excuse for a public gathering like this warmed my heart and made me love the people of Bangladesh. I saw the same sorts of throngs in visits to local museums.

That is one side of Dhaka. Another side is officialdom, which leads back to the main subject of this chapter—the management of the arsenic crisis.

Dhaka has directed virtually all attempts to deal with the arsenic crisis. However great the concern members of the government or of international aid organizations might feel for the rural people who have been poisoned by arsenic in their water, the problem is somewhat abstract to them. The water supply for Dhaka and its suburbs has no arsenic because the tube wells that supply the city are now very deep, bringing water from a stratum where arsenic is not present. Shallower aquifers in the central region that includes Dhaka have been depleted or polluted; avoiding arsenic is an unintended benefit. Dhaka's problem is that the groundwater will eventually run out, and there's been no effort to harvest rainwater from the city's many tall buildings.

Well off natives of Dhaka along with most foreigners live and work in neighborhoods at the north end of the city called Gulshan and Banani. A young woman named Shazia told me her father used to go hunting near her Gulshan office building. Cheetahs roamed the woods just a few decades ago, so fast has this city grown. Now the wide tree lined streets of Gulshan and Banani are full of people who have enough money to live there; additional plush neighborhoods are being constructed farther north into the city's outskirts.

Gulshan and Banani do not have the appearance you might expect of the capital in one of the world's poorest lands. They are chock-a-block with new and attractive apartment buildings, five-star hotels, new hospitals, shiny shopping areas, and bumper-to-bumper air-conditioned vehicles of recent vintage. Rickshaws pedaled by village men in lungis thread among them when gridlock brings the vehicles to a halt. The city is so choked with traffic the government has decreed that shops and businesses must close down one day a week according to a rotating schedule. This will theoretically cut down on traffic in at least one quadrant of the city each day.

Though I frequently had to travel an hour or more across the traffic-choked city to meet and interview people in other sectors, a surprising number of experts were available in Gulshan and Banani—sometimes in walking distance, or within a short auto- or bicycle rickshaw ride. One development expert I met said that Bangladesh suffers from a plague of NGOs—nongovernmental organizations. The foreign aid industry thrives here and is housed in comparative luxury.

The hotels are full of foreign advisors here for shorter stays than the diplomats or consultants who have set up house. They may spend several months a year in Dhaka—usually in the cool winter season—as consultants to the international nongovernmental organizations, or INGOs, and to Bangladeshi businesses.

A couple of such experts were staying at the small guesthouse I visited in 2010. They could have stayed in more expensive high-rise hotels down on the main streets, but they wanted to get away from the bowing and scraping of the fancy hotels to be with our hostess Maya and the casual comfort of home-cooked Bengali food.

Whatever help Bangladesh receives from these many foreigners—and that may indeed be substantial—even most of those advisors would agree that their presence adds to confusion about who is running the country.

In the case of the arsenic crisis, not only were the dimensions of the problem unclear, trying to get a handle on it was complicated by the number of cooks in the kitchen. Various government bureaucracies were involved. International groups were also involved, some of which had helped to build the millions of tube wells to pump up the now dangerous groundwater. The World Bank, UNICEF, and the UK's Department for International Development (DFID) got involved in arsenic "mitigation," along with CARE and large Bangladeshi development organizations like BRAC and Grameen.

The World Bank wanted to support further government research on the magnitude of the problem. Other organizations wanted to focus on household filtering devices. There was some duplication of effort and lack of coordination. There were short-term projects, incomplete efforts, lack of focus and continuity. The work of foreign aid groups is typically short term, and the government that had the longevity to set up long-term structures didn't have the financial resources or enough people working in the various concerned ministries. These problems led to duplication, mismanagement, delays, and even corruption.

Who was really in charge? The government? The foreigners? And why after almost twenty years since the discovery of arsenic in Bangladesh's groundwater does there still seem to be no well-coordinated strategy to solve it? Fortunately the magnitude of the problem seems to be smaller than public health officials like Iftikhar Hussain feared after those first terrifying estimates, which should make the solution easier. Some of the INGO project managers and government officials I met in Dhaka were somewhat mystified by their collective inability to come up with an effective strategy.

The government is not strong. Its agencies do not necessarily stay focused when foreign donor money is to be had for projects dealing with HIV, global warming, or other problems. The sheer number of people who are potential victims remain staggering. And those victims are very dispersed, throughout rural Bangladesh, in small villages, all living much like the people I met with Anwarul Karim in Kushtia. Arsenic is a slow-moving catastrophe in a country where the fast-moving catastrophes in the form of floods and cyclones are frequent. It's easier to direct money to them, while the creeping catastrophes often get pushed aside, as Iftikhar Hussain noted.

The search for cheap, effective devices to remove arsenic from drinking water has not ended. On my last visit to Dhaka I had the opportunity to observe the complexities of launching a new one.

My friend David Sowerwine tried to build bridges in Nepal. His low-tech, inexpensive *tar pul,* or wire bridge, consisted of a gondola suspended from a sturdy cable. It was just the thing for getting across swollen monsoon rivers in the Himalayan hills. People in remote villages could cross otherwise impassable torrents to get to schools, doctors, or festivals, or to take their goods to market. But his efforts got caught up in the aid game as well as in the cross-currents of Nepal's Maoist insurgency. Still, he didn't give up easily. I met him and his wife Haydi toward the end of their fourteen years in Nepal during one of my earlier, shorter stays in Kathmandu. They are both tall and fit; he is balding and she is graying now, but they still have the energy of people twenty years younger.

Trained as a chemical engineer, David got interested in the arsenic issue after turning the *tar pul* business over to his Nepali partner. Some graduate students at Dartmouth College were looking for a challenge. David suggested they might try to create a device that would take arsenic out of water using a little electricity. The theory was that iron electrodes placed in the water would produce iron oxide: rust. The rust in turn would capture the arsenic. Pouring the water through sand would filter out other impurities like bacteria, along with the resulting particles of rust—which ideally would no longer be soluble in water, making the waste harmless. Most important, the water would be free of arsenic and safe to drink.

After just four months' work in 2009, the students succeeded. They created a device that removed virtually all the arsenic in samples of water—below the World Health Organization standard of 10 parts per billion (ppb) and well below the Bangladesh government's standard of 50 ppb. Amounts of arsenic in raw water from tube wells in Bangladesh ranged from that standard up to 200 ppb.

When I was trying to finish this story of the Ganga and her tributaries and planning my last visit to Dhaka in November 2010, David was setting up meetings there. His goal was to develop a prototype of the device so that a manufacturer could market it in rural Bangladesh. David has worked around the world for international businesses as well as development organizations like USAID; he is committed to helping developing countries. But he believes that entrepreneurs will better solve the problems of the developing world than foreign aid will.

Convinced that even the poorest people value something more if they choose it and pay for it, David wanted to explore the possibilities of an effective arsenic-mitigation device that could be manufactured and sold cheaply. He was not interested in profiting from this effort personally; he wanted to promote a good device through a market. He said he would happily give the technology to a business that seemed able to manufacture and distribute such a device without help from an INGO, which so often seemed to undercut attempts to boost private enterprise.

In Dhaka David hoped to get one of the aid organizations interested in supporting a test run of such a device; he also wanted to locate a potential manufacturer to produce and market it commercially. The Dartmouth technology was probably not the most innovative aspect of this endeavor. It was a new idea, but what was more novel was David's intention to go against the grain of the aid game.

Most of the devices currently in use had been given to villagers at no cost. And most employed a patented substance of some sort, so they could not be easily duplicated. Using electricity instead of a patented ingredient meant there was no big secret here. It might be duplicated fairly readily.

David began by going to see one of the devices already in use. I tagged along, always eager to get outside of sprawling South Asian cities.

Our destination, the Comila district, was not a great distance from Dhaka, but the first two hours were spent just trying to get out of the city. Our driver threaded his way through its insane traffic jams, honking all the while. In time we reached the village of Kumarbagh. Leaving our boxy vehicle on the shoulder of the asphalt highway, we walked down a dirt road hemmed in on both sides by lush green trees. We passed two long, one-story buildings from which emanated the din of cloth-weaving machines, and reached a cluster of small homes.

Anuwadha, the matron of one of them, was expecting us; she greeted us outside a house constructed of sheets of corrugated metal. In the kitchen, a bag of rice rested atop a black oil barrel. A rusty portable electric fan hung from the ceiling, attached to a narrow bamboo beam. The base of the fan had been removed and the cage of blades pointed downward. Shelves held tin pots and pans and plastic buckets. A hammock hung against the wall. The only light was from the open doorway. Behind the small kitchen were two additional rooms.

The bright electric green of the portable filtering device we had come to see was so intense it practically illuminated the bare platform where it sat. The little device glowed next to Anuwadha's black shirt and maroon sari. From a tin jug she poured water, which had been pumped from a sixty-foot tube well just outside her house, into the cylinder. A second tin jug sat on the floor beneath the cyclinder's spout.

The water poured out of the little green spout and into the tin jug on the floor in an instant. The beauty of this unit is that the water flows out, cleansed of arsenic, virtually as quickly as it enters the filter. The water passes through sand in the top cylinder to clean out any solids and bacteria, then passes through a layer of pellets in the bottom cylinder to remove the arsenic.

The plastic pellets inside the filter have been coated with a chemical that first bonds with iron oxide—rust—which in turn captures the free-floating arsenic. The chance of an arsenic atom bumping into the surface of one of these pellets as the water passes through the filter is pretty high, David told me. When it does, the arsenic sticks to the iron oxide. Gradually the pellets lose their effectiveness as they get coated. The only way to know when it's time to replace the pellets is to test the treated water for arsenic.

A little over a foot tall and six inches in diameter, this Japanese invention goes by the odd name "Read F." The unit is made of sturdy plastic, has a snug lid, a swinging handle like a bucket's, and a spout from which flows the purified water. Compared to the large Sono unit I had seen near Kushtia a few years earlier, this one is much smaller, sturdier, faster. The technology for this compact unit was invented in Japan; the green plastic housing is now made in Bangladesh but the pellets for the filter still come from Japan. The plastic pellets are the secret of the Read F method. Like the Sono filter, the Read F was one of four devices for filtering arsenic that the government had approved.

Anuwadha told us she likes the filter and would be willing to pay to keep it. David was happy to hear this; it supports his idea that village people without a lot of money will find a way to pay for a product as long as it's within their means. Replacing the patented resin would cost five thousand taka, about sixty-five dollars. The whole unit cost 6,500 taka. A new filter could last three to four years, depending on the concentration of arsenic in the water and the size of the family. The monthly cost of the unit averaged over several years was minimal.

Anuwadha, who had no signs of arsenicosis, had volunteered to be part of a public health study. She was one of four hundred villagers in this area who received the Read F unit in exchange for submitting to regular blood tests. Dr. Pritimoy Das had met us at Anuwadha's to help explain the unit. On our way out of the little house, we met his assistant Helena Akter, a pretty young woman swathed in a black abaya, with only her face and hands showing. She carried a small cooler for blood samples.

This same technology has been adapted so that an entire village can use one filter unit. An hour or so down the road we walked out to another village called

Anuwadha pours arsenic-contaminated water into a filter

Khunta to see a larger version of the Read F filter. The unit was about the size of a large diesel backup generator, housed in an open-sided shed. Children from nearby houses took turns showing us how they pumped water up from a tube well. The water then passed up into the guts of the filter after being sprayed into it from a showerhead to aerate the water. Adding oxygen sped up the process and allowed the arsenic to bind with the chemicals in the pellets faster. The water quickly poured from a spigot on another side of the concrete box that housed the filter.

World Vision Korea, a Christian relief organization, donated units like this to eighteen villages in the district at a cost of three hundred thousand dollars. Khunta has about two hundred families, so more than a thousand people depended on the unit we saw for safe drinking water. The surrounding jungle was so dense we could see only the house closest to the shed. A local man who had worked in Saudi Arabia and Libya for fourteen years donated the plot of land where the well and filter were housed. Now he had a small metal works shop out on the main road.

There had been some cases of arsenicosis in the village. No one we met that day was able to say how many or how serious. David tried to figure out whether it might be feasible for the villagers to pay for replacing the resin inside the filter when World Vision leaves. Villagers had begun setting aside money, but it seemed they had nowhere near enough yet for replacing the pellets of the village-size filter when they wore out.

David was impressed by both Read F filters. "It's a well-engineered product," he said. "The problem is that it's not a real product." It is a donation; so, however nifty the unit, and however marketable in theory, in practice its commercial appeal has not been tested in a market where people might, for example, choose between the Sono and Read F. Though it's slow, the Sono filter might suit some families because it's very cheap. Other families with a little more money might buy the Read F home unit because they can afford it and value its speed and portability.

Nothing like this has so far happened in Bangladesh. Some people were given the Sono filter and used it. Some were given it and used the buckets for something else, either because the water came too slowly or they were not convinced it was worth the trouble if they saw no signs of arsenic disease. Others have been given the small Read F units in exchange for participating in a health study. It's not clear yet what those users may do when the study is over, or when the unit loses effectiveness and the pellets no longer remove the arsenic.

Will there be an infrastructure in place so a family can take the unit somewhere to remove the old pellets and insert new ones, or do any other maintenance the unit might need? Will the family just buy a whole new unit since replacing the pellets is almost 80 percent of the cost of a new unit? And for the villages using the large units donated by World Vision, how can they manage them permanently on their own when World Vision leaves, as it surely will? And what will be done with the toxic residue from these filters?

All these questions would ideally be scoped out in advance for a commercial product. This was the chief problem with aid, as David saw it. "The model doesn't work," he said, talking about the Read F filters we saw. "It's an expensive medium. It's made in Japan, and it's costly to replace. Plus they gave them away. They should not give them away."

"That's the problem with development," said Rajiv Pradhan, the country director of International Development Enterprises (IDE), an organization that aims to steer developing countries like Bangladesh toward creating markets. In Rajiv's bright, functional office in Gulshan, he and David discussed what we had seen in the villages where the Read F unit was in use. Like David, Rajiv—a Nepali with twinkling hazel eyes, a cherubic face, and strong voice—believes the typical development approach of giving tube wells, solar setups, or storm shelters to villages and then walking away is badly flawed. Often there was no maintenance or monitoring after donor money was spent and the project report completed. That could be a waste of money, sometimes leaving facilities that become useless after a while.

"Even if people want to buy the filters, they can't," Rajiv continued. "The whole idea of promotion and the development of the supply chain just isn't there. The focus on projects takes away the incentive for producers." Rajiv thought David's ideas sounded good and wanted to help plan the strategy for getting the device into production.

"NGO prices are not market prices," Rajiv pointed out, "which makes price comparison pointless." David was concerned that too many entrepreneurs had simply adapted to what could be called the "development market," and would have trouble shifting to a real market approach.

Helping poor people while bolstering a market-driven economy might be a good idea, but it was far from easy to implement. Rajiv and David continued to talk about how they might proceed with the Dartmouth students' electrode device; manufacturing and marketing could take from three to five years, perhaps using multiple manufacturers and marketing schemes. First they would have to test the device under all the different conditions in Bangladesh in order to make it an acceptable commercial product. And to do that, they were going to have to find a donor to support research and development.

Thus they would have to straddle the development model and the commercial model because they needed a donor for the research-and-development phase. A donor like the World Bank (which sometimes gives grants for small projects) or one of the national aid organizations like the Danish DANIDA could play the role of venture capitalist. The project would need an "angel investor," someone who would be repaid by seeing the project succeed, not by receiving dollars.

Ironically, a big project backed by big foreign aid money would be much easier to launch than what David had in mind, starting small and growing. David had already talked with a promising manufacturer in Dhaka, but he and Rajiv could not get the manufacturer lined up until the product promised to recover the costs of the fabricator and make him some profit.

This was one hurdle for them to leap. The other was getting government approval. During the rest of the week, David looked for support. He met with people in government, with several INGOs, with one of the big development banks, and with a potential manufacturer. Most found the proposal intriguing, but at the end of the week it wasn't yet clear how the right combination of donor support and entrepreneurial effort was going to get what looked like a good idea off the ground.

One government official in public health was helpful, offering useful advice about how to deal with Bangladeshi bureaucracy. But the head of the agency charged with testing and approving arsenic devices gave David a chilly reception. The official tapped his fingers and fidgeted while David outlined his ideas, as if he already knew what he was going to hear and wanted to get on with it and have tea. "Unfortunately we cannot help," he said, whether with testing in the field or with any other suggestions.

The official was a short-timer in his post, perhaps eager to get back to his own scientific research on arsenic. Perhaps he was tired of foreigners meddling in the problem. Perhaps many people thought they had good ideas and most turned out to be duds. Still, in the back of David's mind, was past experience. He had tried to get approval for implementing projects in other developing countries. Often the only way to make progress in situations like this was to be part of an organization that had enough money to offer substantial support to the bureaucrat's organization. That may not have been the case here at all; but David had encountered the phenomenon often enough to be alert to the signals.

From staff at the World Bank we learned that in the past people had come and started distributing arsenic filters without even telling the government. That led to the government's setting out a formal procedure for approval. They assured David approval was not difficult. Yet when David commented that there was "an awful lot of smoke and mirrors in this game," they readily agreed.

The World Bank staff did not see how they could help David, though his product sounded like a good one in their opinion. Staff at another development organization, Wateraid, said they would be willing to conduct field tests when David got approvals. That group believed there was a general movement away from filtering toward developing alternative sources of water wherever possible.

The Bangladeshi representative of the Read F product had strong feelings about his past dealings with the government. He told David he had waited more than four years for government approval. He tested the technology in the field for three years, but still he could not get a clear yes or no from officials. He said the government wanted the fourteen-thousand-dollar village units given as a donation, at which the Read F people balked. Finally they got approval for the family devices, later for the larger one.

I recalled Dr. Iftikhar Hussain's difficulties getting what might have also been a promising technology approved. In terms of speed and cost his tube well attachment was similar to the small Read F device and supposedly had no residue that would require disposal. The arsenic was bound up in stone pellets that were benign waste; even thrown on the ground they apparently would not contaminate water.

Like David, Iftikhar had also wanted people to pay for their own filters. "People paying for their own is the best way. If it comes from the donor, people don't trust sometimes. Because they think you are trying to do something just to make money, not to help them."

A less determined man might have just decided to forget the whole thing after the discouraging chat with the government official. But David has persevered. He applied for a grant to build the models and test them and remained optimistic about pursuing this project.

The arsenic threat has not been corralled. Perhaps the obstacle to the cheaper solution—going back to ponds and shallow hand-dug wells—was similar to what I had seen in Kathmandu and Bihar. The Melamchi water supply plan and embankment raising in Bihar were ways for someone to make a lot of money. There was money in arsenic too. And there is a general preference in the developed and developing world for technological solutions, perhaps because technology seems more trustworthy, even if it's as simple a technology as the Dartmouth students' electrodes. Thus the solutions tend toward filtering devices and deeper tube wells, not cleaner ponds and simple water disinfection. Still, it's not difficult to understand why people hesitate to take on the challenge of making surface water safe enough to drink in such a populated country with so little effective sewage management.

Perhaps a combined approach—using all the methods available, depending on local circumstances—and focusing concerted attention on the issue once again, could make the arsenic in Bangladesh's groundwater a manageable national liability. In any case, arsenic may turn out to be more manageable than the problem that has taken center stage in the list of threats to Bangladeshis' well-being: global warming.

# 17

# Where the Rivers End

> But here in the tide country, transformation is the rule of life:
> rivers stray from week to week, and islands are made and
> unmade in days. In other places forests take centuries, even
> millennia, to regenerate; but mangroves can recolonize a
> denuded island in ten to fifteen years.
> —Amitav Ghosh, *The Hungry Tide*

It's not yet dawn. The Bon Bibi is anchored. Some members of the crew have risen and one has brought hot water to the dining room. I have some tea to warm me as I lean on the deck rail, looking out over the silent river this December morning. Our guide and my fellow travelers are still asleep below in cozy bunks draped with mosquito nets.

Looking to the nearby shore where we walked yesterday, I see the forest guards' cottages, crumpled by cyclone Sidr's winds and water. All is quiet. And the fishing skiff that anchored near us yesterday evening is silent too. Perhaps the men fished all night and are still sleeping in the morning chill.

I walk around to the other side of the deck. Just a few yards away there's a log floating down the gentle current toward the Bay of Bengal. Or wait. Maybe, it's a crocodile heading obliquely across the wide mouth of the river? No, it must be just a clump of vegetation moving in the slight swells of water. Then the clump sinks slightly at both ends. I feel a surge of excitement.

The shape is closer to me than any of the crocodiles the boat captain tried to approach as they sunned themselves on the shore upriver yesterday. They always slithered back down the sand and into the water before we could get a good look. There's little light now, and still unsure, not trusting my aging eyes, I peer intently at the dark hump. It seems to sink imperceptibly again, until finally it slips under the water altogether. The shape moved as quietly and slowly as leaves floating on the surface of the water might, but clumps of vegetation don't usually vanish underwater quite like that.

Fifteen minutes later our young, long-haired Bangladeshi guide asks me, smiling slyly, if I had my glasses on when I made this sighting. When I insist I did, he's convinced this is a good sign. Apparently this is the first sighting of a crocodile near the Bay since cyclone Sidr the previous month.

The rest of the guests are now up and having tea. In the predawn light our group steps down from the Bon Bibi into a canoe. We ride silently up a narrow channel perpendicular to the river. Our eagle-eyed guide, Emile, is in front; an equally alert young crewmember stands in the stern, rowing and steering the green canoe with figure eight movements of a single paddle. We glide closer to the shore.

A chartreuse snake has draped itself exquisitely over a dead branch like a vine. Another one presents himself as a graceful tannish-gray loop on a branch with leaves. Maybe this snake species, like chameleons, changes color. One of the passengers whispers that they seem to have it backward: the green one should be on the branch with leaves. But maybe the snakes know best.

An extended family of monkeys grooms each other and watches us from the safety of the trees. Wide-eyed deer stare out at us from behind a protective screen of bushes. Birds call and perch and show us their stunning reds, blues, and teals. A pair of kites sits stately in the top branches of a tall tree.

This is the third day of my long-awaited excursion into the Sundarban. After many trips to Himalayan glaciers, finally I am approaching the Bay of Bengal, where all the water joins the sea. The ice and snow that melts, the moisture that falls in the monsoon and does not water crops or evaporate—this is its destination. Unnamed channels here are wider than many named rivers in California and minor rivers are as wide as the wide Missouri. The bottom of the watershed where the rivers coalesce and melt into the ocean is the end of my journey.

The first morning, we left soon after dawn. We had reached Khulna and boarded the small cruise ship after midnight. We traveled the whole lazy day after our short night's sleep, stopping only to pick up the two armed forest guards that are required to accompany all tourist boats into the protected Sundarban, the largest remaining mangrove forest in the world.

We chugged down the Pussur, which branches from the Gorai, past the port of Mongla, past barges transporting wood and bamboo, past cement factories perched at river's edge. Midday we veered off into a channel; as the day faded, only a few dozen one- or two-man fishing skiffs remained on the water with us. Emile said the fishermen stay out for weeks at a time. A larger boat would pass up the river and buy their catch about once a day.

We traveled on down the wide channel until late that night, our companionable group of nine getting acquainted: reading, playing cards, enjoying Bengali home cooking, and even recharging cell phones and laptops as the engine and generator rumbled. We were four Danes, two Slovenians, two Bangladeshis, and me. There were as many crewmembers as passengers. The boat was simple but spotless and well tended. We awoke to find ourselves all alone on the water in the silence of the Sundarban. As a bright orange sun blinked between the thickly ranged trees, we took our first canoe ride into the forest. Families of otters scurried up the slippery bank and into the brush at our approach.

Fishing skiffs on the water near the Sundarban

The Bangla name for the mangroves is *sundar bon*, or beautiful forest. Perhaps the name is also derived from the tall tree that grows here, the Sundari, a species that has developed a mysterious disease in recent decades, most likely because of changes in the ecosystem. There are many other kinds of trees here, providing pulp, timber, and fuel wood.

The Sundarban also yields honey as well as fish and shellfish, sometimes harvested at risk of life, for this is also the home of the Royal Bengal tiger and a refuge of many other species, some like the tiger endangered. When we were out walking one day the forest ranger pointed to some tiger tracks, maybe a day or two old, in the dried mud near a water hole. Emile said he's been coming to the Sundarban for eleven years and has never seen a tiger, only tracks.

The world's highest mountain stands at the top of the Ganges basin, looming above the frigid gray waters of Imja Lake and Tsho Rolpa. The world's largest mangrove forest fans out here at the bottom. The alluvial delta and the mangrove forest evolved long after the Himalaya arose, and it's still growing, fed by the sediments from upstream. Bangladesh might have extended itself out into the bay, were it not for the improbably named Swatch of No Ground. Just off the coast, beyond where islands are still forming, is this sea canyon of unplumbed depth, where much of that troublesome yet beneficent sediment from the Himalaya lands.

But not all of it. Emile took us to an island that didn't exist a generation ago. Now it's covered with trees, a home to deer and tigers. Both species are good swimmers who can come to colonize the new territory and play their role in the new island's food chain.

Given my taste for cascading rivers and big conifers, it took me a day or two to feel the tug of this tightly knit pale green forest of relatively low trees marbled through and through with channels of sediment-laden water. Then I eased into the gray-blue skies, the thick jungle that hugs the land, its edges slick with sticky mud.

With no snow-flanked mountain kissed by the gold and pink dawn to astound me in the morning, I came instead to appreciate the perfection of form and function here. The trees that nourish and shelter myriad wildlife have evolved to tolerate various levels of salinity in different seasons, adapting to the mix of salt and fresh water that ebbs in and out twice daily and makes the soil spongy. The tree roots poke up through the mud so they can grab oxygen when the tide ebbs.

The forest that grows up, palms succeeding grasses, hardwood trees succeeding palms, fixes the new land that the rivers arrange during the monsoon. The forest that springs up here on the verge of the Sundarban stands guard facing the sea, taking the brunt of the occasional cyclone, being blown down perhaps, but protecting the older forest and the farms inland.

There are many threats to this unique and precious ecosystem, this tropical salt marsh. Some come from near: the needs of a growing population for which the ecosystem is a resource base. Some come from far away: a warming planet, warmer seas stirring more cyclones, rising oceans inundating low-lying lands.

The six thousand square kilometers of the Bangladesh Sundarban are only a remnant of the mangrove forests that once ringed the Bay of Bengal, from Thailand down the coast of India, providing a buffer between the sometimes turbulent bay and the exposed coastline. So perfectly knit together are form and function in the mangroves, one could easily see some divine hand at work, devising this natural barrier to protect the land and its creatures when cyclones stirred over the warm waters of the bay and slammed into the coast.

The vast mangrove forests did not survive colonization and a population explosion in South and Southeast Asia. In Bangladesh the Sundarban once reached all the way up to Khulna, where we started our cruise. Then trees were steadily cut and the land was claimed for farms. The cutting likely began at least six hundred years ago; by the eighteenth century the forest was diminished but still twice the size it is now. It was declared a reserve forest and offered some protection in 1875.

More than six million people now live on the edges of the Sundarban, half in Bangladesh, half in and around the Indian Sundarban, the smaller less lush third of the mangrove forest to the west. The steady encroachment of people has led to deterioration of these coastal forests for centuries. Still, as in many magnificent places on the planet, what remains is beautiful and thriving. For how much longer is uncertain.

Cyclone-damaged forest department camp in the Sundarban

The myriad nameless channels here in the Sundarban—where the waters of the Ganga and Brahmaputra join and reach the sea—mirror the mountain streams that feed the Ganga's tributaries. The quiet of the Sundarban echoes the silence of the glaciers. Now both are changing, and we have probably already lost any control over the direction that change will take.

During our short voyage on the Bon Bibi, we saw damage from a recent cyclone. Sidr—called a "super cyclone" because of the speed of its winds—had passed through the area where we traveled in December 2007. After breakfast our first morning deep in the forest, we visited a Forest Department camp. The dock was badly damaged; we had to climb up to it on a rope ladder and then walk carefully down the dock where planks alternated with big gaps. The rangers told us the water swept in and submerged the half dozen cottages in the camp. The water withdrew only fifteen minutes later, but the houses, set on stilts to accommodate floodwaters, were ruined. The group in the ranger village had taken refuge at the top of a little mosque and been safe until the water receded.

Sidr's 250-mile-an-hour winds and storm surge—a wave of water about twelve feet high—had done their worst damage east of where we traveled in the Bon Bibi. Dhaka's daily newspapers were still carrying small front-page articles noting the discovery of another body, or two, or three in that devastated area. Some were found under rubble in a village, some out in the fields and forests. It was already more than a month since the cyclone. The death toll was not as great as the ten to fifteen thousand feared early on, just after the storm, but at least three thousand people died. The lives of two million and more were disrupted. Homes were destroyed, coastal agriculture devastated.

Other articles bemoaned the destruction of a quarter of the mangrove forest, a great loss when the forest is so much smaller than it once was. But at the same time, the trees had a job to do. And at least where we were traveling they seemed to have done that job; they protected the land. Though we saw trees down each time we stopped to walk along the shore, far more were still standing. East of here, where the eye of the storm struck the land, the proportion of standing trees to downed ones was, however, the reverse. It might take twenty years for that forest to grow back.

Back in Dhaka I talked with Nayeem Wahra, who then worked for Save the Children—an international relief organization focusing on the needs of children in disasters. I asked him about the articles that stressed Sidr's damage to the forest. "We should not be so panicked about this happening to the tree," he said. Nayeem agreed the mangroves were doing their job and hoped the Forest Department would leave all the downed trees where they were so the forest could regenerate undisturbed.

He was more concerned about the blunders that sometimes led to increased suffering for people in the well-populated areas outside the protected forest, still near the coast and exposed when cyclones move inland. He told me that the majority of Sidr's casualties were children and that most of the dead bodies were found inside houses, not in the rivers. He attributed the deaths of the children to a combination of local custom and to the planting of shallow-rooted trees not native to the coastal region.

Children generally tend the family's animals. They see that the cows and chickens are sheltered before the storm comes, and thus they stay near the house. The wind comes, a tree falls on the house and kills the child. But according to Nayeem the wrong kinds of trees have been planted in the villages in recent years. "It's not the native trees, the coconut palms and betel-nut trees that fall. We should not blame the cyclone," he said. "Social structure and practice and bad development initiatives are the killers. The deaths at the homestead could be avoided. If we can do proper village planning with local trees, then the trees protect the houses."

Storm shelters may have helped some people during Sidr, but the warnings came more than a day before the storm hit. People went back home, perhaps to care for elderly people, disabled children, or their animals. Some never left their homes, Nayeem said. "A poor woman with some chickens and ducks, she won't leave them until it's too late."

Nayeem does not believe that building more shelters in the coastal area, as international donors want to do, is necessarily the right answer. People may not go to the shelters if they don't feel comfortable in them. Instead, he says, "people need houses that will stand during the disaster. It's not costly." Raising the houses a few meters from the ground could keep them above the storm surge.

For the cost of one cyclone shelter, houses in several villages could be raised and thus protected, Nayeem contends. "But a cyclone shelter can be photographed, you can show it to the donor. So there's a big lobby for cyclone shelters." Native trees growing around the villages are cut to build the cyclone shelters; when the big trees are gone, the birds stop coming, more pesticide is needed for paddy fields, and then the fish go away too.

Good cyclone shelters, raised houses, and better planning of villages may be only the beginning of the kinds of preparations that Bangladeshis living in the coastal region will need to make in coming years. The increasingly warm and turbulent Bay of Bengal will churn up more super cyclones, most experts say. Such storms are just one of the challenges a warming climate will bring to Bangladesh.

"I had a meal with people fourteen hours before the cyclone. Six of them are still missing," said Ahsan Uddin Ahmed, a scientist and researcher, a month after the storm. "It still haunts me." His voice trailed off.

"The tide was low when Sidr came. Otherwise the storm surge could have been eighteen feet and the death toll easily 100,000." Ahsan agreed that people died because the warning was broadcast too early. People went to the shelters, then left before the brunt of the storm.

"This is not the first time a super cyclone has come to Bangladesh," Ahsan said. "In 1991 the coastal areas were devastated because of a super cyclone; 138,000 people died in a matter of minutes along with two million cattle. There was a big storm surge. International media came, then the donors came. We built more than 2,000 cyclone shelters. They were all built by Bangladeshi people."

The shelters were put to good use in 1994, 1997, and 1998, Ahsan says, and saved lives. But by 2007, when Sidr struck, the population had increased while the capacity of the storm shelters remained at the 1993 level. "There was not enough room in the shelters. People stood for hours. Women and the elderly often don't even go to the shelters." Rain made the roads slippery; pregnant women and elderly people did not want to walk on them. "And the shelters have no ramps. Staircases are steep. One woman told my colleague she miscarried. There are no toilets for women. No place for nursing. Women say 'why should we go there?'"

"Our capacity to save human lives during cyclones is much less than we actually need." Ahsan said he would not be surprised if fifty or a hundred thousand people were killed the next time a cyclone comes.

The inadequacy of storm shelters points to a central development issue for Bangladesh: where to invest scarce resources as the population increases and the list of problems this country has to cope with seems to grow longer. "There are floods and droughts and salinity intrusion every year," Ahsan says. "Where to put the money?

"We have invested money in good projects, but the benefit of them is about to be washed away." Ahsan cited a recent study showing that people who had risen from under the poverty line to just above it, fell back again. "That is one-fifth of the total poor now. So no matter how the government tries to help them cross the barrier, holding them above that barrier is extremely difficult because so many things are happening."

A family might lose a crop, or have to borrow money to repair a house, or because someone in the family is sick. "This is the most common way of becoming impoverished. Once or twice you can borrow money. Then the family doesn't know how to pay it back, and lenders ask for high interest. The lenders take the dwelling, so people become landless and even homeless."

Or, a flood or cyclone can leave a family with nothing. "People cannot cope with that shock." Such shocks, combined with poverty, lack of jobs, the highest density of population anywhere on earth, inefficient management and lack of governance create challenges as great as or greater than any the young nation has ever faced.

More money is already going to the social safety net than to development financing, says Ahsan. "We cannot afford this to happen, but it is happening," because so many people need help. To these stresses add global warming. "All our development gains will probably be washed away because of climate change."

Global warming's potential to devastate their nation has been a central concern for researchers in Bangladesh for decades.

Ahsan Uddin Ahmed is one of the Bangladeshi scientists who started thinking about global warming more than twenty years ago, even before the UN's Intergovernmental Panel on Climate Change (IPCC) was formed. Early on, people in Bangladesh wondered why he was concerned about something so remote when the country has many other problems. When he and others talked about the need to prepare for climate change, people in government agencies would respond that climate change was an international problem, something far away; their focus was developing Bangladesh. Ahsan told them they would see the effects in their lifetimes. "Now more people accept this."

Trained in chemistry and atmospheric sciences, he began working on climate change at the Bangladesh Center for Advanced Studies, a think tank in Dhaka. The logistics of getting to work in the city's daunting traffic jams became too much; he had a disabled daughter at home and his wife, an economist, had to commute to the opposite side of town to her university. He started his own small research organization, the Center for Global Change, so that he could continue his work closer to home. He hopes someday he can telecommute and move outside Dhaka where the "air is still clean."

Ahsan and his group collaborate with international academic institutions and nongovernmental groups. They apply regional climate models to explore how hydrology will change in coming years and how human vulnerability will increase as a result.

For scientists like Ahsan, climate change in Bangladesh means a complicated set of causes and effects, shifts and adaptations, over a span of decades as temperatures and patterns of precipitation change. For those of us who read typical news accounts, the idea of rising sea levels typically grabs our attention. This might conjure visions of the Bay of Bengal inching its way up with each high tide, eventually inundating up to a quarter of the country. Low-lying land in places like Bangladesh, Burma, the Philippines, and the Pacific Islands are the most vulnerable to the rising seas caused by global warming.

According to Bangladeshi sources, high tides have been rising about five centimeters a year. Climate models—based on current and projected temperature increases worldwide, along with melting ice and expanding oceans—indicate that this will continue. In the past decade, both the IPCC and the World Bank have published estimates that higher sea levels of a meter or more will affect low-lying countries like Bangladesh by the end of this century. Other sources have predicted even faster rising. A one-meter rise in sea level could completely engulf the Sundarban.

Like the glaciers high above in neighboring Nepal, India, and Tibet, the fate of the Sundarban—indeed of Bangladesh itself—is in dispute. The extent of melting at the top of this immense watershed is difficult to measure; similarly the consequences of global warming here at the bottom cannot yet be quantified and predicted with certainty.

"Thermal expansion" would account for about half the increase in sea levels, because warm water takes up more space than cold water, and ocean temperatures are rising. Melting ice would cause the rest. The melting of Himalayan glaciers would create little noticeable change here, despite predictions to the contrary. Far away in Greenland and Antarctica, where ocean temperatures are rising by degrees of magnitude greater than in South Asia, ice masses vastly larger than Gangotri or Ngozumpa are melting. Just as the industrial gases that have warmed the planet came not from this region but from far away, the unwanted water that may lift the ocean and nibble at Bangladesh's coast also comes from far away.

"Some say a large part of Bangladesh, maybe 30 percent, will be inundated within the next thirty to seventy years. I think that's a misstatement," says Ahsan. "There is no reason to believe that. In the coastal zone, all areas are embanked. Even in the highest tide during the full moon these embankments withstand the swelling of the tide."

Ahsan says the catastrophic scenario of inundation has been repeated and become the accepted view even though it's not based on science. Researchers who made the predictions did not factor in the coastal embankments when they made their calculations. "We analyzed the hydrological model in different coastal districts and identified embankments which could be inundated and those that would not." He says his group would like to analyze all the coastal districts when there is money for such research.

During my first visit to Bangladesh I spoke with the Conservator of Forests in the Khulna district. Shaikh Mizanor Rahman wasn't very worried about the rising seas predicted to eat away at Bangladesh as ice melts far away in Greenland and Antarctica. He said the land would keep growing as it always had. "All silt goes through Bangladesh. New land is coming."

A group of scientists at the Bangladesh Government's Center for Environment and Geographic Information Services (CEGIS) in Dhaka has countered the IPCC predictions of eroding coastlines along the same lines. A billion tons of sediment each year would counteract rising seas. CEGIS has thirty-two years of satellite images showing Bangladesh's landmass has been increasing.

This seemed comforting and made sense to me from my limited understanding. The lower Himalaya were formed by the erosion from the upper Himalaya. Vast amounts of sediment in the mountains will come down the rivers and create new land on the coast. Some will be washed away each year, as it always has been. Land near the coast and the rivers in Bangladesh comes and goes; for generations people have known their land could be swallowed up by a flood. People living on the chars—the large temporary islands in the main rivers—have developed their own way of life. When a char erodes and disappears in a flood, another generally forms and people move to that newly formed island. The people of Bangladesh are known for such resiliency.

Ahsan Uddin Ahmed agreed that in the framework of geological time, the Himalaya could replenish what Bangladesh might lose to rising seas. "So in 10,000 years probably this silt will help form a new part of the delta along the shorelines. But we are talking about something that is happening in two, three, four decades. If we are talking about world climate change and sea level rise, then probably we should be thinking in terms of decades, not millennia."

And during those decades, Bangladesh has some time to prepare. "In the worst-case scenario," Ahsan warns, "if embankments are all broken, then probably a significant portion of Bangladesh *will* be inundated because of sea level rise. But that's wild," he adds. "We hope we won't get there. We will become conscious. We will do something," both to protect the coastline and to protect Bangladesh's economic progress.

Bangladesh's birthrate steadily dropped from the 1970s through the 1990s but has begun to increase again. "Half of the population is below reproductive age. With that huge population pressure we will be in deep trouble—we will still be a 'least developed nation,'" says Ahsan. "Problems are overlapping and deepening. That's why there is urgency. Climate change alone will erode most of our development gains.

"If we spend in the right areas, this doesn't need to happen. But if Bangladesh does not invest in adaptation, then probably our dream of becoming a middle-income country will never be achieved. With our population and pressure on food security, we won't even have enough money to mend the embankments."

Embankments can prevent or at least forestall inundation, but saline water can still do damage near the coast as it moves inland through the rivers. The salinity of groundwater and soil in coastal regions is already a chronic problem in southwestern Bangladesh due to upstream diversions from the Ganga. In the short run, increasing salinity is a threat to agriculture. For people trying to survive and get beyond abject poverty, this is happening now, not in fifty or a hundred years.

In the monsoon, the rivers literally keep the sea at bay. During high tide, salt water presses into the mangroves; but when the rivers are full, their flow pushes it back into the Bay of Bengal. As the rivers subside after the monsoon, salt water pushes farther and farther up the rivers, past the Sundarban, into farmland. As less fresh water flows down the rivers through the delta, sea water can push farther inland more often.

The districts of southwestern Bangladesh, adjacent to and directly north of the Sundarban, are those most affected by the lack of water in the Gorai River during the dry season. Maps showing the saline-affected areas of Bangladesh indicate how greatly this region has been affected by inland flows of seawater. The districts to the east, flushed by the combined flows of the Padma, Brahmaputra, and Meghna, have not been so affected. Even in the dry season there is plenty of sweet water to push the salt water back to the Bay.

The widespread belief that salinity is increasing in the region watered by the Gorai because of global warming frustrates Ainun Nishat. He is now vice chancellor of BRAC University; before he was Bangladesh's country representative in the International Union for Conservation of Nature (IUCN) and a negotiator on the treaty governing water sharing at Farakka. He says salinity has been steadily increasing over the past thirty years chiefly because less water is coming down the Padma. "Due to Farakka there is no flow through the Gorai, no flow to the Sundarban, and salinity is increasing. Now the salinity problem is compounded by climate change, but I am still convinced the main problem is Farakka because sea level rise is not that great—maybe ten centimeters.

"The coastal line is very flat, so that means even with a ten-centimeter rise in sea level, the front coastal line has moved maybe by five kilometers, ten kilometers. It's very flat over there," he told me in his office in Dhaka in 2010. "The problem has been aggravated by sea level rise."

He believes that predictions about loss of land to rising seas are somewhat alarmist. "Sea level rise will not cause inundation. Why should you go underwater if the sea level rise is three feet? I can raise my ground. Salinity is the problem. Food security would be the problem."

"Salinity increases because it pushes in during tides and fresh water is not enough to push it back. The land is still fresh. The waters of the river have become brackish." He says a sufficient quantity of fresh water flowing down the rivers would counteract a minimal encroachment of seawater from the Bay of Bengal. But the farther saline water pushes inland on a regular basis, the greater the likelihood groundwater will become more saline, making it useless for drinking and irrigation. The water could also seep into the soil and make it too saline for crops.

All the water flowing into Bangladesh, which appeared to Indian engineers to be wasted on tiny East Pakistan, has a purpose. It continually flushes the coastal region, keeping the land fresh and the Sundarban from becoming too saline. In the past thirty years, much of this region has been given over to shrimp farming.

A fifty-eight-year-old farmer in Khulna told me that when he was young he could produce an abundance of crops. Now he can't. His small plot of land is productive for only four or five months a year. He has a pond for raising fish. The wild, freshwater species were already gone. He said people in his village get by; they share crops. But the land and water were no longer congenial to cattle, so milk products had begun to disappear.

The more salinity increases, the more land in the coastal region will be lost to farming and given over to shrimp cultivation. Some of the cultivators are small local landholders who switched to shrimp as their soil became too saline for year-round crops. Some are nonresident landlords with large holdings who have bought up small farms and converted them to shrimp ponds. In any case, the shrimp are for export. For the most part, they do not feed rural Bangladeshis. A variety of rice that tolerates salinity has been developed, allowing one crop a year to be produced in the saline soil. Shrimp farming and salt-tolerant paddy were necessary and useful adaptations as the long-term effects of salinity became clear. Still, land has been lost, in a country with limited land that was already struggling to feed itself.

The depleted Gorai River and the increasing salinity of farmland in southwest Bangladesh provided an early warning for the country's global warming specialists. They showed them what would likely happen in other parts of Bangladesh as the climate changed. It would bring longer and warmer dry seasons. As even more water would be needed to offset the effects of rising temperatures on thirsty land and crops, increased withdrawals upstream would mean less water in the rivers downstream and increasing salinity near the coast.

Bill Collis has worked in Bangladesh for more than thirty years and takes a more sanguine view of the agricultural opportunities for farmers in the southwest region. Collis is the country representative for World Fish, an organization that does research to promote fisheries in developing countries. He believes soil salinity in the dry season has been the reality for southwest Bangladesh for a long time. Upstream withdrawals from the Ganga have just made it worse.

"Even in shrimp areas we look for cropping options. Most successful farmers I've seen in Bangladesh have switched from two rice crops a year," says Collis. During the wet season, from July to November, they cultivate freshwater prawns and fish like carp and tilapia and grow vegetables. Then, during the early dry season from December to March, they grow rice. March to July, when the salinity increases, they cultivate saltwater shrimp. Collis says all the farmers he works with are small landholders. Shrimp is a high-value crop that needs more care than absentee landlords can provide. There are 200,000 hectares in shrimp cultivation in Bangladesh, almost half a million acres, mostly in Khulna.

Whether salinity in the southwest began with Farakka or even before, it is still increasing. "I'm looking for data on this," Collis says, "but anecdotal evidence says saltwater is coming a month earlier than it did ten years ago." This means a shorter rice-growing season, impinging on the one crop a year farmers are able to grow. Research on saline-tolerant rice is continuing, which could be useful for farmers in this region if they can afford the expensive seeds for rice that gives good yields.

As flows in the major rivers decline, preserving surface water becomes a challenge. "Wetlands can help," says Collis. "We're trying to preserve them and all the environmental services wetlands provide. Thirty to fifty percent of the permanent wetlands in Bangladesh have disappeared in the last few decades." Wetlands are the source of the small fish that used to be so important in the Bengali diet. People ate the small fish whole, thereby gaining vital micronutrients, says Collis, the lack of which is increasingly becoming the cause of malnutrition in Bangladesh.

Here as in other parts of the world, scientists still can't say with certainty that global warming has caused any particular drought, hurricane, or cloudburst. Evidence may point to climate change, but "we just can't put a tag on all the changes happening around us," says Ahsan Uddin Ahmed.

He has put the tag of climate change on one dilemma, however. Bangladeshi fishermen—who are among those struggling to get above poverty—are seeing the effects of climate change on their work and their income. Sea surface temperatures have warmed in proportion to global temperature increases, leading to more cyclones forming in the Bay of Bengal. Some of them strike land, but some of them stay in the bay and make it more treacherous; and the bay tends to be more turbulent already because of the increase in surface temperature. Ahsan says ten years of data show a doubling of the number of events of turbulent seas. "This year it happened twelve times," he told me in 2007.

Fishermen go out into the bay to fish for two weeks at a time. They typically borrow money for fuel and food for these trips. When warnings about rough seas are broadcast, fishermen have to come back to shore or risk their lives. If they abort a trip, they lose their investment, make no profit, and may end the trip in debt instead of with money to feed their families. If they stay at sea despite the warnings, they might be lucky, escape harm, and make money on the trip. If they are not lucky, they could lose their boats or their lives along with their investment. Some of Sidr's casualties likely fell into this group of frustrated fishermen who had abandoned their ventures too many times already and were desperate to have a profitable trip.

The bodies are rarely found. The men are simply missing. Sensors in fishing boats that will alert the coast guard to a fisherman in danger could be useful, provided there are helicopters to rescue the men, Ahsan says, but more is needed. "Planning is what is needed, starting now. The sons of deceased fishermen will go to sea prematurely and die in the same way. We cannot just let this happen. The government must realize these are human lives, not insects. We must respect people's lives and livelihoods."

Artisanal fisheries alone won't support coastal people, he says. "In an advanced economy we need carpenters, masons, machine operators. Why not start training centers in coastal zones so that sons of fishermen are a priority?"

Such crises are emerging across the economy, which is why Ahsan worries that climate change will wash away the gains Bangladesh's poorest people have made. "These things are happening today. We don't have to wait twenty or forty years to see effects of climate change in Bangladesh."

"Our problems would be smaller if we had one-tenth of our population," says Mozaharul Alam, another of the Bangladeshi scientists who have been paying attention to signs of climate change for the past two decades and have participated in international negotiations.

Researchers like Alam represent Bangladesh at the periodic international meetings on climate change, where agreements about curbing greenhouse gas emissions remain stalled by big-power recalcitrance. These meetings are still opportunities to meet with other "least developed nations," to form alliances and to keep pressing the developed nations for the funds they have promised to help countries like Bangladesh adapt.

Alam began working at the Bangladesh Center of Advanced Studies (BCAS) in 1992, when scientists there were assessing sea level rise and trying to understand how much land Bangladesh might lose. More recent research at BCAS focuses on increasingly erratic weather.

BCAS was chosen to represent the views of Bangladesh at the UN. At the beginning most people assumed that developed countries would have to take responsibility for solving the problems of the developing countries that are most vulnerable to climate change. Since then, researchers in countries like Bangladesh have begun to steer their governments to play a much more active role in solving their own problems.

Alam says the country's coastal region is not the only area of concern. "You hear that Bangladesh is flood and cyclone prone, but we have a drought-prone area and it will become more so." Alam has focused on the drought-prone northwest area, gathering data and interviewing local people. "People in villages who don't know anything about climate change say that the distribution of rainfall is different. Rain is not gentle. The number of days without rainfall has increased. So the rainfall is coming in a shorter period of time. Now the issue is whether that will continue. Most likely it will."

The challenge for researchers once they see such trends is to use the information to help people: to "merge the climate change data and the socioeconomic analysis" of a region to design projects that will help people prepare and cope. Drought will not affect all the people in a region in the same way. A large landholder, a small farmer, a landless family may have very different specific needs even though all of them might need to plan for storing water, diversifying crops, and devising better ways to irrigate them.

Bangladesh is small but its topography is diverse, with many different ecosystems and microclimates, Alam says. Each of them will see somewhat different effects from climate change, and have to devise unique coping strategies: "In terms of doing something, we need local understanding."

As researchers generate data about how climate change is affecting specific locales, global warming is increasingly understood as a multitude of local crises instead of a single looming catastrophe. And researchers like Alam have a better chance of gaining a hearing with government officials when they can show local trends.

Researchers are trying to convince politicians that they must include the realities of climate change in all aspects of governance as they plan for the future. This does not happen easily, because researchers and politicians speak a different language. Politicians think in five-year cycles. Climate change researchers like Mozaharul Alam say adaptation requires at least a thirty-year planning timeline.

"New research is generating information about how climate change will impinge on development in thirty years," he says. "There's the beginning of awareness about this at the policy level." But the country's budget priorities don't yet reflect this awareness. "The primary thrust of the government is poverty reduction, so we try to show that this is also a poverty-reduction issue."

BCAS researchers talk to the environment ministry, the agriculture ministry, and the water resources ministry, says Alam. And they work with local, community nongovernmental groups, which he says play an important role in terms of diversification of livelihood, job creation, and disaster risk reduction.

Alam agrees with those who say additional cyclone shelters will be needed irrespective of climate change because the population is increasing. For them to be well maintained the shelters should serve other purposes in the community, he believes; if they are just for cyclones, they tend to be ignored until they are needed. Sidr did not take the usual track that cyclones have taken in the past, he says, so fewer shelters had been built in the region that Sidr devastated. Cyclones tend to go to the northeast, toward the Chittagong coast.

As late as thirty years ago, a patch of the Bay of Bengal's ring of mangroves could be found in the arm of Bangladesh that reaches to the southeast toward Burma, on the Chittagong coast. This patch is completely separate from the protected Sundarban I visited. Environmental journalist Philip Gain attributes the loss of several thousand lives in the devastating 1991 cyclone to the deliberate destruction of this piece of mangrove forest, called the Chokaria Sundarban. Had it still been standing, it might have protected part of the region near Chittagong where the cyclone struck and which is frequently the destination of cyclones out of the Bay of Bengal.

This eighty-square-kilometer patch of forest and mangrove swamp was decimated for shrimp farming, a tragic mistake funded by the World Bank and the Asian Development Bank in the 1980s. The banks loaned money to the Bangladeshi government to fund the creation of shrimp farms to provide a product for export. The forest was not just a protection against frequent cyclones and shelter for the wild animals that once inhabited the entire coastal region. It was the source of fish and wild shrimp. The mangrove roots that sheltered the aquatic population are gone, yet the shrimp farms did not succeed.

Bill Collis of World Fish is trying to find funding to convert the area back to mangroves. He says the destruction of the Chokaria Sundarban was an unfortunate idea initiated by the Bangladesh government, not the big development banks. "I'm sure they regret it," says Collis. "Shrimp culture doesn't work in mangrove areas for the most part because of soil acidity."

Chemical changes in the soil caused by the years of shrimp farming may complicate the return of mangroves in the Chokaria Sundarban. Yet the effort may signal a growing recognition of the value of mangrove forests to protect the coastline, as a nursery for fish, a source of other valuable products, and a habitat for many species.

The many pressures on the protected Sundarban—from the surrounding human population and from increase in salinity—will continue to erode the ecosystem. The hallmark tree of the forest, the Sundari, has shown signs of stress for many years. These trees have been affected by something called "top-dying disease." Ahsan Uddin Ahmed said changes seem to have begun even before Farakka. Back in the 1920s there was evidence of the "top-dying" disease that has increased in recent decades. The top branches of the tall Sundari trees die while the rest of the tree continues to stand.

The conservator of forests in Khulna, Shaikh Mizanor Rahman, told me the reduced amount of sweet water coming down the Padma will continue that trend and take a toll on the other varieties of trees in the mangrove forest. As he saw it, lacking enough sweet water, the Sundari will either adapt or go extinct. He said this made him sad, but he took the long view: everything changes and adapts. Maybe the Sundari trees will adapt in time to changes, growing less tall because of lack of sweet water and increased siltation. The trees in the Indian Sundarban are less high and lush than those in Bangladesh, he noted. They have adapted to the less favorable conditions there.

"We reach out to local people and try to understand their vulnerability through their own eyes," says Ahsan Uddin Ahmed of the work his group does, especially with coastal populations. "We have to allow vulnerable people to analyze their own vulnerability."

That includes educating the future generation, along with equipping local governments with knowledge and financing. "Only if we do the right things today can we safeguard our development work," says Ahsan. This applies to decisions made at all levels, from the central government down to the household level. Each village will need to assess its own problems, and then with help from local government agencies, prepare its people to cope.

Women in particular can be a resource for information about changes in and around the home and farm, an Indian woman who had done research on rural women told me. As they adapt to climate change, even unconsciously, changes in their work habits can point to shifts in the availability of water and fodder and how crops are growing. Women see changes in the health of animals and people. Eliciting such knowledge from women in a traditional culture will require patience and care so that it can be incorporated into village planning.

As resources grow scarce, climate change could increase gender inequities, a trend that would slow overall progress in Bangladesh. If instead women were drawn into planning, whether for adaptation to gradual climate changes or choices about how to be ready for periodic disasters, an opposite trend might be fed.

In any case, Bangladeshis familiar with past trends in disaster management say previous responses to cyclones and floods will have to change. "Asking for a few million tons of food in an emergency won't work. We have to create our own food reserves." This applies at a regional level as well, says Ahsan. "Policy makers don't think about it unless it's really happening. And those of us living in Dhaka don't think about it. If a crop is gone we can still survive, but those who depend on the crop face starvation." Ahsan would like to see a regional food bank. A quarter of the world's poor live in south Asia. "Who knows when Pakistan or India will need something?"

He conjured an apocalyptic vision unless this kind of preparation starts now, so that people will have food and shelter during emergencies, episodes of seasonal disaster, or the chronic deterioration of their lands and resources that climate change may bring. Otherwise, victims of global warming will leave rural areas for cities in India and Bangladesh; such migrations have come to pass already in Africa.

"Same with our coastal zone. Can we create a fence? They will fight for their survival. They will come to Dhaka. And they are already coming."

Before I left Dhaka in December 2010, I asked for help finding a family recently forced to come to the city. I was looking for "environmental refugees" in the broadest sense, people no longer able to farm and survive in their home village whether due to flood, lost crops, salinity, or cyclone. Hundreds of thousands who fall into this broad category have over time come to Dhaka to live in the slums and shanties that can be found near any of the city's water bodies. There are now four million people in Dhaka's slums.

I had misgivings about asking to be introduced to a "typical victim." Stories about developing countries often start this way, sketching the plight of some poor villager. I did the same thing in Bihar and Madhya Pradesh as well as Bangladesh. Giving voice to the people who are actually suffering a crisis is preferable to talking only to the aid experts, but still I was an intruder. I was asking for help so that I could go peer more closely at the desperation of people like those I might see fleetingly in the streets, whose experiences I might never fully grasp even if I could speak Bangla.

I was put in touch with Ranajit Das, a public health specialist. He worked in an organization that tried to improve access to safe water and sanitation for people living in Dhaka's slums. We took an auto-rickshaw from his office in Shamoli, on the western side of Dhaka, down toward the Buriganga, or "Old Ganga." It flows to the west and south of the megacity and is the receptacle of much of the city's waste.

We reached Lautala Akram Miar slum off Basila Road. Slums may connote tightly packed tenements. Some of Dhaka's poorest people do live in buildings that more closely match that picture. But Lautala Akram Miar was a new slum, its flimsy barracks built right on the floodplain.

We slid down a steep hillock from the road onto the flat expanse of sandy soil. A series of long, narrow structures stood on the plain ahead of us. Refuse dotted the ground: bits of orange peel, remnants of cloth, paper, and cardboard. Some was likely garbage dropped by the nearby residents, but most appeared to be the small pieces of debris a receding flood might leave behind. A scattering of low weeds struggled up through the hard-packed silt.

We walked a few hundred feet across the empty ground. On our left stretched a dark pool of stagnant water, ringed by scum. It was a small inlet of the river. At its edge, the four bamboo legs of a makeshift privy were planted firmly in the murky water. Two lengths of mature bamboo set side by side served as a gang-plank out to the booth of woven bamboo. A stiff oversized sack hung slackly; it would partially cover the entry when pulled inward. Short segments of bamboo were laid parallel to make a rough floor for the booth; a gap between them was the opening through which the members of at least twenty families might defecate.

The riverside barracks we were approaching, constructed of corrugated metal, sheltered multiple families. It was the length of a couple of railroad cars. As we got closer I could see the structure's foundation consisted merely of lengths of bam-boo pounded into the sand, spaced a few feet apart to hold up the sheets of metal. A few square openings in the wall showed that some residents had punched through the metal wall to create a window.

We climbed onto a triangular bamboo ramp that served as a rough staircase up to the hallway of the tenement. The barracks had a double row of rooms with a passage in between, open to the air. We went inside one of the first rooms on the right. Lengths of bamboo held the walls in place and offered support for a few shelves that contained some cooking utensils, a bottle of cooking oil. A mosquito net was caught up in one corner of the room.

Ranajit's staff had selected two families as fitting my criterion—recent envi-ronmental refugees from the coastal area. Minara and Kanchon Mazi were renting this room. Other adults and some children came in to watch us, all sitting in a semicircle opposite Ranajit and me. The women and girls glowed against the flat gray of the metal walls, their saris and kurtas in shades of pink, orange, coral, and purple. A woven bamboo mat with a faded green cloth softened the rough plank floor where the women, children, and men squatted while Ranajit and I sat on a woven mat.

I was not sure which of the children in the room might belong to Minara and Kanchon, who had come to Dhaka with their four daughters and three sons. Kanchon crossed his arms, resting them on elevated knees, his legs folded under a green lungi. His hair was flecked with gray. A pretty little girl with a serious, inquisitive expression sat next to him. The girl's long narrow face resembled his. She was dressed scantily in a boy's white undershirt and underpants, but little orange earrings that looked like miniature teddy bears dangled from her lobes. An older girl in a kurta and pants who looked like Minara sat beside her mother.

During the past monsoon the family's house on Bhola Island, the largest of Bangladesh's coastal islands, east of the Sundarban, had been washed into the river. Bhola is prone to cyclones, but that is not why they had to move. Their home was in Char Fasson, an island within the larger island, surrounded by a river. The land around their house had been eroding for years, and this last monsoon had taken it all.

Bhola is even poorer than the rest of rural Bangladesh. In the past half century the island has lost half its landmass to the water, to erosion from the rivers that surround it to the north, perhaps to increasing turbulence of the seas to the south. The name Bhola means lost place in Bangla. About two million people live on the island. Ainun Nishat told me the offshore waters around Bhola used to be fresh enough to use for irrigating crops year-round; water flowed from the rivers to bathe the island with fresh water, pushing seawater back. Now in the winter months the water on all sides of the island has become too saline to be pumped in for irrigation. If used for irrigation, the water makes the land saline.

The Mazis used to grow two crops of rice a year, along with vegetables. They grew enough food to feed the family year-round. "That life is good," said Ranajit, translating what Kanchon was saying. "They can grow food, they have their own land. Here they have to buy everything, even water. Everything costs a lot. He sometimes pulls a rickshaw, but not every day. It's hard work. Sometimes he works in construction."

Their oldest son came to Dhaka first, then returned for them and brought them here about six months before I visited. The son does day labor, loading and unloading trucks. He lives in another room like this one nearby with his young wife. The whole family eats together.

Their home in Char Fasson, at the south end of Bhola Island, was likely small and made of similarly humble materials—sheet metal and bamboo. But they had their garden, trees, and animals outside their door. Owning their own land and house made them feel proud and independent, Ranajit said. I thought of the green, quiet Bangladeshi villages I had visited. That home must seem like a dream to them, this littered urban floodplain a nightmare.

Ranajit asks him why they didn't just stay in their village. "Here we can rent a house," Kanchon replied. "In the village you have to have your own house, your own land. There is no place to rent."

Kanchon and Minara do not smile. They are not just thin, they look depleted, almost in shock.

"They will never go back," Ranajit says. "They have no land there. They don't know what will happen."

"It is not possible to save money and buy land," says Minara.

The youngest child cries. Minara, thin and silent, nurses him. The children look healthy, better fed than their parents. They went to school in the village, but not here. There is no free school nearby. The children just roamed around the flood-plain all day when they weren't inside the room.

Their oldest daughter, aged sixteen, was working in a garment factory a mile from here; she was the family's main wage earner. The daughter earns 1,600 taka a month. The son brings in one or two hundred taka the days he can find work. The family's income is never more than four thousand taka a month, or fifty-three dollars; twelve hundred taka, or sixteen dollars, pays for the rent on the two rooms. They pay fifty taka a month to take water from the hand pump outside. They eat rice and sometimes fish and try to scavenge vegetables when they can. They don't buy vegetables because they are too expensive.

Ranajit said the cost of food is going up, double what it was a few years ago; but entry-level pay for garment workers has been the same for the past ten years.

When the Mazis arrived during the monsoon, water was standing a foot deep here in this room, Ranajit tells me. Even though the sheet metal barrack rested on bamboo stilts, and the plank floor was several feet above the ground outside, the structure was inundated. And still they lived in it.

Ranajit sees the amazement on my face and laughs a little. "You cannot under-stand. This is their coping mechanism."

No, I could not begin to imagine how they lived in this small room with its flimsy floor under water. Apparently, Ranajit seemed to be saying, somebody would sleep while someone else stood in the room. A hammock could be strung above the water. And there was a narrow shelf above my head about a foot wide. Perhaps a small child or two might sleep there. Still, this discomfort, dirty flood-water invading their skimpy shelter, was almost unimaginable. Next monsoon maybe it will be the same. By then, they hope to move to another place, higher on the floodplain.

I ask their ages. They aren't sure. Ranajit asks Minara how old she was at liber-ation, thirty-nine years before, and concludes she is in her early forties. Kanchon is about fifty. The exact ages of the children are also a little uncertain, ranging from early twenties to toddler.

Another couple and their children crowd into the room to tell us their story. This is a younger family, the couple in their thirties.

The man, Alamgir Hossain, is handsome. He has a mustache and the stubble of a beard. He smiles slightly, quizzically, which makes him look friendly; but his eyes are sad and wary. Alamgir is related to Kanchon Mazi. The two families came to Dhaka together. The same monsoon took their land by the river.

His wife Rina has a job as a housemaid, washing dishes, clothes, and floors, for about three hours a day for a thousand taka a month, or thirteen dollars. She doesn't like the work. She never thought she would be a housemaid; she thought she would have her own house.

Alamgir drives a rickshaw about twenty days a month to make a little money. Like Kanchon, he says he can't do it every day; the work is too tiring.

"That life was settled," Alamgir says of the village. "Life is very uncertain here. Every day you have to work. There is social support in the village, but not here. People are from all over Bangladesh, not from our village."

"There is no social bonding here," says Ranajit. "But they can't go back either. Their land is gone. They have no roots there anymore."

The couple has four children. The small children in the room are laughing and romping while we talk. The older children in both families continue to sit quietly, close to their parents, and watch.

I ask whether Alamgir wants his children to go to school. "They're not thinking about that at this moment," says Ranajit. They're worried about more basic things. Their oldest daughter, a very pretty girl, sits in between her parents with a scarf modestly over her head like her mother's. When she is sixteen she too will go to the garment factory.

They have no assets, only some cooking things here and a few things with a relative back in the village. "They never thought they would come to the city. But the river changed everything," Ranajit tells me. "They were forced to come to earn money, not because they want to. If an emergency comes they have nothing to use for it. They have to fight just for food. They said they have no hope in life. Only God knows what will happen."

More people from their village will join them, Alamgir told Ranajit. The river will keep taking the land.

Minara took a stained white plastic container from the shelf. Inside were green leaves and white paste. Kanchon sliced an areca nut with a well used curved knife. They starting making paan with the medicinal betel leaves for the adults to chew.

As we were getting ready to leave, the younger couple's oldest son came into the room, grimacing from pain. A boy of twelve, he had a bad burn on his arm from a hot water spill earlier that day or the day before. He held his red, blistered arm out in front of him as he squatted on the floor briefly. The arm was shiny; his parents had put some oil on the burn. The boy soon left the room, the discomfort too intense for him to sit still for long.

I asked if they had taken him to a doctor, knowing the answer would be no. I asked Ranajit if it would be all right for me to leave some money for them to take him to a doctor.

I remembered something Nayeem Wahra said about most Bangladeshis: "We only go to the doctor before we die." Ranajit and I left after I slipped Ranajit 500 taka to give to the boy's mother, hoping it might be used to pay a doctor to look at the boy's arm.

We took a rickshaw back into the main part of Mohammadpur so I could catch a taxi to the other side of Dhaka, where the foreigners cluster. A pileup of vehicles and rickshaws in the curve of the narrow road left us immobilized next to a gar-bage dump. Men and women were raking through the contents of dumpsters to find usable items. Flies covered the garbage and swarmed over the road onto us, so thickly I kept my mouth closed until we could move on down the road.

If the worst predictions of global warming and inundated coastlines come to pass, even very slowly, tens of millions of people could be displaced in Bangladesh. They would have to be absorbed somewhere, but it's hard to imagine where they might fit and make lives for themselves. Bangladesh is short of land; most of the rest of the country is already too populated. India doesn't want them. And in a city like Dhaka, public services do not yet function well enough for the people who are already there.

I thought of the Indian water expert who explained the need for the inter-basin transfers of the river-linking plan. It would be easier to move the water to the people, he said, even through expensive canals. That would be preferable to moving the people—or to their having to move en masse because of drought.

As the abundant waters in the Bay of Bengal edge slowly closer to the people of Bangladesh, moving the rising water back into the bay so those people won't have to relocate would require a worldwide effort of unimaginable proportions. Even if developed and developing countries finally agree to the kinds of steep emissions cuts many scientists say are needed right now, warming will continue for a long time before the trend might reverse. Seas may have to rise, and salinity increase, before seas fall again in this or some other century. For a long time to come, the resilient people of Bangladesh will be in the crosshairs of calamity.

# Notes

Except in Bolivia and Ecuador; they have amended their constitutions to grant rights to nature. There is a movement in Turkey to do the same.

Formerly known as Uttaranchal.

The connections, the dance of mountain and cloud and rain and soil, are even more complicated. Plate tectonics pushed up the Himalaya, which block the summer monsoon, forcing it to dump enormous amounts of rain, which in turn washes away millions of tons of sediment.

See Michael Thompson et al. *Uncertainty on a Himalayan Scale* (Kathmandu: Himal Books, 2007), p. xlv.

Dr. Ravi Chopra of the People's Science Institute in Dehradun told me in spring 2011 that "as of this moment we think the government of Uttarakhand has plans for 675 dams of all sizes: 155 of them are 5 megawatt or more." Chopra and his institute provide scientific analysis for groups mobilizing over hydropower. He says there are many alternatives to avoid some of those dams or make them less damaging. Solar power is one. Conservation is another—helping Indians use electricity more efficiently and stemming unnecessary losses. "Losses here in Uttarakhand were 40 percent a decade ago. They are now down to 28–30 percent from eliminating illegal connections. We can also reduce loss by redesigning transmission systems; more skilled technical people can reduce it even more."

Ravi Chopra says blasting for tunnels weakens mountain slopes, but without blasting there are fewer sites available. One alternative is to use the kind of tunnel-boring machines used for the New Delhi Metro. They are expensive, and so large they would be difficult to transport up into the Himalayan foothills, but they could reduce some of the damage. They are being used on a hydropower project on the Alaknanda, funded by the World Bank.

Diana Eck, *Banaras: City of Light* (New York: Columbia University Press, 1999), p. 211.

The names of the three projects were Loharinag Pala, Pala Maneri, and Bhairongathi.

Chapter 2, The Real Poop

27    ***Globally 1.2 billion people***
      United Nations University, Institute for Water, Environment and Health. *Sanitation as a Key to Global Health: Voices from the Field.* 2010.
      "India has more mobile phones than toilets: UN report." *Daily Telegraph*, July 17, 2010.
      Ravi Nessman, "India: Land of many cell phones, fewer toilets," Associated Press, October 30, 2010.

28    ***something in the neighborhood of a billion liters***
      "Ganga pollution reaches alarming levels," *Hindustan Times*, June 5, 2007.
      "India allows 25 billion litres of sewage into its rivers," Times Online, January 12, 2010.
      Chemical effluents are probably even more dangerous; containment and regulations of these are just as weak as of sewage. But sewage is the most visible and has cascaded into rivers in urban areas like Delhi in such quantities that the multiple benefits these great rivers once gave people are completely overwhelmed. The sacred uses, the drinking and bathing and navigation, all are compromised beyond belief though not yet beyond repair.

29    ***But the system I have known and used quite unconsciously ever since I was potty trained is nuts.***
      A century ago, a prescient American president saw the absurdity of using drinking water in toilets. Teddy Roosevelt thought civilized people should be able to come up with a system that made more sense.
      Rose George, *The Big Necessity* (New York: Metropolitan Books, 2008). p. 37.

32    ***Thames is said to have reeked***
      Rose George, *The Big Necessity* (New York: Metropolitan Books, 2008), p. 23.

32    ***Putting an end to this scavenging***
      Most of the information about Bindeshwar Pathak's work comes from an interview with him in January 2010 in New Delhi.
      For more information, see Bindeshwar Pathak, *Road to Freedom: A Sociological Study on the Abolition of Scavenging in India* (New Delhi, 1991).

33    ***Protects groundwater***
      Effluent can seep into the surrounding soil. In dry soils that is fine because the soil microbes will break down all pollutants within a radius of a few meters around the pit in which these processes take place. However, in places with a high groundwater table or places prone to flooding, this does not work well.

34    ***I met a group of women who had been scavengers***
      For portraits of these and other former "scavengers," see Bindeshwar Pathak, ed. *New Princesses of Alwar* (New Delhi, Sulabh International, 2009).
      Further reading:
      Gita Ramaswamy, *India Stinking: Manual Scavengers in Andhra Pradesh.* (Pondicherry: Navayana, 2005).
      Kumar Alok, *Squatting with Dignity* (New Delhi: Sage, 2010).

37    ***nonrenewable resource***
      "What we can say for sure is that urine contains all the major plant nutrients, N, P, K (nitrogen, phosphorus, potassium) and is thus a 'complete' fertilizer. Many farmers in South Asia who rely on inorganic (or chemical) fertilizers only use N, which leads to depletion of the other nutrients." Marijn also notes that phosphate is not just running out, but that the geographical distribution of it is awkward. Approximately 80 percent is in Western Sahara, 10 percent in China, and the rest fairly scattered. "This means that the geo-politics of phosphate may turn very nasty indeed," he says.
      See also:
      Melinda Burns, "The Story of P(ee)," *Miller McCune* magazine, February 10, 2010. And *The Wealth of Waste: the Economics of Wastewater use in Agriculture*, United Nations Food and Agriculture Organization, FAO Water Report 35, Rome, 2010.

40    ***Has relaunched a campaign***
      A campaign begun in 1986 to boost rural sanitation led to nearly ten million latrines by the end of the century. But the population continued to swell, and millions of those

toilets were unused because they had not been designed in consultation with the people who were going to use them. Rose George, *Big Necessity*, p. 176.

For a more detailed discussion of the challenges to ending open defecation in India, see George, *Big Necessity*, Chapter 8.

On one of my last trips to Delhi I saw the new trash bins that had been installed as part of city beautification for the Commonwealth Games. They were cleverly made, one side for biodegradable waste, the other side for nonbiodegradable. But they were not being used properly yet. There were cans in the biodegradable side and trash strewn around outside the non-biodegradable side of the bins. The hardware is there, as it is with toilets. But like rural toilet use, there's the matter of education and habit. I saw the same phenomenon in Kathmandu, where bins donated by the Chinese government were being used for everything but garbage, which sat in piles on the streets. In my youth, American highways were littered with all the stuff Americans tossed from their car windows. Some still do.

41    ***the tracks themselves are the recipients of excreta***

In summer 2011, in an effort both to save train tracks from corrosion and relieve scavengers of the unpleasant work of cleaning the tracks, the Delhi High Court directed the Indian Railways to begin installing "bio-toilets" on trains. Anaerobic bacteria inside the toilets would digest the waste, converting it into water and gas. A pilot project was underway by early 2012; toilets on some long-distance trains had been converted. With something in the neighborhood of 170,000 toilets on Indian trains, this promising effort will likely take quite a while.

46    ***Westerners just want to know that water is safely sanitized.***

A few pilot programs in the United States are now undertaking the challenge of turning sewage into drinking water.

For example, see Randal C. Archibold, "Los Angeles Eyes Sewage as a Source of Water," *The New York Times*, May 16, 2008. See also Randal C. Archibold, "From Sewage, Added Water for Drinking," *The New York Times*, November 27, 2007.

48    ***And the black water can***

A small town in Tennessee, for example, decided the solution to old, failing septic tanks was not piping its sewage to Nashville. The cost was prohibitive. Discharging even treated wastewater into a nearby river was not possible either, because another town took its drinking water from it. Instead, the town of Pegram set up a system in which homes or clusters of homes and businesses could treat the wastewater through sand-gravel systems, after which the treated water was used for subsurface drip irrigation in nearby pastures. *Pipeline*, Fall 2000.

### Chapter 3, Delhi's Yamuna

51    ***There is no water in our flat this morning Mrs Puri.***

Excerpt from *City of Djinns* reprinted by permission of HarperCollins Publishers Ltd. Copyright William Dalrymple 1993.

60    ***red and white radish, green cabbages***

Such vegetables grown near the banks of the Yamuna may no longer be healthy, having absorbed pollutants like lead, mercury, and cadmium carried by the river.

62    ***Archaeological evidence shows***

Nitya Jacob, *Jalyatra*. (New Delhi: Penguin Books India, 2008), p. 14.

65    ***Then there was the problem of leaks***

"DJB officials have pegged the total water loss at over 50 per cent of the total 800 million gallons it is equipped to produce every day, as against an average demand touching 1,100 million gallons per day." *Indianexpress.com*, March 22, 2011.

73    ***Instead of hiring someone again and again***

In March 2011 the DJB announced it had filled up its coffers by settling some long-overdue accounts, uncollected bills that Madhu Baduri had deplored a decade earlier. It was launching an effort to expand water-distribution and sewage systems without any government subsidies. The following month the DJB announced it would begin an extensive survey of customers to find out what they think of the utility and whether they were still opposed to any form of privatization.

Since I began looking into Delhi's water problems, the Delhi Jal Board has taken some steps toward conservation. In 2009 it curtailed the amount of water allocated to large hotels, malls, and other commercial institutions, urging them to begin recycling programs and dual pipe systems. "This city will have to recycle all its water," Manoj Mishra has said. "If it takes care of its groundwater and recycles wastewater, it will have all the water it will ever need." New leadership at DJB seems to be moving in this direction, suggesting the city's 16,000 parks use recycled sewage water. The DJB banned groundwater abstraction by businesses and private individuals in 2009. The Board itself has announced it will begin rainwater harvesting at its own facilities.

78    **Outright corruption**

After several arrests of his underlings, the chief organizer of the Commonwealth Games, Suresh Kalmadi, was arrested in April 2011.


## Chapter 5, The Shrinking Third Pole

110    **outside the polar caps**

Glaciers at the two poles are melting at a rate that has alarmed many scientists—faster than glaciers in high mountain ranges like the Alps and Andes. After the poles, the area that appeared a few years ago to some observers to be melting the fastest was the Himalaya. Now it's more generally believed that the rate of melting in the greater Himalaya is about average.

111    **That same spring, the World Wildlife Fund (WWF) released a report**

*An Overview of Glaciers, Glacier Retreat, and Subsequent Impacts in Nepal, India and China.* WWF Nepal, March 2005. Quotes predictions of various specialists on p. 2.

115    **estimates of the amount of meltwater in the rivers**

According to some estimates, from two-thirds to three-quarters of the water that sustains the arid Gangetic plain until the monsoon begins is melting ice and snow.

An example of such older estimates:

"A significant portion of the low flow contribution of Himalayan rivers during the dry season is from snow and glaciers melt in the Himalayan region. The runoff supplies communities with water for drinking, irrigation and industry, and is also vital for maintaining river and riparian habitat. It is posited that the accelerated melting of glaciers will cause an increase in river levels over the next few decades, initially leading to higher incidence of flooding and land-slides (IPCC, 2001a). But, in the longer-term, as the volume of ice available for melting diminishes, a reduction in glacial runoff and river flows can be expected (IPCC, 1996b, Wanchang et al., 2000). In the Ganga, the loss of glacier meltwater would reduce July-September flows by two thirds, causing water shortages for 500 million people and 37 percent of India's irrigated land (Jain, 2001; Singh et al., 1994)." WWF Report, p. 3.

These days in Nepal there is a serious shortage of electricity, thanks to a ten-year insurgency and bad planning. As the dry season proceeds, the hours of "load shedding" (more or less orderly blackouts) increase from six hours a day to sixteen. The city goes dark, quadrant by quadrant. Kathmandu residents adapt, with candles, small solar lamps, or diesel generators depending on their incomes. Food spoils, businesses close. Only as the snow and ice in the high mountains begin to melt in the spring do the hours of electricity slowly increase until the monsoon puts rain in the rivers and water into the nations' few reservoirs to generate hydropower.

116    **He believed that panic was created unnecessarily**

Dipak Gyawali and Ajaya Dixit. "Water and Science: hydrological uncertainties, developmental aspirations, and uningrained scientific culture." Futures 33 (2001) pp. 689–708. Gyawali thinks the international aid community has given up on science. "I think they're all into social science. You need physical science to tell you how the earth is behaving, how the atmosphere is behaving. There's very little funding for this stuff."

133   **the latter date had been suggested in the late 1990s by a Russian scientist**
      V. M. Kotlyakov, ed. *Variations of Snow and Ice in the Past and at Present on a Global and Regional Scale* (Paris: UNESCO, 1996), p. 66.

134   **Glaciers in the Himalaya are receding. The full quote is:**
      "Glaciers in the Himalaya are receding faster than in any other part of the world . . . and, if the present rate continues, the likelihood of them disappearing by the year 2035 and perhaps sooner is very high if the Earth keeps warming at the current rate. Its total area will likely shrink from the present 500,000 to 100,000 km$^2$ by the year 2035 (WWF, 2005)." IPCC Fourth Assessment Report. Climate Change 2007: Working Group II: Impacts, Adaption and Vulnerability. 10.6.2 The Himalayan glaciers. Intergovernmental Panel on Climate Change, 2008.

      "Flooded Out—Retreating glaciers spell disaster for valley communities." *New Scientist*, June 5, 1999. Noted on p. 2 of 2005 WWF report.

      From blog, *The Carbon Brief*:

      "'Glaciergate' refers to the controversy surrounding the inclusion of an unsubstantiated and inaccurate claim that Himalayan glaciers could melt by 2035 in the 2007 report from the Intergovernmental Panel on Climate Change. The false projection was contained within Chapter 10 of the 938-page Working Group II report on 'Impacts, Adaptation and Vulnerability,' but not in the more widely-read summary for policymakers or the section discussing glaciers in the Working group I report.

      "WWF in turn appear to have taken the projected date from a report in the New Scientist magazine by journalist Fred Pearce. The article was based on an email interview with Syed Hasnain, who was head of the publication in 1999 of the Report on Himalayan Glaciology from the Working Group on Himalayan Glaciology of the International Commission for Snow and Ice. Hasnain said in the interview that the Himalayan glaciers would have melted by 2035, although his published report did not include this date. The International Hydrological Programme report published by UNESCO in 1996 gave an estimate for a significant melting of ice-caps globally as 2350, and this figure may have been misquoted and taken out of context to produce the 2035 estimate."

134   **Asian brown cloud**
      Also known by the less controversial term "Atmospheric Brown Cloud," or ABC.

135   **Recent research suggests**
      "In addition, it is necessary to understand that the location of a glacier's terminus is not a comprehensive assessment of total glacier condition or health. Such measurements represent nothing more than the location of the terminus at a given point in time as it responds to both the dynamics of the ice body and the current climate. For example, if a glacier is noted to be retreating, this simply means that the ice at the terminus is melting faster than the rate at which ice is being supplied to that location by movement (dynamics) of ice from further upslope in the system. It is possible that a glacier may be gaining in total mass from one year to the next, due to increasing amounts of snow arriving at the higher elevations by precipitation, wind deposition, and avalanching, while the terminus, at the lowest elevation, is retreating. Therefore, it should be understood that measurements showing short-term retreat only indicate that the recent climate does not support the extension, or even stability, of the lowermost elevation of a given glacier, and does not define the current conditions controlling the changes in volume over the entire glacier at all elevations."

      Richard L. Armstrong. *The Glaciers of the Hindu Kush-Himalayan Region: A summary of the science regarding glacier melt/retreat in the Himalayan, Hindu Kush, Karakoram, Pamir, and Tien Shan mountain ranges* (Kathmandu: ICIMOD, 2010), p. 2.

Chapter 6, In the Valley of Dhunge Dhara

139   **How did they build something so durable**
      From a collection of interviews with elderly Nepalis, *Water Wisdom* (Kathmandu: Panos South Asia, 2000), pp. 30–31.

139    **leaving the valley dry by around 10,000 years ago**
       Bhim Subba, *Himalayan Waters*. (Kathmandu: Panos South Asia, 2001), p. 30.

142    **The beginnings of this water system**
       Most of the information in this discussion is from personal interviews with Sudarshan
           Tiwari, supplemented by his book *The Brick and the Bull* (Kathmandu: Himal Books, 2002).
           See also his *Temples of the Nepal Valley* (Kathmandu: Himal Books, 2009).

149    **The stupendous festival of Rato Machhendranath**
       Kesang Tseten's wonderful documentary "On the Road with the Red God" captures this yearly
           festival in all its chaos and color, showing the building of the chariot and its yearly journey.

## Chapter 7, Melamchi River Blues

157    **We started an organization**
       *Water Wisdom*, p. 90.

157    **kept the city from growing beyond certain limits**
       From interview with Sudarshan Tiwari

158    **The history of modern Nepal**
       John Whelpton, *A History of Modern Nepal* (Cambridge University Press, 2005).

159    **Must be treated**
       Local NGOs have promoted a simple method called SODIS, solar water disinfection. Clear
           PET bottles are filled with water and set in the sun for six hours. For residents with
           limited income this method works to destroy most of the bacteria and parasites common
           in Kathmandu water. Residents with more income use devices like the Euroguard.

162    **In 1992 Ajaya wrote an article . . . outlining the failures of Kathmandu's water system**
       Ajaya Dixit, "Little Water, Dirty Water." *Himal*, Jan/Feb 1992, p. 8.

164    **Meters would have allowed Kathmandu's water officials to assess how much water
           was actually flowing through the system**
       That something so basic to water management was a stumbling block seems a telling
           example of just how dysfunctional the management had become. Meters are basic
           tools of water management, especially for a system that experiences shortages during
           part of the year. Yet even in California's central valley, including the state capital,
           Sacramento, meters are still not universal. The whole concept of meters has been a
           source of controversy for the past two decades.
       See Peter Gleick, "Smart Water Meters, Dumb Meters, No Meters," *San Francisco Chro-
           nicle*, April 28, 2010.

169    **A further editorial urged the KUKL board to appoint "someone with a vision" as his
           replacement.**
       "The Rising Nepal," May 1, 2011.

## Chapter 8, More River Blues

181    **In the past, there were no latrines**
       *Water Wisdom*, p. 87.

182    **Ratna's guru**
       The Norwegian hydropower engineer Odd Hoftun, who completed the Tinau hydropower
           project in 1978.

185    **But now there's no sand.**
       The riverbeds of both the Bagmati and Bishnumati are mined upstream for sand that is
           used in making concrete for construction.

193    **after their deaths their ashes not be consigned to the filthy Bagmati**
       Cremations are commonly carried out on the edge of the Bagmati River, at the large and
           famous Pashupatinath Temple on the northeast side of town. There is a fairly new
           sewage treatment facility upstream of the temple, the only one in the city that works,
           which makes the water along that stretch of the river somewhat cleaner than it is far-
           ther downstream.

## Chapter 10, The Sorrows of Bihar

215    ***cultivated on these plains since at least 2500 BCE.***
  B. G. Verghese, *Waters of Hope* (New Delhi: India Research Press, 2007), p. 16.

215    ***There was no discussion of the inherent fallibility of embankments***
  One exception was a report several months after the flood on Al Jazeera English, which featured some of the experts cited in these chapters on Bihar.

222    ***Dinesh Mishra had been traveling when I arrived in Bihar***
  Much of the information in this chapter and the two subsequent ones is from interviews with Dinesh Mishra and various other people in India and Nepal, many of whom are quoted by name. That information is supplemented from the books and pamphlets written by Dinesh Kumar Mishra. These include:

  *Trapped! Between the Devil and Deep Waters* (Dehradun: People's Science Institute, 2008).

  *The Kamla, River and People on Collision Course* (Patna: Barh Mukti Abhiyan, 2006).

  *Living with the Politics of Floods* (Dehradun: People's Science Institute, 2002).

  *Story of a Ghost River and Engineering Witchcraft* (Patna: Barh Mukti Abhiyan, 2004).

## Chapter 11, The Koshi's Revenge

227    ***A photo taken as the flood began***
  For some video footage of the flood and its aftermath, see http://vimeo.com/21963072.

228    ***For several years before the flood, the braided channels of the Koshi had been moving east***
  This led to the erosion of the protective spurs that jut out into the river, built to slow the force of the river in spate. That combined with the height of the water created tremendous hydraulic pressure on this earthen wall.

234    ***a revered Hindu pilgrimage site***
  Dipak Gyawali told me that the shrine at Barakshetra is dedicated to the Boar incarnation of Vishnu (baraha), the incarnation that rescued Bhoomi Devi (the earth goddess) from the underwater world. This seems to offer very appropriate symbolism near the *triveni* where the Saptakoshi begins to deliver her floods to the land and people below.

236    ***a fat camera bag slung over his shoulder***
  A selection of Eklavya's photographs of the resilient people of North Bihar was exhibited in Chicago from May 21 to July 31, 2011. *After the Flood: Eklavya Prasad's Photographs of Life in North Bihar, India,* Loyola University Museum of Art.

246    ***These villages are among the 380 settlements***
  There are another sixty-one villages north of the border in Nepal. Thirty-four are below the barrage, the others above it.

248    ***Some activist musahars deplored this***
  "It is official, no rat farming in Bihar." *Thaindian News,* December 7, 2008.

## Chapter 12, The Engineers

252    ***The dam was first proposed***
  Dinesh Mishra notes that though the dam on the Koshi was first proposed in November 1937, at a conference in Patna on floods, only in 1947 was a public announcement made.

252    ***The flood control strategy of the colonial British government in Orissa***
  Rohan D'Souza, *Drowned and Dammed: Colonial Capitalism and Flood Control in Eastern India* (New Delhi: Oxford University Press, 2006), pp. 15, 51.

253    ***A flood in 1927***
  D'Souza, pp. 157–158.

254    ***If there were a dam in Nepal***
  Results of a World Bank study, released in preliminary draft after this chapter was completed, support conclusions of critics who contend that a dam in Nepal on the Koshi would do little to curb potential floods in the flatlands of Nepal and Bihar. *Ganges Strategic Basin Assessment: A Discussion of Regional Opportunities and Risks,* World Bank, 2012.

254    **catapult that poor country into prosperity**

I heard about the vast hydropower potential supposedly locked up in Nepal's mountains from the very beginning of my interest in the country. Almost every news article about hydropower would mention it, as did people who argued that canceling the proposed 201-megawatt Arun III project had been a huge mistake, resulting in the oppressive lack of electricity that was crippling the country. Many politicians still talk about dams in Nepal and selling electricity to India as the nation's salvation. Such hydro schemes are still being proposed, some to be paid for by India. Some even give away water in the process.

Ratna Sansar Shrestha, along with many others, is vehemently opposed to planning dams in Nepal with Indian needs in mind. He is interested in judiciously situated dams that will submerge as little land as possible and displace as few people as possible, and which will primarily benefit Nepal in terms of irrigation water, electricity, and flood control. If there is surplus power, that can be sold to India. And if there is water Nepal does not need during the dry season, that too can be sold to India. "Let's work out the cost of the project: cash cost to implement the project, and non-cash cost in submergence of land and displacement of local populace. Nepal should pay part. India should pay for the downstream benefits of flood control and irrigation as well as power."

Correcting those who say that Nepal's greatest resource is its potential and as yet undeveloped hydropower, Ratna says Nepal's resource is not hydropower; it is not even water. "We are rich not because of water but because of topography and terrain. This allows us head—water falling from a height. And valleys in the mountains allow us to build small dams to store large quantities of flood water during the rainy season without inundating large tracts of land. Our topography allows us to transfer water from one season to another and to control flood in downstream areas." Ratna believes the way to raise the living standards of people in Nepal is to give them more irrigation water during the dry season so they can grow three crops instead of one. "The only way to give them this is with water storage, with dams. But we must build dams with discretion. I am against a dam that will provide water free of cost to India. But to help Nepali farmers, that's a different question."

257    **Many other manageable changes will help**

Dipak Gyawali offers a logical approach to avoiding the risk inherent in transboundary rivers in this region.

"The idea was to evolve an alternative perspective on disasters, and we came up with some principles. One: Disasters are the unfinished business of development. Two: Disasters are a recurring phenomenon, rather than one-time events, and will occur more frequently and intensely due to climate change. Therefore, what happens between disasters is more important than what you do in the immediate aftermath of the disaster. The focus should be on the interim period to re-think development from a disaster risk reduction perspective. Three: You have an extreme physical event, which interacts with social vulnerability—that is what makes for a disaster. You cannot avoid the former, so work on reducing the latter. Four: It is the structures within society and the relationships between different social groups that determine why certain sectors are more vulnerable to disasters than others. Five: Disasters are also a forensic moment, a window of opportunity, for changing the way things are to reduce overall social vulnerability."

"The Right of the River," an interview with Dipak Gyawali. *Himal*, March 2011. The 2012 World Bank study cited above confirms the importance of the variety of "soft" interventions mentioned here.

259    **Trees were cut down, others replanted. The amount of sediment the Koshi brings down has not varied much over the years.**

For a discussion of erroneous beliefs that overcutting of forests in the Nepal Himalaya has led and continues to lead to increased landslides, sediment in rivers, and floods in the lowlands, see Michael Thompson et al. *Uncertainty on a Himalayan Scale* (Kathmandu: Himal Books, 2007).

261    **the great university at Nalanda**

The university at Nalanda, located in Bihar south of present-day Patna, was a center of Buddhist learning from about 400 to 1200 CE. Along with Taxila in present-day Pakistan, it is one of the oldest universities of the world. Much of today's Tibetan Buddhism comes from documents from Nalanda that refugee professors took with them and preserved. Before it became a center of learning, Nalanda was a prosperous town that the Buddha visited.

## Chapter 13, The Garland

264    *The first such concept cropped up in the mid–nineteenth century as the British Raj sought to extend its control over India's productivity*

Sir Arthur Cotton, an engineer with the East India Company, proposed the first of these ideas in 1839. His proposal would have linked rivers in South India for purposes of river navigation. In 1972 Dr. K. L. Rao proposed linking the Ganga with the Cauvery in South India. In 1977 Captain D. J. Dastur proposed a Garland Canal to be joined to a Himalayan canal for irrigation purposes.

267    *Rather than have people move, you move water*

This idea is the basis of water management throughout the United States and the world. California would not be California without it. Los Angeles grew by stealing the water of the Owens River and Valley. The rest of California's water system and the dams in the western United States were based on a concept of moving water to the people. However, it was done in the opposite order to what Verghese has in mind: much of the water was moved first, to attract the people. Most of California's water is in the underpopulated mountainous north. The bulk of the population is in the dry south. By means of dams, inter-basin transfers, and an aqueduct this inconvenience was corrected so that water flows down to Los Angeles and arid agricultural regions in the Central Valley. Water from the Trinity River in the north is transferred to the Sacramento River so it can reach the southern part of the state. The rest of the transfer occurs through the San Francisco San Joaquin Delta, whose levees are vulnerable to earthquakes and climate change and could see a disaster of Bihar, or Katrina, proportions.

The idea of moving water from "surplus" to "deficit" areas in the United States is seen by some people as a solution to the predicted weather extremes climate change may continue to bring. The engineering firm Black and Veatch has developed a plan to siphon one thousand cusecs from the Mississippi River to Colorado, which would require 775 miles of pipe, 110 miles of canal, 85 miles of tunnel, and 7 pumping stations to lift the water up to 7,500 feet. The cost is estimated at $11.37 billion. Pat Mulroy, general manager of the Southern Nevada Water Authority, has proposed moving the "excess" water from the Mississippi River to Southwestern Colorado. "The Ogallala aquifer would be replenished and the needs of municipalities for the next 30-50 years downstream, all the way to California, would be taken care of." (http://www.foxbusiness.com/industries/2011/12/02/shovel-ready-jobs-could-help-relieve-droughts/#ixzz1qMWDGZP9) Peter Gleick of the Pacific Institute calls the plan "foolhardy."

As Texas sizzled and the northeast was swamped in the summer of 2011, a caller on NPR's *Talk of the Nation* proposed—and not entirely in jest—that somehow the excess water in the Northeast be moved to the parched Texas plains.

269    *Ramaswamy Iyer, one of the most respected water experts in India*

See, for example, *Towards Water Wisdom: Limits, Justice, Harmony* (New Delhi: Sage Publications, 2007).

269    *when archrival China was busy assembling the Three Gorges Project*

China has an ambitious and controversial river linking plan of its own, the North South Water Diversion Project, which would siphon water from the upper Yangtze to send to the Yellow River and on to Beijing. It would cost twice as much as the Three Gorges project. "Plan for China's Water Crisis Spurs Concern," *The New York Times*, June 1, 2011.

269    *None of India's states, even those that sometimes suffer floods during the monsoon, are likely to say they have water surpluses they want to get rid of.*

Bihar, for example, was interested in river linking from early on, but did not want to just be a conduit of Himalayan waters to other parts of India. Under the leadership of its Chief Minister Nitish Kumar, detailed plans for intra-state linking of rivers were in progress in 2011. Using language similar to that used at the national level, the project aimed to solve floods and provide irrigation water to dry areas by linking not only some of the smaller rivers in the state, but major rivers like the Koshi and Bagmati.

"Bihar launches ambitious river linking projects," *The Hindu*, June 3, 2011.

270    **the group that works on sustainable development in the Jhansi district of Bundelkhand**

Development Alternatives works in various districts: Jhansi (U.P.), Datia (M.P.), Tikamgarh (M.P.), and Chattarpur (M.P.) districts, with plans to expand its activities in the remaining nine districts of Bundelkhand.

Air Vice Marshal (Ret.) S. Sahni is a senior adviser at Development Alternatives, a specialist in degraded lands, wastelands, and water harvesting. He supplied additional information about the region.

"Bundelkhand from time immemorial had a pastoral economy, which implies that the bulk of the income came from animals in the form of milk, milk products, wool, and meat, along with skin, bones, blubber and other parts from dead animals. In the rest of the country agriculture was the main source of income. Thus the percentage of animal population in Bundelkhand compared to other parts of the country was much higher. Also, the region had nearly 50 percent area under good forest with water bodies in most villages, so there was no shortage of fodder and water, and man and animal lived in harmony with nature. Another feature of Bundelkhand has been that during the monsoon there are generally two to three cloud bursts and long periods of no rain during the SW monsoon (June–September). Bundela kings who ruled the area a little over 500 years back, realizing the vagaries of monsoon, undertook the construction of a large number of tanks as part of a drought mitigation programme. These tanks would get filled up with water during normal monsoon and charge the surrounding wells. The result was that even in a drought year there was adequate water in the wells to get one crop and no shortage of drinking water for humans and animals even during the hot summer months. All this started changing from early 1940 when for the requirement of Second World War large quantities of trees were cut. Also Bundelkhand had a large number of princely states who controlled their forest. During 1947–48 when the princely states were being amalgamated with the rest of the country, there was literally a rape of the forest and today hardly 10 percent of the area of Bundelkhand is forested. Old water bodies have disappeared or silted, and animal population remains very high. Man to animal ratio is 1:1 in Bundelkhand, compared to 1:45 in other parts. Thus animals instead of being an asset have become a liability due to lack of fodder and drinking water. The region is facing acute shortage of water during the last decade, adversely affecting the production of food and fibre and resulting in large scale migration from villages to towns."

Sahni believes the Ken-Betwa link is not the answer. "Bundelkhand area being dominated by the Vindhya Range of mountains is generally sloping, where despite an excellent average rainfall of approximately 900 mm in a normal year during the SW monsoon (June–September) most is lost in runoff. What we need is to absorb every drop falling *in situ*. Check dams where suitable sites are available, other water bodies such as tanks, desilting of old tanks and covering the entire region by a professionally implemented watershed development programme will transform the economy of Bundelkhand in hardly two decades at a pittance of the cost compared to the cost of Ken-Betwa link."

270    **Ashok helped form a committee of experts to allow for some exchange of ideas on river linking**

A compilation of papers by members of this committee explores in detail the pros and cons of river linking, focusing on the Ken-Betwa link.

Yoginder Alagh et al., eds., *Interlinking of Rivers in India: Overview and Ken-Betwa Link* (New Delhi: Academic Foundation, 2006).

275    **The debate began with a question posed by JNU professor of science policy, Rohan D'Souza**

See Rohan D'Souza, "Supply-Side Hydrology in India: The Last Gasp," *Economic and Political Weekly*, September 6, 2003.

276    **He said India could use an institution like Cal-Fed**

The Cal-Fed (California and Federal) Bay Delta program began in 1994. It was a department in the California Resources Agency.

277     ***In recent decades, just these sorts of changes have come about in the neighboring state of Rajasthan***
    See Zac Goldsmith, "Back to the Future in Rajasthan," *The Ecologist*, Vol. 28, no. 4, July/August 1998, for a longer discussion of the revival of rainwater harvesting in Rajasthan. Also, Sunny Sebastian, "The Water Man of Rajasthan," *Frontline*, v. 18, August 2001.

278     ***traditional rainwater harvesting systems had been abandoned in favor of state-sponsored pumping of groundwater***
    The government and aid groups sank tube wells in Rajasthan to give people an alternative to water containing guinea worms. The worms induced a nasty, disabling infection in people.

286     ***Together they had spent ten years studying the tigers in the nearby Panna Tiger Reserve***
    The documentary *Tigers of the Emerald Forest* chronicles their work. Producer Mike Birkhead, *Natural World*, BBC, 2003.

286     ***But others were deliberately being poached for their parts***
    "Panna Poacher Nails Official Lie," Express News Service, July 31, 2005.

## Chapter 14, Susu

289     ***From the story of Ganga's descent to earth***
    Excerpted from a quotation sent to me by Ravindra Sinha from Pilleri, G. The Indian River Dolphin in the Babur Nama Miniatures, the Moral Edicts of King Asoka and Ancient Literature. Legends and Superstitions concerning the Amazon Dolphin. *Investigation of Cetacea*, Vol X: 351–358. 1979.

290     ***The Ganges river dolphin, widely known now as susu***
    This is the name scientists and conservationists now commonly use for the dolphin. Its name in Hindi is *soons*, in Nepali *swangshu* or *swonsh*, in Bengali *shushuk*.

296     ***When he started his research, Sinha could find little information about the susu and thus had to start from scratch***
    "The Ganges River dolphin is extensively mentioned in India's mythological and historical literature. However, it was William Roxburgh, the then Superintendent of the Calcutta Botanical Garden, who wrote the first scientific paper on this species in 1801. John Anderson published a report on its biology, including a distribution map, in 1879 after which there is no record of further scientific work on this animal for the next one hundred years.
    "The first efforts in the 20th century to document the status and threats faced by the Ganges River dolphin were made under the Ganga Action Plans (GAP) I (1985) and II (1991) through research and conservation projects. These provided baseline scientific information about the species, including the fact that habitat degradation, through pollution and reduced water flow, and poaching were threatening its existence."
    R. K. Sinha, S. K. Behera, and B. C. Choudhary, *The Conservation Action Plan for the Ganges River Dolphin, 2010–2020*. Ministry of Environment and Forests, Government of India, 2010, p. 7.

## Chapter 15, Beyond Barrages and Boundaries

305     ***The smaller slice of pizza to the west would include Pakistan and the Indus River, but that's another complicated story***
    India's very name derives from Pakistan's chief river, the Indus (or Sindhu). Pakistan was not pleased when India appropriated the name at partition for the majority Hindu part of the former empire. However, Indians tend to refer to India as Bharat, not by the name most foreigners use.

307     ***India has the resources to build infrastructure and would like to capture and store Nepal's water, but so far the latter has resisted the control that extensive development would entail.***
    As Bhim Subba analyzes this standoff, India and Nepal have never been able to come to agreement, because they are not talking about the real issue.
    "A more valid reason for the impasse can be directly traced to a fundamental flaw in the negotiations: Nepal has been trying to sell electricity, while it is water that India needs.

Given these divergent—if not conflicting—objectives, it is not surprising that efforts so far have failed to satisfy either side.

"The Ganga cannot meet the future demand for water unless its variable natural flow is regulated. The success of an Indian water strategy to meet the growing water demand in Uttar Pradesh and Bihar, the country's most populous states where more than a quarter of the total population live, hinges on Nepal."

. . . "This implies that stored water must have a monetary value, and that this price must be attractive enough for Nepal to find it viable to design and build projects that optimize water storage instead of maximizing power generation only."

Bhim Subba, "Water, Nepal and India," in *State of Nepal*, Kanak Mani Dixit and Shastri Ramachandaran, eds. (Kathmandu: Himal Books, 2002).

307    **But for hydrologists, there is really no true mouth for a river system as vast as the Ganga's. As there are many sources of these great rivers, there are also many outlets.**

Alan Potkin, "Towards Watering the Bangladesh Sundarbans," in *The Ganges Water Diversion: Environmental Effects and Implications*, ed. Monirul Qader Mirza. (Kluwer Academic Publishers, 2004), pp. 166–167.

"To orthodox Hindus, the Bhagirathi-Hooghly is the 'true mouth' of the sacred Ganges, diverted from heaven by the sweat and sacrifice of its namesake, the mythical King Bhagirat, for the benefit and sanctification of mankind. Married finally to the sea and made pregnant by it, *Mata Ganga*, 'Mother Ganges,' birthed holy Sagar Island, where now every January, a Puja festival commemorates those wondrous events. A million pilgrims come then to wade and bathe at Ganga Sagar.

"To hydrologists, there is no 'true mouth' but certainly since the Sixteenth Century A.D., the largest part of the Ganges' flow—having mostly abandoned the old Hooghly outlet— has been delivered to the Bay of Bengal below confluences with the great Brahmaputra/ Jamuna, from Assam and Tibet, and finally through the estuaries of the Meghna, also a major river, draining the monsoon-drenched Meghalaya Hills."

308    **The reason given for diverting the Ganga's waters was that the riverbed near the port of Calcutta had an excess of silt**

The plan may have promised other advantages as well, such as reduced salinity of Calcutta's water supply and rejuvenation of West Bengal's declining economy.

Ben Crow, *Sharing the Ganges* (Dhaka: The University Press Limited, 1995), pp. 62–64.

As Crow also notes: "technical data and technical choices are frequently not neutral; they contain political aspirations clothed in technical language." *Sharing the Ganges*, p. 22.

308    **A way to save the port from further deterioration**

The port of Calcutta was the chief eastern port during British rule, providing a commercial outlet for all the inland, landlocked regions in the eastern part of South Asia, including Nepal and Bhutan as well as the western states of India. The Ganga herself was once a shipping lifeline for Indian states from Uttar Pradesh down to West Bengal. But as the Ganga shifted into the Padma, abandoning the Bhagirathi—and as the river below Calcutta kept silting up—this shipping has steadily declined.

A new deep water port has since been built south of Calcutta at Haldia, near the mouth of the Hooghly.

308    **The idea of such a diversion had been kicked around during British rule, but it didn't go anywhere**

As early as 1852 the East India Company had concerns about Calcutta's port silting up. Sir Arthur Cotton, a leading company engineer, proposed diverting the Ganga into the Hooghly. Cotton had also proposed the first of the river-linking ideas for India in 1839.

308    **At the time of partition, the potential site for such a barrage was placed in Indian territory**

"Before India and Pakistan existed, but only by a few weeks, lawyers acting for the fledgling states had crossed swords in a tussle for territory which would allow the victor to control the Ganges. This tussle, out of which the new state of India gained the district of Murshidabad and the site for the Farakka barrage, was only a slight murmur almost lost in the hubbub of historic events. But it is here that a narrative of the Farakka dispute must start." Crow, p. 76.

308   *The chief engineer of West Bengal, Kapil Bhattacharya, opposed Farakka*
      Crow, pp. 49–51. See also Manisha Banerjee, *A Report on the Impact of Farakka Barrage on the Human Fabric* (New Delhi: South Asia Network On Dams, Rivers and People, November 1999), pp. 9–10.

309   *When government officials in Pakistan learned of the plan to build a barrage at Farakka*
      Miah, M. Maniruzzaman, *Hydro-Politics of the Farakka Barrage* (Dhaka: Gatidhara, 2003), p. 13. For a discussion of negotiations between Pakistan and India even before the formal announcement about Farakka as well as after, see B. M. Abbas, *The Ganges Water Dispute* (Dhaka: University Press Limited, 1984). Abbas was one of the chief negotiators for Pakistan and Bangladesh.

309   *most of the body's member states were disinclined to take sides in the dispute*
      As Crow notes, member states didn't want to take sides against India, and most of them were also upper riparian states themselves. Crow, p. 111.

313   *As India sees it, the water is somewhat wasted on Bangladesh*
      "In the context of using Ganges water to combat saline intrusion in Bangladesh, Indian negotiators have argued that it is wasteful to allow water to flow into the sea. By the same argument, it is wasteful to use scarce dry season water in flushing Calcutta Port." Crow, p. 73.

314   *They had the Meghna, the Brahmaputra, and so on to choose from*
      Ramaswamy Iyer summed up the Indian perspective—not his own—this way: "The Indian view was that Bangladesh still had enough water. There was so much water going into Bangladesh in proportion to the amount of agricultural land dependent on it. One third of Bangladesh is dependent on the Ganga. We had no alternative. Bangladesh, as the lower riparian, is certainly entitled to a share of that water. There was substantial water after the diversion, but a lot of it goes to sea, it is not being used."

315   *But after the 1970s the region collapsed because of the rise in salinity*
      Studies funded by the World Bank support such changes in the water quality, especially near the industrial center of Khulna, Bangladesh's third-largest city. Crow, pp. 128–148.

316   *The growing demand for irrigation water in both India and Bangladesh would have led to conflict eventually*
      Crow, p. 19. He points out that most major river developments have multiple purposes (irrigation, flood control, etc.). Farakka was pretty much a one-issue intervention.

316   *may have limited later discussions*
      Crow, pp. 19, 74, 75.

317   *the amount of water that agriculture requires and wastes looms as a larger threat for the future*
      Crow notes that as China and India become more and more industrialized, the amount of water industry needs may in time exceed that used in agriculture.
      Ben Crow and Nirvikar Singh, "The Management of International Rivers as Demands Grow and Supplies Tighten: India, China, Nepal, Pakistan, Bangladesh." *India Review* 8:3, 2009, pp. 307, 311.

318   *Diverting water from the Brahmaputra is proposed as a benefit to Bangladesh*
      The potential advantages for Bangladesh have been explored in great detail already. See Anik Bhaduri and Edward Barbier, "Linking Rivers in the Ganges-Brahmaputra River Basin: Exploring the Transboundary Effects." International Water Management Institute, Conference Papers, 2008. pp. 373–395.

319   *The bank has organized forums in different parts of the region and urged the various nations to share information*
      "The objective of the Ganges Strategic Basin Assessment (SBA) is to build knowledge and promote dialogue on the risks and opportunities of cooperative management in the basin. The centerpiece of this regional research is the development of a set of nested hydrological and economic river basin models that can be used to examine alternative scenarios across a range of Ganges futures. At present there is no common knowledge base to explore options and facilitate cooperative planning in the basin. The SBA models will be of adequate reliability and detail to facilitate an informed discussion,

and help focus efforts towards better, more cooperative management of the river system. It is hoped that these models can be shared with riparian countries who could then build on and utilize these models in the future."

"The Ganges Strategic Basin Assessment," The World Bank Group, Briefing Note, March 8, 2010.

319   ***Ben Crow and a colleague proposed in 2009 that ICIMOD might even play a larger role in the region***

Crow and Singh, p. 330.

320   ***Claudia says research on the watershed is continuing; but based on what the team has seen so far, the "traditional wisdom" about dams in Nepal to save India and Bangladesh from floods is not accurate***

In addition, the team suggests there are better means for storing water than behind big dams in Nepal.

"A similar scale of water storage could be gained unilaterally and at much lower cost from better groundwater management. There is great untapped potential for enhanced conjunctive use of groundwater and surface water in some parts of the Ganges basin. Increased strategic and sustainable use of groundwater resources could utilize natural underground storage in the basin on a scale comparable to the full suite of Nepali dams considered in the Ganges SBA."

On the subject of floods, the bank's Ganges basin team notes that better sharing of information among the three countries would be more useful than dams. In the past, India has not been forthcoming with such information.

"From a regional flood perspective, gains are likely to come from cooperation on data—not dams. Regional efforts focusing on large scale upstream infrastructure strategies for systemwide flood management will not be effective. Regional flood management gains could be great, however, if the Ganges countries cooperate on data collection, monitoring, forecasting and warning systems. Flood responses will need to be more localized, i.e., embankment asset management, drainage systems, land zoning, safe havens and evacuation plans."

"The Ganges Strategic Basin Assessment," The World Bank Group, Briefing Note, March 8, 2010.

320   ***But the bank's own analysis indicates that water stored in Nepal could be of some benefit downstream***

South Asia Water Initiative Multi-Donor Trust Fund Annual Report FY10. World Bank, 2010, p. 19.

321   ***Cordial discussions have lately focused on less volatile issues***

"Draft Deal on Teesta Water Sharing Finalized," *The Financial Express*, June 7, 2011.

"Indian Home Minister Visits Dhaka in August," *The Independent*, June 23, 2011.

321   ***so remote it has been only partially explored and that only by the most intrepid adventurers***

See Ian Baker, *The Heart of the World: A Journey to Tibet's Lost Paradise* (New York: Penguin Books, 2004).

321   ***Chinese engineers would like to tap that power***

"China's Himalayan Plan: Dam on Brahmaputra," *Hindustan Times*, May 26, 2010.

"China Plans to Divert Tibet's River: Feasible?" *The Tibet Post International*, June 23, 2011.

"Push for new dams across Brahmaputra as China faces drought," *The Hindu*, June 11, 2011.

321   ***the flows in the Brahmaputra after the river reaches India will remain about the same***

It isn't clear yet how much of the water in the Brahmaputra comes from the upper reaches of the river and could be affected by dams. Climate change will probably alter the amount. The majority of the water in the river comes from the eastern Himalayan foothills, the wettest region of South Asia. Most of this area is in India.

## Chapter 16, Poisoned Blessings

323   ***When rivers flood, two-thirds of the land may be covered by water***

Bangladesh has 250 rivers; 54 are shared with India. Only one major one originates and ends in Bangladesh, the Meghna, which merges with the Padma just west of the Bay of Bengal.

324    **became associated with pictures of starving people**

Henry Kissinger was the U.S. secretary of state when the new nation of Bangladesh was born, and he is said to have dismissed it as a "basket case." The label stuck for decades. Kissinger opposed Bangladeshi statehood, not wanting to weaken Pakistan at a time when U.S.-India relations were not as friendly as they have since become. Pakistan did fine at the outset without its eastern arm, while Bangladesh struggled. Even though Bangladesh remains one of the poorest nations in the world, it's not entirely clear that Pakistan is a whole lot better off despite its greater landmass and military power.

324    **where chemical processes release it into groundwater**

See Jennifer Weeks, "Charles Harvey: Water Detective," *Miller-McCune*, August 17, 2010. "Putting together all of these puzzle pieces, including chemical analyses of water from the rice fields, flood-control ponds and underground reservoirs, Harvey and his students concluded that water from the man-made ponds was seeping into the ground, carrying organic carbon with it. Once the organic material was deep underground, bacteria broke it down, using iron oxide and releasing arsenic."

Current research on the source of arsenic in Bangladesh's groundwater points to the tremendous amount of carbon in the soil from a combination of organic wastes—human, animal, and vegetable. Water carrying the carbon seeps deep into the ground, carrying food for bacteria. Some of this makes its way into the anaerobic domain where bacteria can change arsenic from an insoluble form to a more soluble one. In an oxygen environment, iron oxide tends to capture and lock up the arsenic. Down in the depths of the soil, 100–150 feet down, oxygen has been used up by bacteria that need it and below a certain level the environment is anaerobic. In that environment there are many bacteria that live without oxygen and can consume energy in the iron or arsenic, thereby releasing the more soluble form of arsenic.

324    **But no one knew anything about this until 1993.**

Arsenic was discovered to be a problem over the border in West Bengal some years earlier in the 1980s, but this was not widely known in Bangladesh.

325    **Possibly water in only a quarter of the tube wells in Bangladesh is contaminated**

According to UNICEF statistics, a total of 4.7 million tube wells in Bangladesh have been tested and 1.4 million of those were found to contain arsenic above the government drinking water limit of fifty parts per billion. Nearly forty thousand people showing the skin lesions symptoms characteristic of arsenicosis had been identified. *Arsenic Mitigation in Bangladesh*, UNICEF report, Updated October 2008.

329    **Apparently this is the reason Sono buckets were often found abandoned or used for other household chores**

Dr. Linda Smith, an American geologist, has had very good results supervising use of the Sono filter in Nepal, where arsenic was discovered in 1999. USAID and UNICEF had established tube wells there twenty years before. Working in the Nawalparasi district of the Nepal *tarai*, where some people were very seriously ill already, she founded a group called Filters for Families. The developer of the Sono filter allowed her to take the technology from Bangladesh. Filters for Families buys the CIM in Bangladesh, then assembles the filter in Nepal. Dr. Smith says the arsenic levels in this region are extremely high, up to 1,000 ppb. No one knows why. Tests on the water filtered by Sono have shown it is working well here. Possibly the kind of follow-up and monitoring that Filters for Families does explains why these filters continue to work well. The filters are a short-term solution, Dr. Smith says. In the next decade she hopes to see a more permanent one. She wants to work on wetland restoration, for example.

330    **Eventually Arsen-X fell out of competition for reasons that remain unclear**

I was never able to get a clear picture of what happened with this device, one that sounded so promising. A product with a similar name is now in use in the United States, which has some pockets of arsenic contamination, especially in the west.

333    **there's been no effort to harvest rainwater from the city's many tall buildings**

There may be some improvement soon. In August 2011 planners in Dhaka announced they would change building codes to require new buildings to include rainwater harvesting systems.

334     ***Some of the NGO project managers and government officials I met in Dhaka were some-***
        ***what mystified by their collective inability to come up with an effective strategy***

"However it is deeply disturbing that the sum of all government, non-government and
private sector efforts to date has not been effective in delivering safe affordable water
supply alternatives on a national scale to address the pressing needs of arsenic affected
communities."

NGO Forum for Drinking Water Supply and Sanitation and WaterAid Bangladesh. *Arsenic*
*2002: An Overview of Arsenic Issues and Mitigation Initiatives in Bangladesh*, 2003, p. 2. At
this point (ten years into the crisis), according to the report, it was still unclear how
many victims there were, how many tube wells were contaminated, and the combined
efforts of government and NGOs left Bangladesh "still a long way from a comprehensive
national response to the crisis."

Bill Collis of World Fish commented on this issue.

"We didn't realize how serious it was. Twenty to forty thousand people die of diarrhea;
only a few die from arsenic. Now we find out it causes complications with other diseases. Sometimes I don't get it. It's not that hard to solve. Either you get a filter, or a
deep well. Another issue is the impact on soil fertility. It declines in areas with long-
term arsenic issues. We have solutions for this. Raised beds. Allowing soils to go aerobic. Let water dry out. Or grow something else whose uptake is less. We have solutions.
The question is being serious about it. There's the cost issue too."

334     ***The government is not strong.***

I covered a groundwater contamination crisis over several years in California. MTBE was
a cheap and widely used additive in the state's gasoline. Gasoline manufacturers
claimed it helped the newly formulated lead- and benzene-free gasoline burn more
thoroughly. A few years into widespread use of this additive, it was found to be a nasty
groundwater pollutant. Some old underground gas storage tanks had leaked into
groundwater, poisoning people's wells.

Just how dangerous a toxin MTBE was and how harmful to people's health would not be
clear without more studies. But MTBE was a nasty substance that ruined water and was
extremely difficult to remove from it. It generally made the water so distasteful there
was little question of people drinking it anyway. The other components of the leaked
gasoline moved more slowly through the ground and were easier to remove from the
water they reached and polluted.

Small gas station owners declared bankruptcy as they realized their liability for cleaning up
the contaminated groundwater. It took many years to get a handle on this problem in
California. Lawmakers were not quick to act to ban the substance. Gas companies argued
the health dangers were not so serious. Besides, they said at first, there were no alternative
additives they could use to meet air quality standards. They proposed better, double-
hulled tanks to contain the gasoline. Environmentalists refuted all this. Eventually MTBE
was banned as other states started to discover they shared California's problem.

I make this comparison because the muscle of California's environmental laws and the weight
of many state agencies were available, along with money, to solve the problem more quickly
had there not been so many conflicting interests to reconcile: oil companies, gas station
owners, health agencies, water boards. Assessing blame for the contamination and responsibility for the cleanup, like in Bangladesh, took up time and energy in the beginning. What
might have been done and what was done were subject to a string of impediments.

The problem of arsenic in Bangladesh's water is far more complicated than getting a handle
on the mess created by a gasoline additive in California. Even when a government *is*
strong, laws can be evaded, backs scratched, cleanup initiatives stalled. In fairness to the
agencies and organizations that have not yet solved this problem, it has many facets.

335     ***The theory was that iron electrodes placed in the water***

In David Sowerwine's words: "The theory is that the electrode's iron dissolves in the
water, quickly being captured by oxygen to form insoluble iron oxide molecules. By
good fortune, as these iron oxide molecules rapidly clump together, they also 'ensnare'

the arsenic which they encounter. Pouring the water through a sand bed filters out other impurities such as bacteria, along with the arsenic passengers entrapped in the rust crystals, which 'stick' nicely to the sand. Both biological contaminants and arsenic are trapped, leaving the water safe to drink."

338 **It's not clear yet what those users may do when the study is over**
David wondered whether renting would be a good approach for the large filter. He proposed taking the units back after the study. "If you want to keep it you have to pay for it," he suggested, "but they don't have the rental model yet."

339 **Helping poor people while bolstering a market-driven economy might be a good idea, but it was far from easy to implement**
I learned from another source that some development organizations had recognized that their development efforts were not turning out well. Money was distorting development, they concluded. But what could they do? They had to spend money to keep the superpowers happy. The general belief had long been that the business of making money should not be combined with development and poverty alleviation. Some members of the aid organizations realized that no longer made sense, so they developed new guidelines. But change was slow in coming.

341 **Still, it's not difficult to understand why people hesitate to take on the challenge of making surface water safe enough to drink in such a populated country with so little effective sewage management**
Nayeem Wahra and others have in recent years organized something called a *nodi mela*, a river celebration. It is being held each April at the time of the Bangladesh new year, and organizers want to get young people in particular involved in protecting rivers. Each year they will stage a fair in a different part of the country, on different rivers. The organizers work to involve universities, government agencies, and civic groups. "People signed a promise to protect their own rivers," says Nayeem Wahra. "They promised not to put any garbage in their own river and to stop others from doing so—to keep it clean, discourage insecticide."

## Chapter 17, Where the Rivers End

345 **The Bangla name for the mangroves is sundar bon**
Another possible source for the name Sundarban is *samudra ban*, or sea forest in Bengali.

346 **It was declared a reserve forest and offered some protection in 1875.**
In 1997 UNESCO and the Bangladesh government declared the Sundarban a World Heritage site.

350 **Or, a flood or cyclone can leave a family with nothing**
As I finish writing this chapter, a million people in Bangladesh have been displaced by floods, some of them just recovering from the cyclone of 2009 that destroyed half a million homes.

351 **In the past decade, both the IPCC and the World Bank have published estimates that higher sea levels of a meter or more will affect low-lying countries like Bangladesh**
The World Bank estimates a 10 cm, 25 cm, and 1 m rise in sea level by 2020, 2050, and 2100. This rise would inundate 2 percent, 4 percent, and 17.5 percent of the total landmass of the country.
Sarwar, G. M., and Khan, M. H. (2007). "Sea Level Rise: A Threat to the Coast of Bangladesh," *Internationales Asienforum*, v. 38, no. 3–4, 2007, pp. 376–378.
*Bangladesh, Climate Change and Sustainable Development,* World Bank, Report No. 21104-BD, December 2000.
Climate scientist James Hansen, head of the NASA Institute for Space Studies, has predicted that virtually all of Bangladesh's population will become environmental refugees by the end of this century.

352 **He said the land would keep growing as it always had**
This would seem to presuppose sediments not being held back by dams or barrages. Increasingly clever ways of releasing sediments from behind dams have been developed,

but dams may inhibit the flushing action of some rivers. The force of the monsoon would have to deliver sediments to the bottom of the watershed to build new land.

352    **Bangladesh has some time to prepare**

Recently Australia (CSIRO) has entered into a partnership with Bangladesh to study its water security. Called the Bangladesh Integrated Resources Assessment study, the two-year study will help Bangladesh update its National Water Management Plan. Also, the Netherlands and Bangladesh are planning to exchange research on managing floods.

352    **Embankments can prevent or at least forestall inundation**

Embankments may also give people a false sense of security, as they have in parts of Bihar.

353    **The districts of southwestern Bangladesh, adjacent to and directly north of the Sundarban, are those most affected by the lack of water in the Gorai River during the dry season.**

At partition, Pakistan got Khulna and the eastern part of the Sundarban in exchange for Murshidabad and what would become the future site of Farakka. "So we got part of Sundarban but our political people did not understand what they were going to lose," Nayeem Wahra told me. "There will be no Sundarban if there is no Ganga. So they started shifting Ganga and our Sundarban is dying. It's not just global warming. It's lack of foresight of our political leaders." The leaders of Pakistan negotiated with India to keep water sources for West Pakistan, but neglected the Ganga and East Pakistan, he says.

353    **He is now vice chancellor of BRAC University**

BRAC, originally known as Bangladesh Rehabilitation Assistance Committee when it was founded in 1972, is one of the world's largest international relief and poverty alleviation organizations. Its university was established in 2001.

354    **increasing salinity near the coast**

Reba Paul of the Bangladesh Water Integrity Network (formerly with Global Water Partnership and Wateraid) gave me some examples of increased salinity in the coastal zone. She said the main power station in Khulna must collect fresh water to cool its boilers by sending a barge upstream to get fresh water. In recent years the barge must travel farther and farther upstream to find suitably fresh water. She added that mathematical models have shown that the salinity front will likely move about 60 km north in the next century. Such salinity intrusion will affect agriculture, drinking water supply, and health for the people living in the coastal zone.

355    **People ate the small fish whole, thereby gaining vital micronutrients, says Collis, the lack of which is increasingly becoming the cause of malnutrition in Bangladesh**

The urbanization that is absorbing 1 percent of Bangladesh's agricultural land each year is also driving changes in agriculture, says Collis. Urbanization has changed diets in Bangladesh and that is driving changes in agriculture. People in urban areas tend to eat less rice and more meat and vegetables. Commercial species of fish are aimed at urban markets. And urban markets require large amounts, "ten tons at a go," not the small catches traditional fishermen can provide.

World Fish is working on nutritional issues in Bangladesh. Collis says half the population is malnourished in some way or other. For some that means absolute calorie deficiency. For others it's a deficiency of micronutrients, not protein deficiency.

"People need some high quality protein but we can get most protein from rice or maize or wheat. The issue in Bangladesh is micronutrients. We used to have a very good diet here. 'Fish and rice makes a Bengali.' People would grow their rice and some lentils and go out and catch a handful of small fish in the evening. They eat them whole so they're extremely high in micronutrients—vitamins A, iron, zinc, calcium—all highly bioavailable."

The loss of agricultural land has contributed to the loss of the small fish that were so important for poor people's diets. The fish thrived in farming areas that were also wetlands. This is the land that is being lost to urbanization, says Collis. "We can't stop this. There's no land use planning. So we are trying to include these small fish in our aquaculture systems to increase yields."

357    **This eighty-square-kilometer patch of forest and mangrove swamp was decimated for shrimp farming, a tragic mistake funded by the World Bank and the Asian Development Bank in the 1980s**

Philip Gain criticizes the aid Bangladesh has received from the big multilateral banks on several counts. "All the assistance comes in the form of project loans, not standard loans. These are soft loans; bureaucrats love it. If they are standard loans there are more obligations and accountability." He says Bangladesh pays back more each year than it receives in fresh loans. And foreign companies are hired to come in and do the projects the banks fund, so the loan money really goes to them instead of staying in Bangladesh. "Soft loans are instruments to influence politics and political elites in this country. This is how they operate; they are in business. Bangladesh is a shareholder, but the banks are controlled by the developed countries. They have the cash, the efficiency. I don't blame anyone, but these are the mechanisms in place. If we get one dollar it will go back in the form of four dollars somehow. That is one estimate."

358    ***Each village will need to assess its own problems, and then with help from local government agencies, prepare its people to cope.***

To ensure safe water after inundation Ahsan suggests one or two deep tube wells in each area that could have pumps placed on them in an emergency to provide water for people. And people can be taught how to purify their own water in emergencies. Water treatment facilities should be maintained by local people.

Outbreaks of diarrhea typically follow floods. People get sick and then get treatment. Prevention is better than cure, says Ahsan. He notes World Bank statistics in 2007 showing the cost of diarrhea is more than seven hundred million dollars a year in Bangladesh. "With one-fourth that amount we could solve the problem in five years. In pilot areas, very flood-prone areas, we tried working with local governments and local people. We demonstrated methods of water purification and they took over. There were only six cases of diarrhea in 110,000 people. The cost was minimal. It's so simple. Empower the local people. The basic ingredients for purifying water are available everywhere. Organize it through science classes in local schools. Give prizes for students."

359    ***Women in particular can be a resource for information about changes in and around the home and farm***

Dhanashri Brahme shared an unpublished paper with me. She undertook an exploratory study, Gender Dimension of Climate Change Adaptation, in collaboration with WOTR (Watershed Organisation Trust) during her sabbatical leave from the United Nations Population Fund. The views contained in the document are those of the author and not of UNFPA or the UN and its other agencies.

359    ***And those of us living in Dhaka don't think about it.***

Nayeem Wahra is critical of the Dhaka-centric tendencies in delivering aid after disasters like Sidr, which would need to be changed to deal with climate change.

"At first the army goes by helicopter and throws things to the people. Journalists go first too and take pictures. Only if you think they are not human beings do you do this. When we distribute non-food items we should give them the things and tell them where they are from. We should give them the chance to say 'no, these are not useful.' Treat them as dignified people, with respect. In the first week it's OK, but it continues. All the relief agencies are now using helicopters. The cost of one trip is enough money for food for a whole month for a village. People from Dhaka want to go back quickly so they can sleep there."

Nayeem says reviving road and water communication should be paramount; local markets must be reactivated as soon as possible. Instead, he says, in the past aid groups have kept distributing things like bamboo and water all the way from Dhaka, because a project proposal calls for such relief. Better to just give money to people, not goods from Dhaka, he says.

# Glossary

| | |
|---|---|
| Aniconic | an image of a deity that does not take the form of a human or animal, such as a stone that is revered as a holy object |
| Avalokiteshvara | the Buddhist embodiment of compassion; also known as Chenrezig, the patron deity of Tibet |
| Baba | father in Hindi, but also an honorific for gurus and sages |
| Bahal | a courtyard |
| Baoli | a stepwell in North India |
| Barrage | a low dam or weir, chiefly used to divert some of a river's water into irrigation channels |
| Biogas | gaseous fuel produced by the breakdown of organic matter; chiefly made up of methane and carbon dioxide |
| Bodhisattva | in Buddhism, one who has vowed to achieve enlightenment for the benefit of all beings |
| Brahmin | the highest caste in Hindu culture |
| Chang | a kind of beer made from barley, millet, or rice; drunk in Nepal and Tibet |
| Chapatti | an unleavened flatbread made of wheat flour |
| Chettri | the second Hindu caste in Nepal, originally the "warrior" caste |
| Choko, jutho | terms referring to ritual purity or impurity |
| Chorten | a religious structure, like a stupa but generally smaller and frequently seen near trails; may hold offerings or prayers, and at one time chortens were reliquaries |
| Chowk | typically an intersection where two or more streets or lanes meet, often a shopping area |
| Chuba | a long wraparound dress worn by women in Tibet and Sherpa women in Nepal; usually made of wool |
| Cumecs, cusecs | measurements of the amount of water flowing in a river; cubic meters per second and cubic feet per second, respectively |

| | |
|---|---|
| Dacoit | bandit |
| Dal | a lentil soup eaten in India and Nepal. In Nepal *dal bhaat* refers to a complete meal that includes dal, rice, curries, and vegetables. |
| Dalit | the name typically used now to denote members of what was once called the untouchable caste in Hindu culture |
| Devanagari | the alphabet used by Hindi, Bengali, Nepali, and other Sanskrit-derived languages |
| Dhindo | a porridge eaten in Nepal; can be made of millet or potato or a mixture of grains |
| Dhoti | traditional South Asian garment for men; a large cloth wrapped around the waist and between the legs with the end tucked into the back of the waist. A dhoti resembles trousers but is made of unsewn fabric; it typically drapes below the wearer's knees. |
| Dunge dhara | in Nepal, a water spout carved from stone, frequently in the form of an animal head |
| Eco-san toilet | Also known as a urine-diverting toilet, it contains separate receptacles for urine and feces and uses little or no water. The urine and feces are available to be used for fertilizer after some processing. |
| Euroguard | a home water purifier |
| Ganesh | the elephant-headed god of the Hindu pantheon; the god of new beginnings, remover of obstacles, son of Shiva and his wife Parvati |
| Gangajal | water of the Ganges River; considered holy by many Hindus |
| Ghat | a broad flight of steps leading down to a holy river like the Ganges; used for bathing or puja. Some ghats are the site of funeral pyres. The western and eastern ghats of India are mountain ranges. |
| GLOF | glacial lake outburst flood |
| Gompa | a Tibetan Buddhist monastery; a meditation hall |
| Gray water | wastewater from showers, sinks, and laundry; does not contain sewage |
| Himal | mountain |
| Hindi | the Sanskrit-derived Indo-European language spoken in much of North India |
| Hiti | in Kathmandu and other parts of Nepal, a sunken step well generally made of red brick. One or more dunge dhara or stone water spouts typically protrude from the walls. |
| IPCC | Intergovernmental Panel on Climate Change, the UN scientific body tasked with compiling and reviewing information about global warming; established in 1988. |

| | |
|---|---|
| Jal | Sanskrit word for water; used in Hindi, Nepali, and other South Asian languages. *Paani* is more commonly used in Nepali, pani in Hindi. |
| Khata | a white or yellow silk scarf placed around the neck of travelers in Nepal and Tibet; also used on festive occasions |
| Khola | river |
| Kurta suruwal | knee-length tunic with pants worn by women in Nepal, typically in bright colors; similar to the Indian salwar kameez. |
| Lakshmi | Hindu goddess of wealth; wife of the god Vishnu |
| Lungi | traditional South Asian garment for men; resembles a skirt and covers the man from the waist down. It's made by wrapping a cloth around the waist and securing it with a knot. |
| Makara | a mythical creature sometimes resembling a crocodile; the vehicle of the Ganga. Dunge dhara often take this form. |
| Mala | a garland of marigolds used to greet visitors and in ceremonies. The word *Malla* refers to a dynasty of rulers in the Kathmandu Valley. |
| Mandala | a circle, symbolic representation of the universe; a concentric diagram used in Hindu and Buddhist art and ritual and an aid to meditation |
| Mandir | temple |
| Mani wall | Prayer stones, or mani stones, inscribed with mantras are frequently assembled into long structures in the middle of pathways or at the entrance to Tibetan Buddhist villages. |
| Maoists | refers to members of Nepal's Communist Party, NCP-Maoist, who instigated a ten-year insurgency. In 2006 they abandoned their civil war and entered into a peace process with other political parties, which led to the ouster of the last Shah king. |
| Mughals | Muslim emperors who controlled most of the Indian subcontinent in the sixteenth, seventeenth, and eighteenth centuries |
| Nagas | mythical half-human serpents believed to cause rain and fertility, they protect water bodies |
| Namaste | a common greeting and farewell in India and Nepal, usually accompanied by palms placed together. *Namaskar* is a somewhat more formal version of the same greeting. |
| Newar | indigenous people of the Kathmandu Valley and the artisans who created the distinctive art and architecture of the region |
| NGO, INGO | nongovernmental group. Sometimes NGO is used to denote a nonprofit group that operates in a single country, while |

|  | INGO is the term used to denote the large international aid groups like CARE, Oxfam, IUCN, and World Vision. |
| Nola | a mountain spring |
| Paani | common word for water in Nepali and other South Asian languages |
| Pahad, pahadi | foothills of the Himalaya in Nepal. Pahadi is a person from this region, typically from one of the Mongoloid ethnic groups who have lived in the Himalaya for centuries and speak Tibeto-Burman languages. |
| Paratha | a fried flat bread eaten in India, often stuffed with vegetables |
| Penstock | an enclosed pipe that delivers water from a river to hydraulic turbines |
| Puja | Hindu religious ritual performed at home or at a temple |
| Prayag | a sacred river confluence |
| Pokhari | pond in Nepali |
| Rabi | crop planted after the monsoon and harvested in winter |
| Raj | refers to the era of British rule in India |
| Ram, rama | an avatar of the god Vishnu, hero of the Ramayana |
| Rishi | a seer or inspired poet; also an honorific for gurus |
| Sadhu | a wandering mendicant, often a devotee of Shiva |
| Sari | a long strip of unstitched cloth draped on the body; worn with a blouse by women in South Asia both as formal and daily wear |
| Sauni, sauji | proprietress, proprietor in Nepali |
| Sherpa | the population of Tibetan Buddhists who live in the valleys below Mount Everest in Nepal. Their ancestors came from Tibet. The term *sherpa* is often mistakenly used as a synonym for trekking guide or porter, because the first men to help Western climbers were from the ethnic group in Nepal known as Sherpa. |
| Shiva, shiva lingam | For many Hindus, Shiva is the central god, known as the creator and destroyer. Mt. Kailash is his home; the abstract form of the lingam is often used to represent his generative power. |
| Soft loan | a low-interest loan typically provided by a government agency or development bank, often with a long repayment period |
| Stupa | an ancient form of Buddhist monument, typically a domed structure in South Asia |
| Stepwell | a well or pond with descending steps so people can easily reach the water |
| Subcontinent | frequently used to refer to India, which is a large landmass separate from the rest of the Asian continent; can include adjacent countries like Nepal and Bangladesh |

| | |
|---|---|
| Tal | lake |
| Tamang | one of the largest of the Mongoloid ethnic groups living in the foothills of the Nepal Himalaya |
| Tarai | the plains stretching east to west at the bottom of the Himalayan foothills in Nepal; the nation's largest agricultural area |
| Tika | a paste typically made from red powder mixed with water; placed in the middle of the forehead as a blessing |
| Topi | a brimless hat or cap worn by men in Nepal |
| Tsampa | flour made from roasted barley; eaten as a porridge in Tibet and the Himalaya |
| Tsho or tso | lake in the Tibetan language |
| Tube well | a well bored with a steel tube; used to access water in aquifers |
| Vedas | the oldest Hindu scriptures, consisting of hymns to the gods and chants to be used in important religious rites; the scriptures were likely passed down orally for thousands of years before being written down |
| Vishnu | along with Shiva, one of the major gods of the Hindu pantheon. Whereas Shiva is known as the creator and destroyer, Vishnu is known as the preserver and protector. He appears in a variety of avatars, including the fish, the boar, the hero Rama, and the god Krishna |
| Yak | a large shaggy bovine native to the Tibetan plateau, similar to bison; most of the yaks in Nepal are the result of crossbreeding with cows |

# Select Bibliography and Further Reading

Abbas, B. M. *The Ganges Water Dispute*. Dhaka: The University Press Limited, 1982.

Agarwal, Anil and Narain, Sunita, eds. *Dying Wisdom*. New Delhi: Centre for Science and the Environment, 1997.

Agarwal, Anil, Narain, Sunita, and Khurana, Indira, eds. *Making Water Everybody's Business*. New Delhi: Centre for Science and Environment, 2001.

Alagh, Yoginder K., Pangare, Ganesh, and Gujja, Biksham, eds. *Interlinking of Rivers in India*. New Delhi: Academic Foundation, 2006.

Albinia, Alice. *Empires of the Indus*. London: John Murray, 2008.

Babu, Suresh and Seth, Bharat Lal. *Sewage Canal: How to Clean the Yamuna*. New Delhi: Centre for Science and Environment, 2007.

Basham, A. L. *The Wonder That Was India*. London: Picador, 2004.

Baviskar, Amita, ed. *Waterlines*. New Delhi: Penguin Books India, 2003.

Bista, Dor Bahadur. *Fatalism and Development: Nepal's Struggle for Modernization*. Patna: Orient Longman Ltd., 1991.

Crow, Ben. *Sharing the Ganges*. Dhaka: The University Press Limited, 1997.

Dixit, Ajaya. *Basic Water Science*. Kathmandu: Nepal Water Conservation Foundation, 2002.

D'Souza, Rohan. *Drowned and Dammed: Colonial Capitalism and Flood Control in Eastern India*. New Delhi: Oxford University Press, 2006.

Gain, Philip. *The Last Forests of Bangladesh*. 2nd ed. Dhaka: Society for Environment and Human Development, 2002.

Frater, Alexander. *Chasing the Monsoon*. New Delhi: Penguin Books India, 1991.

George, Rose. *The Big Necessity: The Unmentionable World of Human Waste and Why It Matters*. New York: Henry Holt, 2008.

Gyawali, Dipak. *Water in Nepal*. Kathmandu: Himal Books, 2001.

Hollick, Julian Crandall. *Ganga: A Journey Down the Ganges River*. Washington, DC: Island Press, 2008.

Iyer, Ramaswamy. *Towards Water Wisdom: Limits, Justice, Harmony*. New Delhi: Sage Publications, 2007.

Jacob, Nitya. *Jalyatra: Exploring India's Traditional Water Management Systems*. New Delhi: Penguin Books India, 2008.

Lahiri-Dutt, Kuntala and Wasson, Robert J., eds. *Water First*. New Delhi: Sage Publications, 2008.

Miah, M. Maniruzzaman. *Hydro-Politics of the Farakka Barrage*. Dhaka: Gatidhara, 2003.

Mishra, Dinesh Kumar. *Trapped! Between the Devil and Deep Waters: The Story of Bihar's Kosi River*. Dehradun and New Delhi: People's Science Institute and South Asia Network on Dams, *Rivers and People*, 2008.

Mishra, Dinesh Kumar. *Living with the Politics of Flood*. Dehradun: People's Science Institute, 2002.

Mishra, Pankaj. *Temptations of the West*. London: Picador, 2006.

Novak, James J. Bangladesh: *Reflections on the Water*. Dhaka: The University Press Limited, 1993.

Pearce, Fred. *When the Rivers Run Dry*. Boston: Beacon Press, 2006.

Rist, Gilbert. *The History of Development*. New Delhi: Academic Foundation, 2002.

Sharma, Sudhindra, et al., eds. *Aid Under Stress*. Kathmandu: Himal Books, 2004.

Solomon, Steven. *Water*. New York: HarperCollins, 2010.

Subba, Bhim and Pradhan, Kishor, eds. *Disputes Over the Ganga*. Kathmandu: Panos Institute South Asia, 2004.

Subba, Bhim. *Himalayan Waters*. Kathmandu: Panos Institute South Asia, 2001.

Thompson, Michael, Warburton, Michael, and Hatley, Tom. *Uncertainty on a Himalayan Scale*. Kathmandu: Himal Books, 2007.

Tiwari, Sundarshan Raj. *The Brick and the Bull*. Kathmandu: Himal Books, 2002.

Verghese, B. G. *Waters of Hope*. New Delhi: India Research Press, 2007.

Whelpton, John. *A History of Nepal*. Cambridge: Cambridge University Press, 2005.

# Index